Manipulating the Sun

Nuncius Series

Studies and Sources in the Material and Visual History of Science Series

VOLUME 13

The titles published in this series are listed at *brill.com/nuns*

Manipulating the Sun

Picturing Astronomical Miracles from the Bible in the Early Modern Era

Edited by

Julia Ellinghaus and Volker Remmert

BRILL

LEIDEN | BOSTON

Cover illustration: Wenzel Jamnitzer, *Measuring Instrument*, 1578, Bibliothèque de l'Observatoire, Paris – *Moon-Side* (for surveying), with permission of The Library of the Observatoire de Paris

The Library of Congress Cataloging-in-Publication Data is available online at https://catalog.loc.gov
LC record available at https://lccn.loc.gov/

Typeface for the Latin, Greek, and Cyrillic scripts: "Brill". See and download: brill.com/brill-typeface.

ISSN 2405-5077
ISBN 978-90-04-47188-7 (hardback)
ISBN 978-90-04-67765-4 (e-book)
DOI 10.1163/9789004677654

This book is printed on acid-free paper and produced in a sustainable manner.

Contents

PART 3
The Horologium Ahaz and Joshua Stopping the Sun in Art and Architecture in Specific Locations

Figures

Notes on Contributors

Julia Ellinghaus
studied history of art, with a focus on iconography and Northern European art of the 17th and 18th century. After her PhD she was research fellow at the Gemäldegalerie Alte Meister in Kassel and worked at various museums afterwards. Since 2018 she has been research assistant at the Interdisciplinary Centre for Science and Technology Studies (IZWT) at the University of Wuppertal, where she works on the project "Iconography of the imagery ON scientific instruments of the early modern period."

Henrique Leitão
is Senior Researcher (*Investigator Principal*) at the Centro Interuniversitário de História das Ciências e da Tecnologia (CIUHCT) at the University of Lisbon, and head of the Departamento de História e Filosofia das Ciências at the Faculty of Sciences, University of Lisbon. His research interests are in the history of the exact sciences in the 15th to 17th centuries (mathematics, astronomy, navigation science, physics, etc.), with a special, but not exclusive, focus on Portugal.

Evonne Levy
an art historian, is Distinguished Professor of Early Modern Art at the University of Toronto.

Walter Melion
is Asa Griggs Candler Professor of Art History at Emory University in Atlanta, where he directs the Fox Center for Humanistic Inquiry. He is author of three monographs and a critical edition of Karel van Mander's *Foundation of the Noble, Free Art of Painting*, co-author of two exhibition catalogues, editor or co-editor of more than twenty-five volumes, and has published more than eighty articles. Melion is series editor of Brill's *Studies on Art, Art History, and Intellectual History* and *Lund Humphries' Northern Lights*. He was elected Foreign Member of the Royal Netherlands Academy of Arts and Sciences in 2010.

Andrew Morrall
is professor of early modern art and material culture at the Bard Graduate Center, New York. He has written widely on the arts of early modern Northern Europe, the Reformation and the arts, early modern collecting, craft and Kunstkammer, intersections of art and science, theories of ornament, and the

material culture of the early modern home. He is currently writing a book-length study of South German craftsmen in the era of the Kunstkammer.

Brent Purkaple
is a historian of science who is currently a visiting professor of history at Grand Valley State University in Allendale, Michigan. His research interests are in the area of early modern science, book history, and the history of science and religion.

Volker Remmert
has been trained as mathematician and historian. He teaches history of science and technology at the University of Wuppertal. His research interests are in the history of early modern science and in the history of mathematics in Germany in the 19th and 20th centuries.

Denis Ribouillault
is a historian of early modern art, landscape and gardens, and a professor at the Université de Montréal. He received prestigious fellowships (Villa Médicis, Villa I Tatti, Dumbarton Oaks) and is the author of numerous articles, edited volumes and a monograph on the villas and gardens of Renaissance Rome (*Rome en ses jardins. Paysage et pouvoir au XVIe siècle*, Paris, INHA-CTHS, 2013). His new monograph *The Villa Barbaro at Maser. Science, Philosophy, and the Family in Venetian Renaissance Art* is forthcoming with Harvey Miller Publishers, London.

Introduction

Julia Ellinghaus and Volker Remmert

In 1543 Nicolaus Copernicus's book *De Revolutionibus Orbium Coelestium* was published. While his argument for the mobility of the earth was gradually accepted by most astronomers by the mid-seventeenth century, it caused considerable critique among theologians and exegetes. As they saw it, many passages from the Bible could be read as supporting the stability of the earth and the mobility of the sun.[1] Didn't Psalm 104:5 state that "it [= the earth] should not be moved for ever"; and didn't Ecclesiastes 1:4–6 stress that the "Earth abideth for ever. The Sun also ariseth, and the Sun goeth down, and hasteth to his place where he arose: and there rising againe, compasseth by the South, and bendeth to the North"? In his 1570 commentary on the *Sphere* by Johannes de Sacrobosco, the most popular textbook on Ptolemaic astronomy during the Middle Ages and the Renaissance, the Jesuit astronomer Christopher Clavius (1558–1612) referred to these two passages and concluded: "What could be clearer?" Clavius went on to mention the two biblical miracles standing at the centre of this volume: "And again, it is recounted among the miracles that God sometimes causes the Sun to go back or to stand still altogether."[2] These two miracles, the sun reversing its course in II Kings 20:8–11/Isaiah 38:8 and the sun standing still in Joshua 10:12, stand out among the astronomical miracles in the Bible precisely because they involve the motion of the sun.

1 For details see the classic by Reyer Hooykaas, ed. and trans., *G. J. Rheticus' Treatise on Holy Scripture and the Motion of the Earth with translation, annotations, commentary and additional chapters on Ramus-Rheticus and the development of the problem before 1650* (Amsterdam: North-Holland Publishing Company, 1984); cf. Kenneth J. Howell, *God's Two Books. Copernican Cosmology and Biblical Interpretation in Early Modern Science* (Notre Dame, IN: University of Notre Dame Press, 2002); Volker Remmert, "'Our Mathematicians Have Learned and Verified This': Jesuits, Biblical Exegesis and the Mathematical Sciences in the Late Sixteenth and early Seventeenth Centuries," in *Nature and Scripture in the Abrahamic Religions*, 2 vols., ed. Jitse M. van der Meer and Scott H. Mandelbrote (Leiden: Brill, 2008), 665–691; also the collection by Pierre-Noël Mayaud, *Le conflit entre l'Astronomie Nouvelle et l'Écriture Sainte aux XVIe et XVIIe siècles*, 6 vols. (Paris: Champion, 2005).

2 Christopher Clavius, *In sphaeram Ioannis de Sacro Bosco Commentarius* (Rome: Victorium Helianum, 1570), 247f: "Quid clarius dici poterat? […] Rursus inter miracula refertur, quod Deus aliquando Solem aut retroduxit, aut prorsus consistere fecit." On the context see Volker Remmert, *Picturing the Scientific Revolution: Title Engravings in Early Modern Scientific Publications* (Philadelphia: St. Joseph's University Press, 2011), 19–21.

 | DOI:10.1163/9789004677654_002

Four astronomical miracles are featured in the Bible: the sun standing still in Joshua 10:12, the sun returning in its path in II Kings 20:8–11/Isaiah 38:8, the eclipse of the sun at Christ's crucifixion (Matthew 27:45, Mark 15:33, Luke 23:44f), and the appearance of the Star of Bethlehem (Matthew 2:2–11).[3] Neither the eclipse nor the appearance of the Star of Bethlehem had a bearing on exegetical discussions of terrestrial motion. But from an exegetical perspective the two miracles involving the motion of the sun were crucial to the discussion of worldviews.

The Book of Joshua recounts how the Israelites conquered the land of Canaan and how, in the conflict with the Amorites, God's help had been required:

> Then spake Joshua to the Lord in the day when the Lord delivered up the Amorites before the Children of Israel, and he said in the sight of Israel, Sun, stand thou still upon Gibeon; and thou, Moon, in the valley of Ajalon. And the Sun stood still, and the Moon stayed, until the people had avenged themselves upon their enemies (Joshua 10:12f).[4]

The sun standing still over Gibeon is one of the most spectacular miracles in the Bible. It was used as a standard argument against the Copernican system ever since the middle of the sixteenth century, and the scene of Joshua stopping the sun, being an ideal subject for visual representation, had been widespread in art since the Middle Ages and well into the eighteenth century.[5]

In the Second Book of Kings, the prophet Isaiah promised the dying King Hezekiah that he would recover and live and reign a further 15 years. Hezekiah, not being convinced, requested a sign from God:

> And Isaiah said, This sign shalt thou have of the Lord, that the Lord will do the thing that he hath spoken: shall the shadow go forward ten degrees, or go back ten degrees? And Hezekiah answered, It is a light thing for the

3 Edward Walter Maunder, *The Astronomy of the Bible. An Elementary Commentary on the Astronomical References of Holy Scripture* (London: T. S. Clark & Co., 1908), 351.

4 Unless otherwise indicated, all citations of the Bible in this volume follow the King James Bible (https://www.kingjamesbibleonline.org/).

5 Cf. Theodor Ehrenstein, *Das Alte Testament im Bilde. Ein Illustrationswerk mit über 2000 Abbildungen von altchristlichen, mittelalterlichen und neuzeitlichen Kunstwerken* (Vienna: Kende, 1923), 441–466; Engelbert Kirschbaum S.J., ed., *Lexikon der christlichen Ikonographie*, 8 vols. (Freiburg: Herder, 1968–1976), here vol. 2 (1970), col. 436–442 (Jürgen Paul); Andor Pigler, *Barockthemen*, 3 vols. (Budapest: Akadémiai Kiadò, 1974), here vol. 1, 113–114.

> shadow to go down ten degrees: nay, but let the shadow return backward ten degrees. And Isaiah the prophet cried unto the Lord: and he brought the shadow ten degrees backward, by which it had gone down in the dial of Ahaz (II Kings 20:9–11).

This miracle was often referred to as the Horologium Ahaz because most commentators thought that the measure of the sun's retreat referred to a sundial. This passage, too, became a standard argument against the Copernican system and was likewise often depicted.[6] Because the miracle of the sun moving backwards was as convincing an argument as the halting of its movement over Gibeon, exegetes often mentioned the two together. Both miracles could easily be presented visually, and consequently their prominence extended well beyond the fields of biblical exegesis or astronomy.

This volume aims to discuss these two passages, the development of related imagery, their role as anti-Copernican arguments (in text and image), their reception, their treatment in the mathematical sciences, and their various cultural layers, with a focus on the history of art and the history of science in the sixteenth and seventeenth centuries. It stems from a workshop organised by the research project *Iconography of the Imagery on Early Modern Scientific Instruments*.[7] The project specifically analysed the imagery on early modern scientific instruments, such as astrolabes, sundials, surveying instruments, etc. It aimed for the first time at a systematic analysis of the multifaceted visual worlds on the instruments, interrogating their role in the various contexts of the adorned instruments. The imagery points to quite a few significant topics, for instance, mediation and illustration of knowledge, in particular by picturing the usability of the instruments, or the role of instruments as patronage artefacts with specific pictorial programmes, or visual statements of specific positions in theoretical debates, such as the Copernican question.[8] As

6 On this, see Klaus Maurice, "Sonnenuhr," in *Lexikon der christlichen Ikonographie*, ed. Engelbert Kirschbaum S.J. (Freiburg im Breisgau: Herder, 1972), 4:col. 180–181, and Hans Bardtke, "Ezechias," in *Lexikon der christlichen Ikonographie*, ed. Engelbert Kirschbaum S.J. (Freiburg im Breisgau: Herder, 1968), 1:col. 714–715; Michael Scholz-Hänsel, *Eine spanische Wissenschaftsutopie am Ende des 16. Jahrhunderts. Die Bibliotheksfresken von Pellegrino Pellegrini im Escorial* (Münster: Lit, 1987), 151f; René Taylor, "Architecture and Magic. Considerations on the 'Idea' of the Escorial," in *Essays in the History of Architecture Presented to Rudolf Wittkower*, ed. Douglas Fraser, Howard Hibbard and Milton J. Lewine (London: Phaidon, 1967), 109.

7 The project was funded by the German Research Foundation (2018–2022).

8 Cf. Julia Ellinghaus and Volker Remmert, "Visual Worlds on Early Modern Scientific Instruments: Types and Messages," in *Reassessing Epistemic Images in the Early Modern World*, ed. Ruth Noyes (Amsterdam: Amsterdam University Press, 2023), 153–174.

illustrations of the two miracles are repeatedly featured on instruments, the question of whether anti-Copernican undertones might automatically have been associated with them, or whether there were other messages they were intended to convey, became key in the project. Thus, the point of departure of this volume is twofold. On the one hand, it further explores the visual worlds on early modern scientific instruments. On the other hand, it aims to construct a better understanding of the relationship between biblical arguments and biblical exegesis and the Copernican debate in the sixteenth and seventeenth centuries. At the intersection of these, we find a vast field of possible visual and material repercussions and reverberations of the two miracles of the sun reversing its course and the sun standing still.

The contributions to this volume are presented in three thematic sections, dealing with the two miracles (1) in early modern printed material, (2) on early modern scientific instruments, and (3) in art and architecture in specific locations.

In the first section, *The Horologium Ahaz and Joshua Stopping the Sun in Early Modern Printed Material*, the two miracles are discussed in the context of biblical readings and early modern science.

Walter S. Melion sets the stage in his chapter "The Prayer of Joshua and the Dial of Ahaz: Paratactic Scriptural Illustration in the Lempereur, Vorsterman, and Van Liesveldt Bibles." Focussing on these three illustrated Bibles published in Antwerp in the 1530s and 1540s, he shows that the woodcuts are for the most part figural, namely either historical scenes or portraits of the authors of the various biblical books. In stark contrast, the diagrammatic *Dial of Ahaz*, positioned at the start of Isaiah 38, is a decidedly allegorical image. Melion's in depth discussion of the various exegetical connotations and interpretations of the miracle of the retrograding sun allows for a clear perception of how well-informed reader-viewers would have understood and perceived the miracles of Joshua and Hezekiah in the pre-Copernican world, putting due emphasis on the divine significance of the sun's motion.

Henrique Leitão in his chapter "Mathematical Certainty and Biblical Inerrancy: Pedro Nunes and the Retrogradation of Shadows at the Dial of Ahaz" reconstructs a debate that has almost faded out of history. The Portuguese cosmographer Pedro Nunes (1502–1578) subjected the miracle of the retrograding sun to a strict mathematical analysis in 1537 (and again in 1566). Nunes proved in a purely mathematical way that under specific conditions the retrogradation of the sun would not be a miracle, but would in fact naturally occur twice a day. As one might expect, such a counter-intuitive claim, not based on any actual observation of the phenomenon, raised a number of problematic questions. In the context of a highly contentious relationship between biblical exegesis and the mathematical sciences that Copernicanism triggered from the late sixteenth century and which culminated in the Galileo trial of 1633, it is significant that Nunes proposed a natural (mathematical) explanation of a biblical miracle while still allowing that the miracle had actually happened. Leitão describes how this confrontation between mathematical certainty and biblical inerrancy generated (often heated) discussions that lasted until the late nineteenth century. Pushing this point a bit further, one might even argue that Nunes was among the first to deliver astronomers and mathematicians power over the interpretation of the Bible.

Brent Purkaple's chapter, "Appealing to Ahaz: The Jesuits and the Sundial of Ahaz in the Era of Copernican Astronomy," is situated exactly at the crossroads between biblical exegesis and Copernican astronomy as seen from the perspective of the Jesuits. He closely analyses the various references to the Horologium Ahaz in the publications of early modern Jesuit astronomers and mathematicians such as Christopher Clavius, Johannes Smogulecki, Giovanni Battista Riccioli, Mario Bettini, and Athanasius Kircher. Purkaple traces how discussions of the Sundial of Ahaz transcend the confines of publications on gnomonics and regularly surface in texts on geometry, optics, and natural magic. He shows that while the story of the retrograding sun could have obvious implications in the context of the Copernican debate, in many cases this aspect did not take centre stage. Jesuit authors were uniform in their understanding of this biblical event as a miracle and an important sign of divine influence on the world, emphasising the literal textual meaning. Thus, the story of the Horologium Ahaz could be seen in the light of its implications regarding the motion of the earth, but it was an important topic in itself among Jesuit astronomers and mathematicians, not least because it nicely illustrated that the tradition of gnomonics stretched back to biblical times.

The three chapters of this section clearly show that the Horologium Ahaz is a point of intersection where various threads cross: history and prophecy, exegetical and mathematical traditions, musings about the nature of time as well as schematic representations of divine signs. Copernicanism is but one thread in this dense web. At the same time, as the chapters by Leitão and Purkaple show, in focussing on the Horologium Ahaz, a correlation between scientific replication of the miracle and biblical exegesis emerges in the second half of the sixteenth century.

The second section, *The Horologium Ahaz and Joshua Stopping the Sun on Early Modern Scientific Instruments*, focuses on material and visual representations of the two miracles at the intersection of patronage strategies and artefacts and early modern scientific instruments.

In his chapter "Turning Back the Sun. Christoph Schissler's 'Horologium Ahaz' as 'Kunststück,'" Andrew Morrall focusses on a specific instrument, a sundial by the Augsburg instrument maker Christoph Schissler (c.1531–1608) that could re-create the biblical miracle of the retrograding sun by actually making the sun's shadow travel backwards. Morrall conceives of Schissler's sundial not only as an instrument, but also as a remarkable work of art or, as Schissler himself put it, a 'Kunststück.' In Schissler's sundial the boundaries between instrument and artwork become indistinguishable. Morrall uses the concept of 'Kunststück,' denoting a specific class of artwork, characterised by a high degree of technical, conceptual and thematic ingenuity, to analyse the sundial in the context of Schissler's oeuvre. The instrument could serve not only to demonstrate the miraculous effect of light refraction, but could at the same time be at the centre of the early modern parlour game of decoding iconographic contexts, and showing off mathematical knowledge and biblical lore, be it in a Kunstkammer, an instrument collection, or in the study of an astronomer.

Julia Ellinghaus and Volker Remmert concentrate on a group of instruments in their chapter "Manipulating the Sun in Depictions on Early Modern Scientific Instruments." The chapter examines early modern scientific instruments bearing depictions of the biblical miracles of Hezekiah's life prolonged and Joshua stopping the sun, either solo or combined. These depictions adorn devices of different types and purposes, though all of them could be used to measure time: a paper sundial by Georg Hartmann (1561), a torquetum by Johannes Praetorius (1568), the refractive sundial by Christoph Schissler (1578), that Andrew Morrall discusses in detail, a measuring instrument by Wenzel Jamnitzer (1578), and an astronomical clock by Hans Gruber (1583). The analysis of the ramified pictorial programmes of these instruments brings to light the different meanings underlying the scenes, which are closely tied either to the way the objects operate or to the purchaser or addressee, showing that none of them is explicitly linked to any kind of anti-Copernican message.

Both chapters stress that the highly adorned instruments in question have specific roles as patronage artefacts, that are not necessarily or easily decodable. In particular they are artworks in themselves, which can often hardly be reduced to their functions as scientific instruments. As prestigious and representative objects their pictorial programmes are designed to inspire discussion, but not provoke serious controversy. Thus, it does not come as a surprise that there are no traces of the Copernican debate in the pictorial programmes of the instruments discussed.

The third section, *The Horologium Ahaz and Joshua Stopping the Sun in Art and Architecture in Specific Locations*, leads us into rather grandiose settings for staging the two miracles, namely the Medici Villa at Pratolino and Il Gesù, the Jesuit mother Church in Rome.

Denis Ribouillault in his chapter "Stairway to Heaven. The Marvellous Staircases of the Medici Villa at Pratolino" draws our attention to staircases serving as sundials. In doing so he takes up the common notion among early modern (and earlier) exegetes, that the Horologium Ahaz had in fact been a staircase, and that the miracle had been that the shadow cast on the staircase travelled backwards by ten steps. Ribouillault discusses the stairs on the south façade of the Villa Medici of Pratolino, built at the end of the sixteenth century, that served as sundials. He argues that some aspects of the configuration of the palatial complex of Pratolino were rooted in medieval and early modern visual ideas about the Sundial of Ahaz. Ribouillault also points to the contemporary interpretation of the staircase by the philosopher Francesco de' Vieri in his *Discorsi delle maravigliose opere di Pratolino e d'Amore* of 1586, who, informed by Aristotle and Plato, understood it in the context of meteorological science and the destiny of the soul. This contextualisation reveals the Horologium Ahaz at Pratolino as a striking instance where the miracle was entangled in a complex web of interpretations, traditions, and connotations, but devoid of anti-Copernican undertones.

In contrast, the Copernican question takes centre stage in Evonne Levy's chapter "Joshua Stopping the Sun in the Gesù: A Hypothesis." The chapter takes up the Jesuit thread from Brent Purkaple's chapter, discussing a fresco planned for the Jesuit mother Church in Rome in the 1670s, which was never realised. The Jesuit General Giovanni Paolo Oliva (1600–1681) had intended to commission Jacques Courtois (1621–1676), the renowned Jesuit battle painter, to create a monumental fresco in the most prominent spot of Il Gesù depicting the miracle of Joshua stopping the sun as a battle scene. As Levy demonstrates, the Joshua scene carried various important undertones within the Jesuit order and its extensive pictorial strategies because the figure of Joshua himself was seen as an Old Testament type for Ignatius of Loyola, founder of the Society of Jesus, as well as providing genealogy for the name of Jesus, for which the Society was named. Thus, it stood to reason that the Jesuits would accord Joshua a place of honour in Il Gesù as Ignatius's temple. However, the original plan was ultimately discarded and replaced with the more universal theme of the Adoration of the Lamb. In reassessing the context of the original plan, Levy argues that, in view of the anti-Copernican connotations of the Joshua scene, one can hardly ignore the Jesuits' role in the debates about Copernicanism and in the trial of Galileo in 1633. She proposes that it is hard to imagine that the Joshua scene, in all its complexity, would not have been taken as a clear sign of the Jesuits' full support for the literal truth of the Bible, and the stability of the earth.

By way of conclusion we wish to highlight that the chapters in this volume show that the miracles of the sun standing still and of the sun reversing its course had many facets, and that their role as anti-Copernican arguments (in text and image) did not play a key role in the early modern period outside of astronomical discourse. We also found that the miracle of the Horologium Ahaz seems to be shrouded in more layers of interpretations and repercussions than the miracle of Joshua.

Joshua stopping the sun over Gibeon was often seen to represent the power that God could confer on a virtuous ruler, as is the case with two of the instruments discussed by Julia Ellinghaus and Volker Remmert (Jamnitzer's measuring instrument and Gruber's astronomical clock). At the same time, he stood as a paragon of obedience and as such was highly significant in Jesuit culture, where he was closely linked symbolically to both Ignatius of Loyola and to the name of Jesus, as Evonne Levy explains in her chapter. As it was easy to imagine or depict Joshua in a grandiose setting, as a powerful military man performing the miracle, it is not surprising that he could also easily have played a leading role in the Copernican debate.

In contrast to the Joshua scene, the miracle of the Horologium Ahaz inspires, in a way, a more subdued and more complex, less extravagant and spectacular, yet stimulating contemplation of many subjects, such as the nature of time itself. In its diagrammatic representations, as discussed by Walter S. Melion in his chapter, the miracle of the sun reversing its course became an allegory of the prolongation of time. Melion's analysis of the illustrations in the Lempereur, Vorsterman and van Liesveldt bibles shows that the diagrammatic Horologium Ahaz operated as an exegetical prompt to the reader-viewer, to draw parallels amongst different verses in the bible. From the exegetical point of view the miracle counted as a divine sign, not to be doubted, and a sign of God's activity within the world, in particular among Jesuit authors, as Brent Purkaple emphasises in his chapter. In the world of early modern intellectual parlour games, the Horologium Ahaz provoked reflection on the method of function and the construction of sundials, as we see in various examples such as Christoph Schissler's refractive sundial (Andrew Morrall), Georg Hartmann's paper sundial (Ellinghaus/Remmert), and the staircases at Pratolino (Denis Ribouillault). In this more technical context, reference to the Horologium Ahaz clearly drew a line of tradition for gnomonics stretching back into the Old Testament. At the same time, the miraculous recovery of King Hezekiah, while rooted in his godliness, was only made possible by Isaiah's medical expertise. Thus the miracle entered the repertoire of legitimisation strategies of early modern physicians, as the pictorial programme of Praetorius's torquetum illustrates (Ellinghaus/Remmert). Finally, the miracle highlighted the divine grace (God gives 15 further years of governance to Hezekiah) that God grants the virtuous and pious ruler, and, more generally, helped to legitimise the doctrine of the divine right of kings.

Acknowledgements

We would like to express our gratitude to all the contributing authors for having embarked on this project with us and for their generosity in sharing their expertise and material. Thanks also to Samantha Leong for her diligent and patient editorial work on this volume. We are also grateful for the support of the German Research Foundation. Finally, our thanks go to the editors of the Nuncius series for taking up our proposition to include this volume in the series.

PART 1

The Horologium Ahaz and Joshua Stopping the Sun in Early Modern Printed Material

CHAPTER 1

The Prayer of Joshua and the *Dial of Ahaz*

Paratactic Scriptural Illustration in the Lempereur, Vorsterman, and Van Liesveldt Bibles

Walter S. Melion

Abstract

The woodcuts that mark key passages in the three most prominent illustrated Bibles published in sixteenth-century Antwerp – Martin Lempereur's *La saincte Bible* of 1530 and 1534, Willem Vorsterman's *Tgeheele oude ende nieuwe Testament* of 1532, 1533/34, 1542/43, and 1545, and Jacob van Liesveldt's *Den Bibel met grooter neersticheyt gecorrigeert* of 1532, 1534, 1535, and 1536 – consist for the most part of figural scenes, with the startling exception of the diagrammatic *Dial of Ahaz*, positioned at the start of Isaiah 38. Whereas the majority of woodcut illustrations in these Bibles eschews allegory, showing either historical scenes or portraits of the authors of the various biblical books, the *Dial of Ahaz* is decidedly allegorical. The image of two sundials shown side-by-side invites sustained exploration of the exegetical meanings that nest within this complex visual allegory about the nature of divine agency. This curious image – part mimetic in that it purports to depict dial faces, part diagrammatic in that it consists mainly of geometrical forms, linear marks, and numerical signs, part exegetical in that it conjugates a string of biblical texts, and wholly allegorical in its compilatory method of construction and argument – constitutes one pregnant example of how preachers and their congregations, and the exegetes trained in the kinds of reading codified by the Glossist and Nicholas de Lyra, would have reckoned with the divine significance of the sun's motion, when viewing it through the lens of allegory. The *Dial of Ahaz* thus offers a glimpse of a pre-Copernican hermeneutics of the sun's relation to time.

The Distinctive Form and Function of the *Dial of Ahaz* in the Lempereur, Vorsterman, and Van Liesveldt Bibles

The woodcuts that mark key passages in the most prominent illustrated Bibles published in sixteenth-century Antwerp – Martin Lempereur's *La saincte Bible* of 1530 and 1534, Willem Vorsterman's *Tgeheele oude ende nieuwe Testament* of 1532, 1533/34, 1542/43, and 1545, and Jacob van Liesveldt's *Den Bibel met*

 | DOI:10.1163/9789004677654_003

grooter neersticheyt gecorrigeert of 1532, 1534, 1535, and 1536 – consist for the most part of figured scenes, with the startling exception of the diagrammatic *Dial of Ahaz*, positioned at the start of Isaiah 38 (Figures 1.1–1.3).[1] Aside from their title-pages, which incorporate an allegory of the two laws, the majority of illustrations in these editions eschews allegory, showing either portraits of the authors of the various biblical books or historical scenes.[2] On the contrary, the *Dial of Ahaz*, as will momentarily become evident, is decidedly allegorical. Briefly put, allegory, defined as 'other speaking' (from *allos*, 'other,' and *agoria*, 'speaking'), refers to the use of one thing to signify something variously distant from it: a triangle for the Holy Trinity, for instance, or a pair of scales to symbolise impartial justice. Allegory relies on signifying images, although the notion being signified can be an abstraction. The discursive context – the allegory's circumstances and co-referents – help to determine what the thing visualised actually means. Self-evidently, a symbolic form can have multiple

---

1 I want to thank Prof. Dr. Volker Remmert and Dr. Julia Ellinghaus for their invaluable editorial suggestions.

On the 1532, 1533/34, 1542/43, and 1545 editions of the Van Liesveldt Bible, see Bart Rosier, *The Bible in Print: Netherlandish Bible Illustration in the Sixteenth Century*, vol. 1 (Leiden: Foleor, 1997), 23–24, 73–74, and, for checklists of illustrations, 196–200 (1532), 200–201 (1534), 208 (1535). On the 1532, 1533/34, 1542/43, and 1545 editions of the Vorsterman Bible, see Rosier, *The Bible in Print*, 1:22–23, 1:75, and, for checklists of illustrations, 1:189–196 (1532), 1:200 (1533/34), 1:245–255 (1542/43), and 1:259 (1545). On the 1530 and 1534 editions of the Lempereur (De Keyser) Bible, see Rosier, *The Bible in Print*, 1:19–20, 1:79–80, and, for checklists of the illustrations, 1:178–183 (1530) and 1:210–204 (1534). On marginal glosses and other paratextual elements in the Liesveldt and Vorsterman Bibles, see Wim François, "The 'Glossed Bibles' of Jacob van Liesvelt and Willem Vorsterman (1532–1534ff.)," *Journal of Early Modern Christianity* 6, no. 2 (2019): 147–184; and, more generally, on the readership of these Bibles, see Wim François, "Bible Production and Bible Readers in the Age of Confessionalization: The Case of the Low Countries," in *Lay Readings of the Bible in Early Modern Europe*, ed. Erminia Ardissino and Élise Boillet (Leiden: Brill, 2020), 191–216. On the exegetical function of the woodcut images in the Vorsterman and Lempereur Bibles, see Walter S. Melion, "Scripture for the Eyes: Bible Illustration in the Sixteenth-Century Low Countries," in *Scripture for the Eyes: Bible Illustration in Netherlandish Prints of the Sixteenth Century*, ed. James Clifton and Walter S. Melion, exh. cat., Museum of Biblical Art, New York; Michael C. Carlos Museum, Emory University (London and New York: D. Giles and Museum of Biblical Art, 2009), 18–21. The *Dial of Ahaz* in the Lempereur, Vorsterman, and Van Liesveldt Bibles ultimately derives from medieval cosmological diagrams, on which see Kathrin Müller, *Visuelle Weltaneignung: Astronomische und kosmologische Diagramme in Handschriften des Mittelalters* (Göttingen: Vandenhoeck & Ruprecht, 2008), 29, 80–89.

2 On the title-page of the Van Liesveldt Bible, see Melion, "Scripture for the Eyes," 26–27. On the title-page of the Lempereur Bible, see James Clifton, "A Lutheran Image on the Title-Page of the Last Bible without a Confessional Label," *Ephemerides Theologicae Lovanienses* 84 (2008): 69–86.

meanings, and deciding which meaning is the most pertinent in any given situation requires the interpreter to be informed, judicious, and skillful.

Typical of the author portraits is the full-length image of Joshua, armed and standing before the Israelite encampment, that introduces *Dat boeck Josua*. The compilatory image placed at the hinge between Chapters 4 and 5 of 1 Kings[3] is an example of the historical type: moving from background to foreground, it depicts Philistines slaughtering Israelites at the battle of Aphec (1 Kings 4:10); the judge and high-priest Heli falling dead from his stool when Benjamin brings word that the ark has been captured (1 Kings 4:13–18); the destruction of the god Dagon in his temple at Azotus, where the ark was held captive (1 Kings 5:1–4); the plague of mice that beset the Azotians while the ark was amongst them (1 Kings 5:6); and the temple threshold whereon the priests feared to tread after the ark had shown its power (1 Kings 5:4–5). The specificity of the image, and its close correlation with the text, are characteristic of almost all the images in these bibles: for example, the stump of Dagon's severed neck appears to show flesh and bone because 1 Kings 5:2–3 treats the idol as if it were the living god himself; and the god's head and the palm of one hand lie by the threshold because 1 Kings 5:3 says that it was here that they fell severed, adding that henceforth the temple priests refused to tread upon the doorsill. The extended caption, printed just above, further insists on the narrative imbrication of image and text, attaching the woodcut to the protracted sequence of events that ensues in Chapter 5, 6, and 7: "The Philistines bring the ark into their land and place it in the temple of their idol, which breaks and falls; the ark is transported from city to city, whose inhabitants are punished and terrified."[4]

By contrast, the image of two sundials that prefaces Isaiah 38 exceeds the scriptural text in that it describes an object whose lineaments the prophet barely particularises, and only in passing (Figures 1.1–1.3). The object in question is the eponymous *Dial of Ahaz*, set up by King Hezekiah's apostate father,

---

3 Van Liesveldt, Vorsterman, and Lempereur use chapter divisions that occasionally differ from the King James Version (KJV) – for example, 1 Kings here corresponds to 1 Samuel in the KJV. In this essay I refer to the chapters with the names used by Liesveldt, Vorsterman, and Lempereur. In cases where the KJV is quoted but the chapter names differ, the KJV verse reference is given in the footnotes.

4 Jacob van Liesveldt, ed., *Den Bibel met grooter neersticheyt gecorrigeert, ende op die canten gheset den ouderdom der werelt, ende hoe lange die gheschiedenissen ende historien der Bibelen elck int zijn voor Christus gheboorte gheweest zijn* […] (Antwerp: Jacob van Liesveldt, 1535), fol. M iiij recto: "Die Philistinen brenghen die Arcke in haer lant ende sy stellen die in haers Afgods tempel, die welcke valt af ende breect, die Arcke wort van deene stadt totter anderder ghevoert, die inwoonders worden ghestraft, ende seer verscrict."

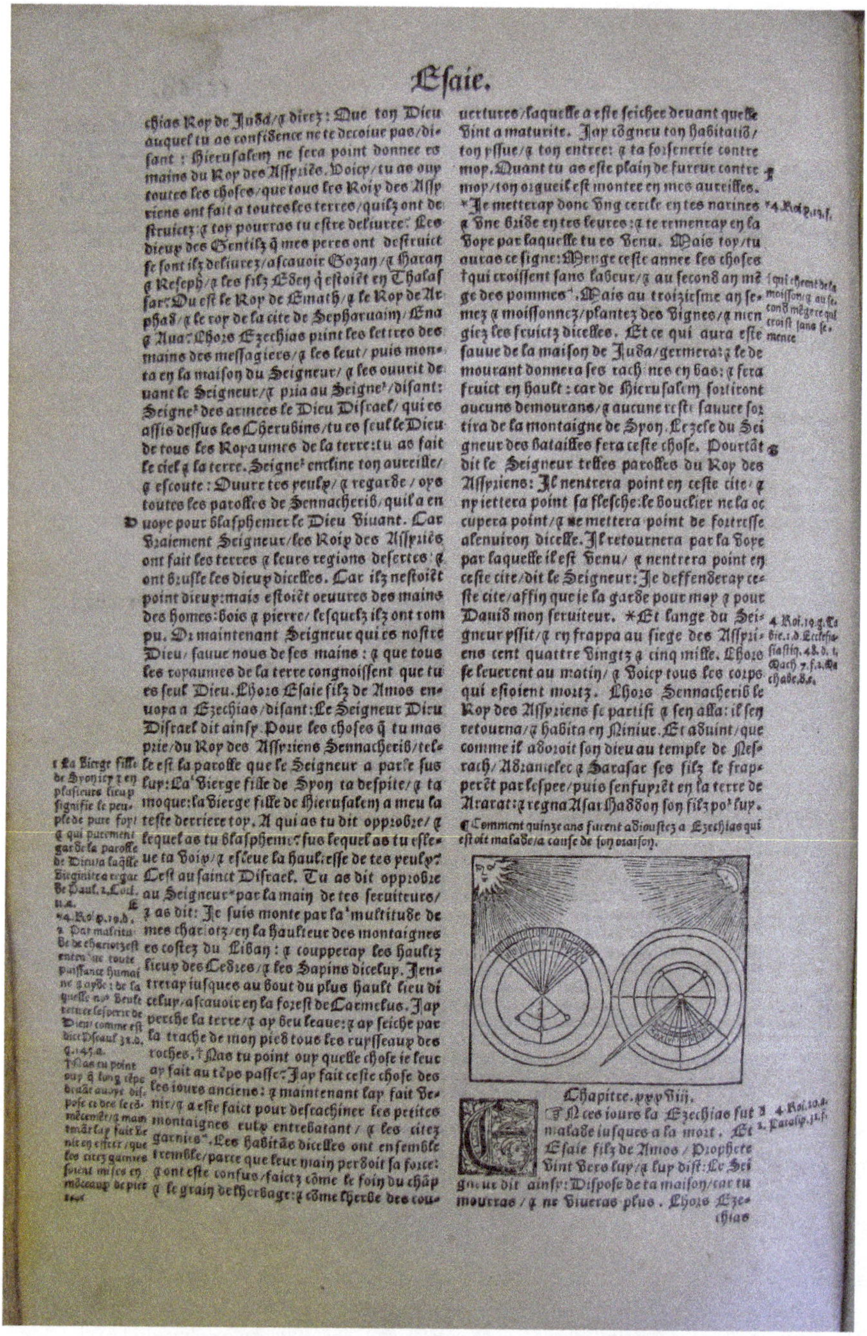

Esaie.

chias Roy de Juda/⁊ direz: Que ton Dieu auquel tu as confidence ne te decoiue pas/disant: Hierusalem ne sera point donnee es mains du Roy des Assyriés. Voicy/tu as ouy toutes les choses/que tous les Roix des Assyriens ont fait a toutes les terres/quilz ont destruictz: ⁊ toy pourras tu estre deliure? Les dieux des Gentilz q̄ mes peres ont destruict les ont ilz deliurez/ascauoir Gozan/⁊ Haran ⁊ Reseph/⁊ les filz Eden q̄ estoiēt en Thalassar? Ou est le Roy de Emath/⁊ le Roy de Arphad/⁊ le roy de la cite de Sepharuaim/Ana ⁊ Aua? Lhors Ezechias print les lettres des mains des messagiers/⁊ les leut/ puis monta en la maison du Seigneur/ ⁊ les ouurit deuant le Seigneur/⁊ pria au Seigneur/disant: Seigneur des armees le Dieu Disrael/ qui es assis dessus les Cherubins/tu es seul le Dieu de tous les Royaumes de la terre: tu as fait le ciel ⁊ la terre. Seigneur encline ton aureille/ ⁊ escoute: Ouure tes yeulx/ ⁊ regarde / oys toutes les parolles de Sennacherib/quil a enuoye pour blaspheme le Dieu viuant. Car vrayement Seigneur/les Roix des Assyriés ont fait les terres ⁊ leurs regions desertes: ⁊ ont brusle les dieux dicelles. Car ilz nestoiēt point dieux: mais estoiēt oeuures des mains des hōmes: bois ⁊ pierre/ lesquelz ilz ont rompu. Or maintenant Seigneur qui es nostre Dieu/ sauue nous de ses mains: ⁊ que tous les royaumes de la terre congnoissent que tu es seul Dieu. Lhors Esaie filz de Amos enuoya a Ezechias/disant: Le Seigneur Dieu Disrael dit ainsy. Pour les choses q̄ tu mas priee/du Roy des Assyriens Sennacherib/celle est la parolle que le Seigneur a parle sus luy: La vierge fille de Syon ta despite/⁊ ta moque: la vierge fille de Hierusalem a meu la teste derriere toy. A qui as tu dit opprobre/⁊ lequel as tu blasphemé? sus lequel as tu esleue ta voix/⁊ esleue la haultesse de tes yeulx? Cest au sainct Disrael. Tu as dit opprobre au Seigneur par la main de tes seruiteurs/ ⁊ as dit: Je suis monte par la multitude de mes chariotz/en la haulteur des montaignes es costez du Liban: ⁊ coupperay les haultz lieux des Cedres/⁊ les Sapins diceluy. Jentreray iusques au bout du plus hault lieu diceluy/ascauoir en la forest de Carmelus. Jay perche la terre/⁊ ay beu leaue: ⁊ ay seiche par la trache de mon pied tous les ruysseaux des roches. Nas tu point ouy quelle chose ie leur ay fait au tēps passe? Jay fait ceste chose des les iours anciens: ⁊ maintenant lay fait venir/⁊ a este faict pour destachiner les petites montaignes entre entrebatant/ ⁊ les citez garnies. Les habitās dicelles ont ensemble tremble/parce que leur main perdoit sa force: ⁊ ont este confus/faictz cōme le foin du chāp ⁊ le grain de lherbage: ⁊ cōme lherbe des couuertures/laquelle a este seichee deuant quelle vint a maturite. Jay cōgneu ton habitatiō/ ton yssue/⁊ ton entree: ⁊ ta forsenerie contre moy. Quant tu as este plain de fureur contre moy/ton orgueil est montee en mes aureilles. *Je mettray donc vng cercle en tes narines ⁊ vne bride en tes leures: ⁊ te remenray en la voye par laquelle tu es venu. Mais toy/tu auras ce signe: Menge ceste annee les choses qui croissent sans labeur/⁊ au second an menge des pommes. Mais au troiziesme an semez ⁊ moissonnez/plantez des vignes/⁊ mengiez les fruictz dicelles. Et ce qui aura este sauue de la maison de Juda/germera: ⁊ le demourant donnera ses rachines en bas: ⁊ fera fruict en hault: car de Hierusalem sortiront aucuns demourans/⁊ aucune reste sauuee sortira de la montaigne de Syon. Le zele du Seigneur des batailles fera ceste chose. Pourtāt dit le Seigneur telles parolles du Roy des Assyriens: Il nentrera point en ceste cite/ ⁊ ny iettera point sa flesche: le bouclier ne la occupera point/⁊ ne mettera point de fortresse alenuiron dicelle. Il retournera par la voye par laquelle il est venu/ ⁊ nentrera point en ceste cite/dit le Seigneur: Je deffenderay ceste cite/affin que ie la garde pour moy ⁊ pour Dauid mon seruiteur. *Et lange du Seigneur yssit/⁊ en frappa au siege des Assyriens cent quattre vingtz ⁊ cinq mille. Lhors se leuerent au matin/ ⁊ voicy tous les corps qui estoient mortz. Lhors Sennacherib le Roy des Assyriens se partist ⁊ sen alla: il sen retourna/⁊ habita en Niniue. Et aduint/que comme il adoroit son dieu au temple de Nesrach/ Adramelec ⁊ Sarasar ses filz le frapperēt par lespee/puis senfuyrēt en la terre de Ararat: ⁊ regna Asarhaddon son filz pour luy.

¶ Comment quinze ans furent adioustez a Ezechias qui estoit malade/a cause de son oraison.

Chapitre. xxxviii.

EN ces iours la Ezechias fut malade iusques a la mort. Et Esaie filz de Amos / Prophete vint vers luy/⁊ luy dist: Le Seigneur dit ainsy: Dispose de ta maison/car tu mourras / ⁊ ne viueras plus. Lhors Ezechias

FIGURE 1.1 *Dial of Ahaz*, in Martin Lempereur, ed., *La saincte Bible en Francoys, translatee selon la pure et entiere traduction de Sainct Hierome, derechief conferee et entierement revisitee selon les plus anciens et plus correctz exemplaires* (Antwerp: Martin Lempereur, 1534), fol. 295v
PHOTO: WALTER MELION, REPRODUCED WITH PERMISSION FROM UNIVERSITEITSBIBLIOTHEEK, BIJZONDERE COLLECTIES, LEIDEN

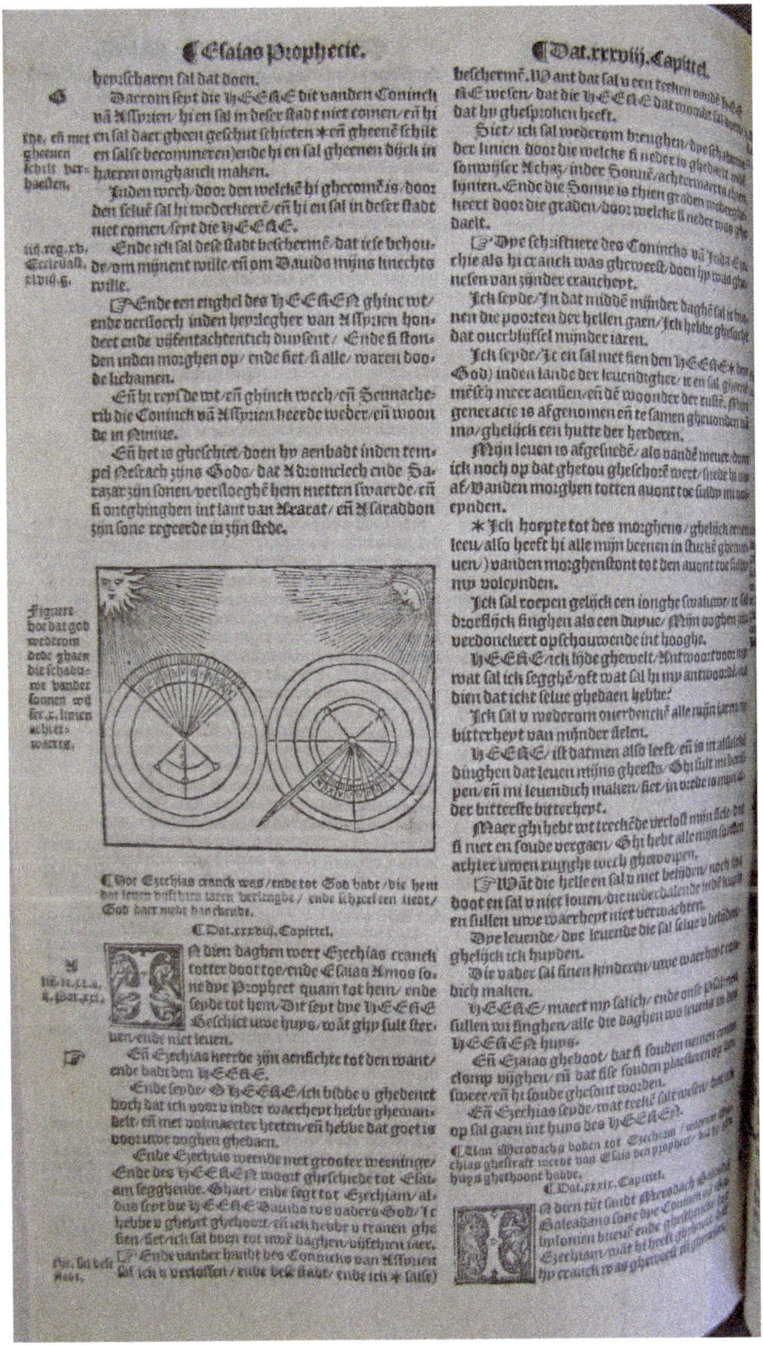

FIGURE 1.2 *Dial of Ahaz*, in Willem Vorsterman, ed., *Den Bibel. Tgeheele Oude ende Nieuw Testament met grooter naersticheyt naden Latijnschen text gecorrigeert* (Antwerp: Willem Vorsterman, 1532), fol. 72v
PHOTO: WALTER MELION, REPRODUCED WITH PERMISSION FROM UNIVERSITEITSBIBLIOTHEEK, BIJZONDERE COLLECTIES, LEIDEN

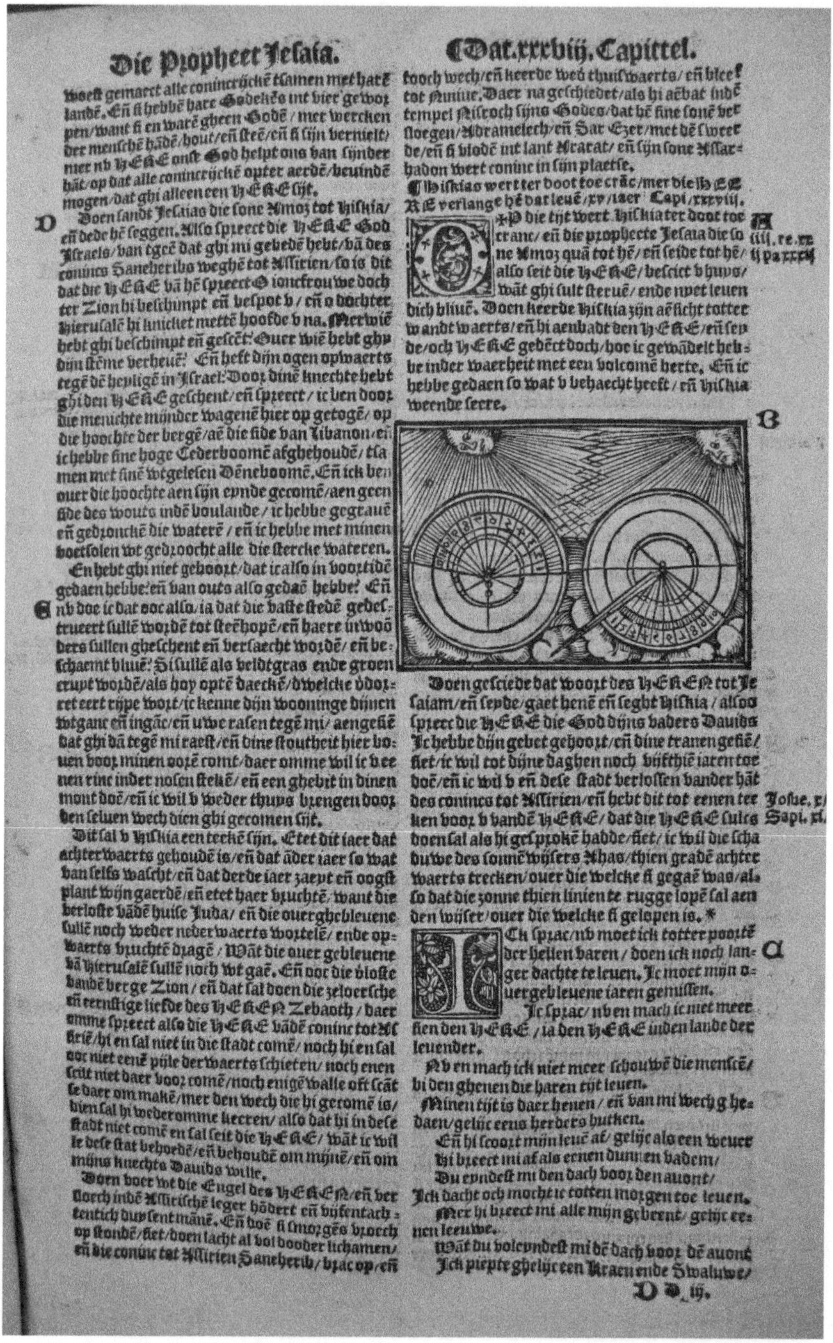

FIGURE 1.3 *Dial of Ahaz*, in Jacob van Liesveldt, ed., *Den Bibel met grooter neersticheyt gecorrigeert, ende op die canten gheset den ouderdom der werelt, ende hoe lange die gheschiedenissen ende historien der Bibelen elck int zijn voor Christus gheboorte gheweest zijn* [...] (Antwerp: Jacob van Liesveldt, 1535), fol. Ddiijr
PHOTO: WALTER MELION, REPRODUCED WITH PERMISSION FROM UNIVERSITEITSBIBLIOTHEEK, BIJZONDERE COLLECTIES, LEIDEN

on the surface of which God signifies, through the retrograde shadow cast by the backward-moving sun, his good intentions toward Hezekiah, Jerusalem, and the Israelites. In Isaiah 38:6–8, God promises to move the sun's shadow ten degrees back upon the dial's face, thus compelling the sun first to retrace its steps, and then, when its forward motion is restored, to re-traverse them; as the Van Liesveldt translation puts it:

> Then Isaiah was visited by the word of the Lord, who told him to go forth and say to Hezekiah, Thus speaks the Lord God of your father David. I have heard your prayer, and seen your tears; behold, I will increase your days by fifteen years, and I will save you and this city from the hand of the king of Assyria. And consider this a sign from the Lord, that he shall do as he has spoken: Behold, I shall draw the shadow of the Dial of Ahaz ten degrees backward from the place it had reached, so that the sun shall again pass through the ten lines it has traversed on the dial.[5]

Serving double duty as a chapter head, the print's inscription distills the evidentiary significance of the prophetic sign: "Hezekiah fell deathly ill, but the Lord extended his life by fifteen years."[6] In turn, as the prophet avows in Isaiah 38:6, this extension of Hezekiah's reign guarantees his people deliverance from the Assyrians and fifteen years of peace and prosperity. The image, then, must be construed as an analogical allegory: the solar miracle, discernible from the clock-face, whereby ten elapsed hours are restored to a day, lengthening it by ten hours (or alternatively, by twenty), is implicitly compared to the miracle of the king's preternatural recovery; as God prolongs the day, so he adds fifteen years to the king's life, presently set to lapse; and so, too, the security of Jerusalem and the kingdom of Judah, formerly threatened with destruction, is restored *in extenso*. In sum, the image of God's sign stands for his gift of a future

5 Van Liesveldt, ed., *Den Bibel met grooter neersticheyt gecorrigeert*, fol. Dd iij recto: "Doe gesciede dat woort des Heren tot Jesaiam, ende seyde gaet henen ende segt Hiskia, also spreect de Here die God dijns vaders Davids. Ic hebbe dijn gebet gehoort, ende dijne tranen ghesien, siet, ic wil tot dijne daghen noch vijfthien iaren toe doen ende ick wil u ende dese stadt verlossen vander hant des conincks tot Assirien [marginal gloss: Josue x. Sapi. xi.], ende hebt dit tot eenen teeken voor u vanden Here, dat die Here sulcs doen sal als hi gesproken hadde, siet, ic wil die schaduwe des zonnen wijsers Ahas, thien graden achterwaerts trecken, over die welcke si ghegaen was, also dat die zonne thien linien te rugghe loopen sal aen den wijser, over die welcke si gheloopen is." 'Gesciede,' which I have translated 'was visited,' could more literally be rendered 'befell.'

6 Van Liesveldt, ed., *Den Bibel met grooter neersticheyt gecorrigeert*, fol. Dd iij recto: "Hiskias wert ter doot toe cranc, mer die Heere verlanghe hem dat leven, xv, iaer."

to his anointed king and people, and, by the same token, it stands warrant for his omnipotent power over time and destiny, and for his merciful benevolence toward supplicants whose tears bear witness to their true remorse.[7]

The print's allegorical form and function are made apparent by the fanciful conceit of showing the sundial twice: seen from directly above, as if from a divine vantage point, the *Dial of Ahaz* in the Van Liesveldt Bible takes the form of two discs hovering over (or amongst) bright clouds, and yet illogically tipped up, as if parallel to the dusky, striated horizon behind them (Figure 1.3). The sun appears twice because the two disks refer to disjunct solar miracles that were, however, connected in the exegetical tradition. Exegesis in the strictest sense is a branch of theology concerned with the systematic interpretation of Scripture. The production and dissemination of printed Bibles in Latin and various vernacular languages in the later fifteenth- and throughout the sixteenth century, together with the provision of scripturally-based sermons and homilies to congregants, and the steadily increasing supply of printed meditative treatises, made simplified forms of exegetical reading available to a wider public. Carefully selected for the purpose of facilitating such reading practices, the images in the Van Liesveldt, Vorsterman, and Lempereur Bibles draw attention to key passages. The *Dial of Ahaz*, precisely because it diverges

7 Soon after, starting around 1570, as Volker Remmert has tellingly shown, a consensus of Jesuit exegetes and astronomer-mathematicians would use the episode from Isaiah 38:6–8, in conjunction with 4 Kings 20:11 and the related miracle of the stopping of the sun in Josuah 10:12–13, to attest the omnipotent power of God and reaffirm the geocentric model ostensibly cast into doubt by Copernicus; see, Volker Remmert, "'Whether the Stars Are Innumerable for Us?': Astronomy and Biblical Exegesis in the Society of Jesus around 1600," in *The Word and the World: Biblical Exegesis and Early Modern Science*, ed. Kevin Killeen and Peter J. Forshaw (New York: Palgrave Macmillan, 2014), 157–173, esp. 158–164. In 1570, as Remmert further demonstrates, Christopher Clavius's *In sphaeram Ioannis de Sacro Bosco commentarius* cites both miracles, along with Psalm 19:4–6, Psalm 104:5, and Ecclesiastes 1:4–6, to argue against the "rectilinear or circular motion" of the earth. And in 1612, the two miracles, now joined by the covenantal rainbow from Genesis 9:12–13 and the new star that leads the Magi to the newborn Christ in Matthew 2:9–10, are incorporated into the allegorical title-page of Clavius's *Opera mathematica*; here they form part of an argument, both exegetical and astronomical, that implicitly equates geocentrism with mathematical truth and construes Copernican heliocentrism as heretical; see Volker Remmert, *Picturing the Scientific Revolution: Title Engravings in Early Modern Scientific Publications* (Philadelphia: St. Joseph's University Press, 2011), 31–39. As illustrated by Johann Leypoldt, designer of the title-page, the format of all four scenes, unlike the Antwerp images discussed in this article, is historical rather than allegorical, even though they are ultimately subsumed into a larger allegorical-exegetical construction. By contrast, the Dial of Ahaz imagery within my purview is explicitly allegorical.

from the narrative conventions and illustrative forms evident in most of the other images, alerts the reader-viewer to its heightened exegetical function.

Positioned at the start of Isaiah 38:4, "Then came the word of the LORD to Isaiah, saying, Go, and say to Hezekiah," the image in question, more than merely illustrating the miracle prophesied in Isaiah 38:7–8 – "And this *shall* be a sign unto thee from the LORD [...]; Behold, I will bring again the shadow of the degrees [...] ten degrees backward" – operates as a kind of comparative gloss, a visual *collatio* that correlates the Isaian passages to the miracle described in Joshua 10:13: "So the sun stood still in the midst of heaven, and hasted not to go down about a whole day" (Figure 1.3). Subdivided into ten numbered segments ordered left to right, the wedge appearing on the lower half of the disk at right depicts the ten degrees across which the sun moved backward. The gnomon's style rests flush against the first retrograde hour-line, and the lines that follow are inscribed counterclockwise to indicate that the sun reversed course. Subdivided into twelve numbered segments or hour lines ordered right to left, the upper half of the disk at left is lit by the sun shining directly above the vertex of the clock face, which has been inverted. As Manfred Kästner points out in his study of earlier printed images of the *Dial of Ahaz*, one of which, Hans Holbein the Younger's, prefaces Isaiah 38 in the so-called *Zwinglibibel* (Zurich: Christoph Froschauer, 1531), the inverted clock face, though dysfunctional as a timepiece, its numbers mirror-reversed, serves perfectly to signify the miraculous stilling of the sun in Joshua 10 (Figure 1.4).[8] In fact, Holbein's image, which he adapted from the woodcut illustrating Isaiah 38 in the *Biblia cum concordantiis veteris et novi testamenti*, printed by Jacob Sacon in Lyons for Anton Koberger of Nuremberg, was the probable source of the reversed print in the Van Liesveldt Bible editions of 1532 and later.

Just as the Van Liesveldt Bible contains marginal glosses directing the reader to consult 4 Kings 20 and Joshua 10, so the image positioned as a chapter head collates visual allusions to both miracles. This is to say that the image functions like an exegetical prompt, urging the reader-viewer to draw parallels amongst the respective verses from Isaiah 38, 4 Kings 20, and Joshua 10, reading them as analogues or, better, as 'common places.' Its position at the start of the sequence of verses wherein God instructs the prophet to tell Hezekiah that

8 Christoph Froschauer, ed., *Die gantze Bibel der ursprünglichen Ebraischen und Griechischen waarheyt nach / auffs aller treüwlichest verteütschet* (Zurich: Christoph Froschauer, 1531). Manfred Kästner, *Die Icones Hans Holbeins des Jüngeren. Ein Beitrag zum graphischen Werk des Künstlers und zur Bibelillustration Ende des 15. und in der ersten Hälfte des 16. Jahrhunderts*, vol. 1 (Heidelberg: Esprint, 1985), 306–307. My thanks to Dr. Julia Ellinghaus for calling this reference to my attention.

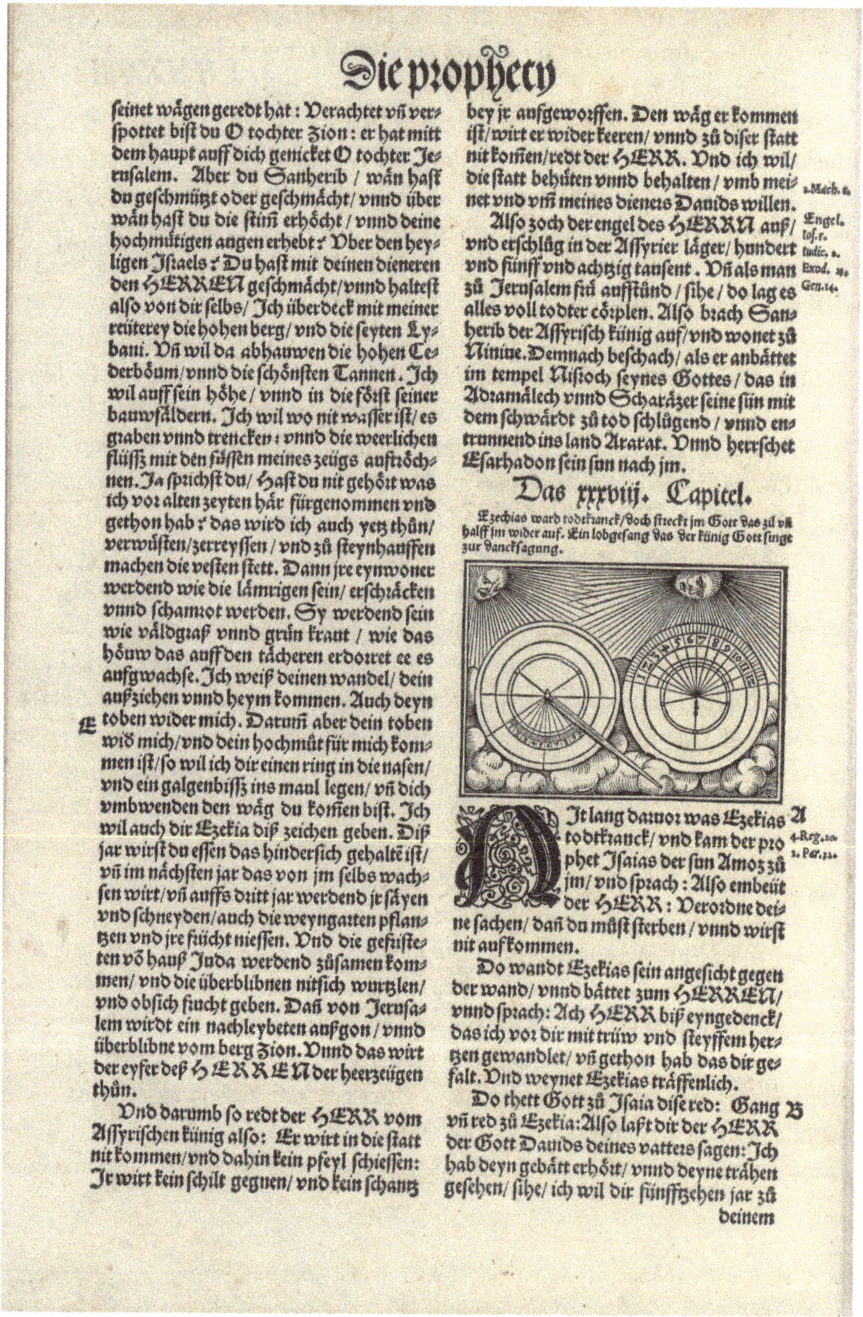
Die prophecy

seinet wägen geredt hat: Verachtet vñ ver-
spottet bist du O tochter Zion: er hat mitt
dem haupt auff dich genicket O tochter Je-
rusalem. Aber du Sanherib / wän hast
du geschmützt oder geschmächt/ vnnd über
wän hast du die stiñ erhöcht / vnnd deine
hochmütigen augen erhebt? Über den hey-
ligen Israels? Du hast mit deinen dieneren
den HERREN geschmächt/vnnd haltest
also von dir selbs/ Ich überdeck mit meiner
reüterey die hohen berg/ vnd die seyten Ly-
bani. Vñ wil da abhauwen die hohen Ce-
derböum/vnnd die schönsten Tannen. Ich
wil auff sein höhe / vnnd in die först seiner
bauwfäldern. Ich wil wo nit wasser ist/ es
graben vnnd trencken: vnnd die weerlichen
flüssz mit den füssen meines zeügs auftröch-
nen. Ja sprichst du/ Hast du nit gehört was
ich vor alten zeyten här fürgenommen vnd
gethon hab? das wird ich auch yetz thůn/
verwüsten/zerreyssen / vnd zů steynhauffen
machen die vesten stett. Dann jre eynwoner
werdend wie die lämrigen sein/ erschräcken
vnnd schamrot werden. Sy werdend sein
wie väldgraß vnnd grün kraut / wie das
höuw das auff den tächeren erdorret ee es
aufgwachse. Ich weiß deinen wandel/ dein
außziehen vnnd heym kommen. Auch deyn
E toben wider mich. Darum̃ aber dein toben
wid mich/vnd dein hochmůt für mich kom-
men ist/so wil ich dir einen ring in die nasen/
vnd ein galgenbissz ins maul legen/ vñ dich
vmbwenden den wäg du kom̃en bist. Ich
wil auch dir Ezekia diß zeichen geben. Diß
jar wirst du essen das hindersich gehaltẽ ist/
vñ im nächsten jar das von jm selbs wach-
sen wirt/vñ auffs dritt jar werdend jr säyen
vnd schneyden/auch die weyngarten pflan-
tzen vnd jre frücht niessen. Vnd die gefriste-
ten võ hauß Juda werdend zůsamen kom-
men/ vnd die überblibnen nitsich wurtzlen/
vnd obsich frucht geben. Dañ von Jerusa-
lem wirdt ein nachleybeten außgon / vnnd
überblibne vom berg Zion. Vnnd das wirt
der eyfer deß HERREN der heerzeügen
thůn.
Vnd darumb so redt der HERR vom
Assyrischen künig also: Er wirt in die statt
nit kommen/vnd dahin kein pfeyl schiessen:
Jr wirt kein schilt gegnen/ vnd kein schantz
bey jr aufgeworffen. Den wäg er kommen
ist/wirt er wider keeren/ vnnd zů diser statt
nit kom̃en/redt der HERR. Vnd ich wil/
die statt behůten vnnd behalten / vmb mei- 2.Mach. 8.
net vnd vm̃ meines dieners Dauids willen.
Also zoch der engel des HERRN auß/ Engel.
vnd erschlůg in der Assyrier läger/ hundert Ios. 5.
vnd fünff vnd achtzig tausent. Vñ als man Iudic. 6.
zů Jerusalem frů auffstůnd / sihe / do lag es Exod. 3.
alles voll todter cörplen. Also brach San- Gen. 14.
herib der Assyrisch künig auf/vnd wonet zů
Niniue. Demnach beschach/ als er anbättet
im tempel Nisroch seynes Gottes / das in
Adramälech vnnd Scharäzer seine sün mit
dem schwärdt zů tod schlůgend / vnnd en-
trunnend ins land Ararat. Vnnd herrschet
Esarhadon sein sun nach jm.

Das xxxviij. Capitel.

Ezechias ward todtkranck/doch streckt jm Gott das zil vñ
halff jm wider auf. Ein lobgesang das der künig Gott singe
zur dancksagung.

[woodcut]

N It lang darvor was Ezekias A
todtkranck/ vnd kam der pro 4.Reg. 20.
phet Isaias der sun Amoz zů 2. Par. 32.
jm/ vnd sprach: Also embeüt
der HERR: Verordne dei-
ne sachen/ dañ du můst sterben / vnnd wirst
nit aufkommen.
Do wandt Ezekias sein angesicht gegen
der wand/ vnnd bättet zum HERREN/
vnnd sprach: Ach HERR biß eyngedenck/
das ich vor dir mit trüw vnd steyffem her-
tzen gewandlet/ vñ gethon hab das dir ge-
falt. Vnd weynet Ezekias träffenlich.
Do thett Gott zů Isaia dise red: Gang B
vñ red zů Ezekia: Also laßt dir der HERR
der Gott Dauids deines vatters sagen: Ich
hab deyn gebätt erhört/ vnnd deyne trähen
gesehen/ sihe/ ich wil dir fünfftzehen jar zů
deinem

FIGURE 1.4 Hans Holbein the Younger [?], *Dial of Ahaz*, in Christoffel Froschauer, ed., *Die gantze Bibel der ursprünglichen Ebraischen und Griechischen waarheyt nach / auffs aller treüwlichest verteütschet* (Zürich: Christoffel Froschauer, 1531), fol. LXXXVIIIv
PHOTO: ZENTRALBIBLIOTHEK, ZURICH, VIII BIS 35, HTTPS://DOI.ORG/10.3931/E-RARA-17064 / PUBLIC DOMAIN MARK

he is about to receive a *signum* (sign) of divine favor – "Hoc autem tibi erit signum a Domino quia faciet Dominus verbum hoc quod locutus est" ("And this shall be a sign to thee from the Lord, that the Lord will do this word which he hath spoken") – indicates that the pictorial subject is explicitly allegorical. The print, instead of narrating the miracle witnessed by Hezekiah, portrays the means and method whereby divine power and intention were made patently discernible to him. Indeed, the source of this image, not only here but also in such precedents as the *Biblia cum concordantiis* and the *Zwinglibibel*, was a key exegetical treatise of the late middle ages, the *Postilla super totam Bibliam* (Postils on the Whole of the Bible) of the Franciscan exegete Nicholas de Lyra.[9] The print's two sundials illuminated by two suns, in what they show, viz., the nature of a divine *signum*, and how they show it, viz., by means of a diagrammatic abstraction, can be seen to emerge from a discursive register that is both exegetical and allegorical. The empyreal placement of the two dials amongst clouds, in the vicinity of the sun, further suggests that the image's manner and meaning are allegorical.

The Van Liesveldt Bible, the text of which was first published in 1526, predates the Vorsterman and Lempereur Bibles, but no Liesveldt edition before 1532 illustrates the *Dial of Ahaz*. The version of the image I have been discussing follows the slightly earlier one in the Lempereur Bible of 1530, but emends it by cleaving closely to Holbein's *Dial of Ahaz* in the Froschauer Bible of 1531. The versions in the Lempereur and Vorsterman Bibles, likely printed from a woodblock shared by the two publishers, derive instead from Sacon-Koberger's *Biblia cum concordantiis* of 1512 (or one of its later editions).[10] Although the sundials in all three versions take essentially the same form, Lempereur and Vorsterman yet show a number of significant divergences from Liesveldt: the change from Froschauer's Arabic to Sacon-Koberger's Roman numerals, the shift in direction of the numbering (with the left-hand dial now moving left to right, and the right-hand dial moving right to left), the substitution of the moon for one of the suns, the tangential positioning of the two disks, and the removal of the clouds (Figures 1.1–1.2). Van Liesveldt's radiant twin suns, each

9 On the importance of Nicholas of Lyra to this illustrative tradition, see Kästner, *Die Icones Hans Holbeins des Jüngeren*, 1:305–306, and my discussion *infra* of Anton Koberger's 1481 edition of the *Postilla*. Further on the *Postilla*, see Edward A. Gosselin, "A Listing of the Printed Editions of Nicolaus de Lyra," *Traditio* 26 (1970), 399–426; Philip D.W. Krey and Lesley Smith, eds., *Nicholas of Lyra: The Senses of Scripture* (Leiden: Brill, 2000); and Gilbert Dahan, ed., *Nicolas de Lyre franciscain du XIVe siècle exégète et théologien* (Paris: Institut des Études Augustiniennes, 2011).

10 On Lempereur's reliance on the *Biblia cum concordantiis*, see Rosier, *The Bible in Print*, 1:19.

aligned with one of the two dial faces, serve to demonstrate that God stayed the sun the space of one day for Joshua and moved it backward then forward ten degrees for Hezekiah. This being said, only the dial at right, marked with ten degrees, can be the *Dial of Ahaz*, who reigned long after Joshua's death and is mentioned for the first time in 4 Kings 16:2 (Figure 1.3).

Vorsterman and Lempereur, on the other hand, utilise the sun and the moon, neither positioned at the vertex of either disk, to convey one temporal phenomenon, whereas the two sundials, the Roman numerals on which appear to track the sun's motion across ten degrees, portray another (Figures 1.1–1.2). The logical conclusion to be drawn from the presence of both heavenly bodies is that the image represents the jointly solar and lunar miracle described in Joshua 10:13: "And the sun stood still, and the moon stayed, until the people had avenged themselves upon their enemies." As the passage of time remains synchronous with the mutual relation between such heavenly bodies as the sun and the moon, so the cessation of time applied equally to both of them. However, the fact that the two dials end with the number 'X' corresponds not with Joshua 10 but with Isaiah 38:8 – "Behold, I will bring again the shadow of the degrees [...] ten degrees backward" – and 4 Kings 20:11: "and he brought the shadow ten degrees backward, by which it had gone down in the dial of Ahaz."[11] In fact, the situation is even more complex, for if the intervals of degree are counted, one discovers nine in the dial at right (the tenth presumably covered by the gnomon) and twelve at the left, which corresponds to the articulation of the two dials in Van Liesveldt: seen in these terms, one dial would refer to the solar miracle of Hezekiah, the other to the miracle of Joshua, with the sun and moon clearly alluding to Joshua.

The two disks touch to intimate how closely connected, indeed identical, they are (if one identifies both as numbering 'X'). The doubling of the *Dial of Ahaz* and the dials' inverted mutual orientation, with one flipped in relation to the other, and the numerical sequence moving left to right (at left), then right to left (at right), serve artificially – by means of two diagrams that stand, somewhat impossibly, for the scribed surface of one and the same sundial – to demonstrate that the sun, having advanced ten degrees per usual, reversed course at God's command, before resuming its normal forward motion. That the two disks are differently scribed, with concentric circles incised at different intervals, and with the numbered hour-lines and quarter-hours set farther out or in, implies that the miracle involved distinct motions of the sun, which

11 2 Kings 20:11 in the KJV. I read the repetition of 'X' three times on the dial at left as a hyperbolic device that insists on the presence of ten degrees only, even though (see *infra*) twelve not ten intervals are actually shown.

behaved unlike itself, even if its normative motion and anomalous countermotion remained methodical and measurable. What the image endeavors to represent is not an actual sundial but notionally a prolongation of time that testifies to God's power over the sun, time, and the course of human affairs.[12] This image, too, in other words, though it alludes to gnomonics and appears to portray a scientific instrument, incorporates instrumental anomalies that signal its allegorical status. Rather than straightforwardly illustrating a biblical episode, it alludes, as I shall now try more fully to explain, to a network of exegetical readings of Joshua 10:11–14, 4 Kings 20:1–11, and Isaiah 38:1–8, that interpet these passages allegorically and complementarily.

Allegory in the Lempereur, Vorsterman, and Van Liesveldt Bibles

The allegorical character of the *Dial of Ahaz* grants it a singular status within the Van Liesveldt Bible, which, apart from the title-page, contains no other explicitly allegorical images (Figure 1.3). In the Vorsterman Bible, again excepting the title-page, there is one additional allegory: Psalm 52, "Dixit insipiens," opens with the image of a fool riding a hobbyhorse, holding a pinwheel instead of a lance, wearing a feathered bonnet in place of a helmet, and barking as mindlessly as the dog racing beside him. Round about, four children closely watch his antics, perhaps preparing to follow his example. As the fool is the *onwijse* (unwise man) "who says in his heart that there is no God," so they are the *menschen kinderen* (children of men), who neither show understanding

12 The dual faces of this Dial of Ahaz type are distantly reminiscent of the Nuremberg vicar-printmaker Georg Hartmann's portable printed paper sundials, which respond in turn to the ivory diptych dials for which the city's instrument makers were famous; the lower face would often incorporate a compass, while the upper face contained the dial. On Hartmann's sundials, see Suzanne Karr Schmidt, *Interactive and Sculptural Printmaking in the Renaissance* (Leiden: Brill, 2017), 205–240, esp. 225–240; on his sundials and those of Johannes Müller, alias Regiomontanus, see Suzanne Karr Schmidt, *Altered and Adorned: Using Renaissance Prints in Daily Life*, exh. cat., Art Institute of Chicago (New Haven: Yale University Press, 2011), 73–82, esp. 77–82. On Hartmann's *Bowl of Ahaz Sundial*, which purports to reproduce, not as a divine but as an artisanal effect, the backward passage of the sun, see Suzanne Karr Schmidt, "Georg Hartmann, *Bowl of Ahaz* Sundial," in *Prints and the Pursuit of Knowledge in Early Modern Europe*, ed. Susan Dackerman, exh. cat., Harvard Art Museums, Cambridge; Mary and Leigh Block Museum of Art, Northwestern University, Evanston (New Haven: Yale University Press, 2011), 296–297, cat. no. 69.

nor attempt to seek God.[13] The lighthearted tone of the image exemplifies the empty-headed godlessness forcefully denounced by the psalm, which warns against the corruption of humankind before the coming of Christ: "For God hath scattered the bones of them that please men."[14] The image, seen in this light, rather than illustrating Psalm 52, instead epitomises the heedless corruption that leads the Psalmist to fulminate against the failure to know and fear God: "They have not called upon God, and they have trembled for fear of that which was not to be feared [viz., everything with the exception of God]."[15] In Lempereur's *La saincte Bible*, a second allegorical image, exegetically quite complex, appears in *La Prophetie de Esaie*. *Christ in the Winepress*, Eucharistic blood pouring from his side wound, illustrates Isaiah 63:3–7, the triumphant prophecy of the Passion, wherein Christ, by submitting to the pressure of the winepress, is seen simultaneously to vanquish his foes and redeem his followers, exact vengeance upon the enemies of Israel and bestow tender mercies upon his people, the house of Jacob: "I have trod the winepress alone, and of all the people, not one was with me. I have trampled them in my fury and tread them down in my wrath. [...] For the day of vengeance is in my heart, and the year of retribution has come. [...] I shall recall the mercy of the Lord, the praise of the Lord for all the things that the Lord has bestowed upon us."[16] A marginal gloss draws a parallel between the prophet's references to vengeful *fureur* and *ire*, and the redemptive *force* and *courage* praised in Psalm 76:15–16, as if to emphasise that the sacrifice of Christ converts the Lord's severity into tender mercy: "Thou art the God that dost wonders. Thou hast made thy power known amongst the peoples: with thy arm thou hast redeemed thy people, the sons of Jacob and of Joseph."[17]

13 Willem Vorsterman, ed., *Den Bibel. Tgeheele Oude ende Nieuw Testament met grooter naerstichheyt naden Latijnschen text gecorrigeert* (Antwerp: Willem Vorsterman, 1532), fol. 9r: "Die onwijse dacht, seggende in zijn herte, daer en is gheen God."

14 Vorsterman, ed., *Den Bibel*, fol. 9r: "Want God heeft verstroeyt die ghebeenten vanden ghenen die die menschen behaghen."

15 Vorsterman, ed., *Den Bibel*, fol. 9r: "Sie en hebben God niet aengheroepen, ende si hebben ghinder van vreese ghebeeft, daer niet te vreesen en was."

16 Martin Lempereur, ed., *La saincte Bible en Francoys, translatee selon la pure et entiere traduction de Sainct Hierome, derechief conferee et entierement revisitee selon les plus anciens et plus correctz exemplaires* (Antwerp: Martin Lempereur, 1534), fol. 303r: "Jay presse le pressoir seul: et de toutes gentz ny a pas ung home avec moy. Je les ay presse en ma fureur: et les ay foulle en mon ire. [...] Car le iour de vengeance est en mon cueur: et lan de ma retribution est venu. [...] Je auray recordation des miserations du Seigneur: et de la louenge du Seigneur, pour toutes ses choses que le Seigneur nous a rendu."

17 Lempereur, ed., *La saincte Bible en Francoys*, fol. 235r: "Tu es le Dieu qui fais les merveilles. Tu as fait que ta puissance a este congneue es peuples: tu as rachete en ton bras ton peuple, les filz de Jacob et de Joseph."

Like these two allegories, the *Dial of Ahaz* is more emotionally and exegetically complex than the majority of images illustrating the three bibles I am discussing (Figures 1.1–1.3). But like the Old Testament images populating them, it answers to the publishers' calls to pay close attention to the books of the patriarchs and prophets who pondered God and foresaw Christ. Van Liesveldt makes his case thus:

> Seeing that the ignorance and simplicity of many men ascribe little or no worth to the text of the holy scriptures of the Old Testament, considering them to have been given to the Jews alone, of no concern to us, spent and having come to naught, taken for nothing more than immoral tales and histories. This in spite of the fact that all throughout them, Christ is attested and promised; and that Christ is wound therein, as if in a winding sheet; and that we shall find him there, as Christ himself bears witness in John 5[:39]: "Search through Scripture, for it testifies to me."[18]

Vorsterman declares that his editors, keen to "correct errors of interpretation that have taken root in the hearts of Christian persons," have consulted the Hebrew text, inserting Dutch translations from it, in the form of judicious glosses, "by way of an alternative" to the Vulgate ("in een maniere van alter atien").[19] And Lempereur, with reference to the Hebrew, Greek, and Chaldaic sources of the canonical books of the Old Testament, counsels his readers, that he has glossed the principal *tropus, ou parolle figurative* (tropes or figurative

---

18 Van Liesveldt, ed., *Den Bibel met grooter neersticheyt gecorrigeert*, fol. A + i verso: "Aenghesien dat die ongheleertheyt ende simpelheyt van veel menschen, den text der heyligher scriftueren des ouden Testaments, seer cleyn achten, ende als niet daer af en houden meynende dat alsulcken schriftuere alleen den Joden ghegeven si, ende ons niet aen en ghae, ende dat het oude Testament heel wt ende te nieute si, ende dattet alleen niet dan voor slechte geschiedenissen ende historien te verstaen en is. Niet teghenstaende dat ons Christus gheheelick daer door bewesen ende belooft wort. Ende dat oude Testament is ons als die wendeldoecken daer Christus in gewonden leyt, ende daer we hem oock in vinden sullen, also ons Christus ooc selve betuycht, Joannis int v. seggende. Ondersoect die scriftuere, want die selve geeft ghetuygenisse van mi."

19 Vorsterman, ed., *Den Bibel. Tgeheele Oude ende Nieuwe Testament*, fol. A vj recto: "Hierom, op dat een yghelic, dye in die heylighe schrift naerstich is, ende die begheert te weten, wat den Hebreeuschen text in heeft, ende oock na dat ons bevolen was den Bibel, tot profite der ghemeynder christenen menschen, om die dwalinghe wter herten te keeren (die welcke nu, Got betert, zeer heeft gheregneert) te corrigeren, om te doen drucken, so salmen de Hebreeuschen text, int duytsche ghestelt, in een maniere van alteratien, opten cant vinden, oftmen daer claerder verstant wt mocht hebben, ende niet om eenighe nieuwe opinien, te genereren."

terms); whenever obscure passages are encountered, one must try to understand the context in which they appear, and also clarify their meaning "by concordance with [passages whose meaning is] more evident," searching through the matrix of Scripture as a whole.[20] What I now hope to suggest is that the *Dial of Ahaz* was designed to function as a node through which multiple scriptural strands intersect, the nature of God is made manifest, and the presence of Christ is discerned *in nuce*.

Derivation of the *Dial of Ahaz* from the Exegetical Tradition

The *Dial of Ahaz* derives not from the tradition of biblical illustration but from that of exegetical commentary: the source images first appeared in Anton Koberger's edition of the *Postilla super totam Bibliam*, published in Nuremberg on 22 January 1481 (Figure 1.5). The two dial faces were adapted from the double illustration of Nicholas de Lyra's exposition on 4 Kings 20:9, where Isaiah questions Hezekiah, asking which species of sign he prefers: "And Isaiah said to him: This shall be the sign from the Lord, that the Lord will do the word which he hath spoken: Wilt thou that the shadow go forward ten lines, or that it go back so many degrees"?[21] The first image, above, depicts a hypothetical model of the *Dial of Ahaz* as it would have looked were two lines scribed for each hour: "There follows a figure according to the first opinion whereby two lines make one hour."[22] The second image, below, depicts an alternative model,

20 Lempereur, ed., *La saincte Bible en Francoys*, fol. v verso: "mais fault ques les passages obscurs soient clarifiez par autres concordans et plus evidentz. [...] Secondement fault necessairement regarder les parolles precedentes bien conferees a celles qui suyvent, en diligemment scrutinant de qui, ou a qui elles parlent, en quel temps et pour quelle cause: tellement que scavoir les circonstances des escriptures nest point petite introduction en icelles. [...] Mais en ce lieu avons notez les plus principaulz passages et lesdictes manieres de parlers (que nous appellons Tropus, ou parolle figurative) avec les diverses acceptions de plusieurs motz."

21 Bonifatius Fischer and Robert Weber, eds., *Biblia sacra: iuxta Vulgatam versionem* (Stuttgart: Deutsche Bibelgesellschaft, 1994): 4 Regum 20:9: "Cui ait Esaias hoc erit signum a Domino quod facturus sit Dominus sermonem quem locutus est vis ut accedat umbra decem lineis an ut revertatur totidem gradibus." Cf. Bishop Challoner, ed., *The Holy Bible, translated from the Latin Vulgate and diligently compared with the Hebrew, Greek, and other editions in divers languages* (Fitzwilliam, NH: Loreto, 2013), 360.

22 Anton Koberger, ed., *Postilla super totam Bibliam* (*cum expositionibus Guillelmi Britonis et additionibus Pauli Burgensis et correctoriis editis a Matthia Doering*) (Nuremberg: Anton Koberger, 22 January 1481), n.p. [Regum quartus. Cap. xxj. 'In diebus illis']: "Sequitur figura secundum primam opinionem in qua du[a]e line[a]e faciunt unam horam."

FIGURE 1.5 *Two dial faces*, from Anton Koberger, ed., *Postilla super totam Bibliam* (*cum expositionibus Guillelmi Britonis et additionibus Pauli Burgensis et correctoriis editis a Matthia Doering*) (Nuremberg: Anton Koberger, 22 January 1481)
PHOTO: WALTER MELION, REPRODUCED WITH PERMISSION FROM UNIVERSITEITSBIBLIOTHEEK, BIJZONDERE COLLECTIES, LEIDEN

in which each hour is scribed by one hour-line: "The second figure according to the second opinion whereby each line makes one hour."[23] Additionally, the first image is inscribed with a reference to Lyra's discussion of the solar miracle performed at Joshua's behest in Joshua 10: "But whether that day under Hezekiah was more protracted than the one in the time of Joshua, has been discussed [with respect to] Joshua 10[:13: 'And the sun stood still, and the moon stayed, until the people had avenged themselves upon their enemies']."[24] The inscription beneath the second image elaborates upon the first: "It should be noted that these two figures were fashioned after the method posited with respect to Joshua 10, namely, that during reign of Hezekiah, the day was prolonged at the time of the winter solstice, and was thus shorter than the prolongation of the day under Joshua."[25] The two Dials of Ahaz in Koberger's *Postilla* codified in printed form the many appearances of the two clocks in glossed manuscript bibles, such as the so-called *Bible of Jean, Duc de Berry*, produced in France between 1388 and 1390 (Figures 1.6–1.7).[26] Starting with Koberger, these illustrative diagrams became standard in printed editions of the Bible with the canonical glosses: for example, in the *Biblia Sacra cum glossa ordinaria*, published in Paris in 1590, they supplement Isaiah 38; in the Venice edition of 1603, *Bibliorum Sacrorum cum Glossa ordinaria*, one pair of woodcuts is appended to Isaiah 38, and a different pair amplifies 4 Kings 20 (Figures 1.8–1.9).[27] The

23 Koberger, ed., *Postilla super totam Bibliam*: "Figura secundum secundam opinionem in qua qu[a]elibet linea facit unam horam."

24 Koberger, ed., *Postilla super totam Bibliam*: "Utrum autem ista dies sub Ezechia fuerit maior, quam illa qu[a]e prolongata fuerit tempore Iosue, dictum fuit Iosue 10. cap."

25 Koberger, ed., *Postilla super totam Bibliam*: "Notandum quod ist[a]e du[a]e figur[a]e fact[a]e sunt secundum unam viam quam posuit Iosue decimo[,] scilicet quod dies prolongata sub Ezechia fuerat in solstitio hyemali, et sic brevior die prolongata sub Iosue."

26 See Vatican City, Biblioteca Apostolica Vaticana, MS Vat. lat. 50, https://digi.vatlib.it/view/MSS_Vat.lat.50, specifically 191v and 192r. On this manuscript, see Kristian Jensen, ed., *Incunabula and their Readers: Printing, Selling, and Using Books in the Fifteenth Century* (London: The British Library, 2003), 236 note 81.

27 See Jacques de Cuilly, François Feuardent, and Jean de Dadré, eds., *Biblia Sacra, cum glossa ordinaria* [...] *et postilla Nicolai Lyrani*, 6 vols. (Paris: Jean-Baptiste Regnauld, 1590); and Jacques de Cuilly, François Feuardent, and Jean de Dadré, eds., *Bibliorum Sacrorum cum Glossa ordinaria iam ante quidem a Strabo Fulgensi collecta: nunc autem novis, cum Graecorum, tum Latinorum Patrum expositionibus locupletata*, 6 vols. (Venice: [s.n.], 1603). On the exegetical form and function of the *Glossa ordinaria*, see Lesley Smith, *The 'Glossa Ordinaria': The Making of a Modern Bible Commentary* (Leiden: Brill, 2009); on early modern editions of the *Glossa*, see Karlfried Froehlich, "The Printed Gloss," in *Biblia Latina cum glossa ordinaria: Facsimile Reprint of the Editio Princeps Adolph Rusch of Strassburg 1480–81*, (Turnhout: Brepols, 1992), 1:xii–xxvi; Froehlich, "Walafrid Strabo and the *Glossa ordinaria*: The Making of a Myth," *Studia Patristica* 28 (1993): 192–196; Froehlich, "An Extraordinary Achievement: The *Glossa ordinaria* in Print," in *The Bible as Book: The*

differences in the appearance of the two dial-faces in the Van Liesveldt, Vorsterman, and Lempereur Bibles derive from the two alternatives presented by Lyra, as illustrated in Sacon-Koberger's *Biblia cum concordantiis* where they are combined and conflated: in Van Liesveldt's *Dial of Ahaz*, the two disks represent two clock faces, one related to Joshua 10, the other to Isaiah 38 and 4 Kings 20; in Vorsterman's and Lempereur's versions, they become two views of the same clock. What had been mutually exclusive models of the *Dial of Ahaz* become simultaneous if incongruous views of a single dial face.

Complementary to Lyra's alternative views of the clock in Koberger's *Postilla* is the diagram of the clock face which accompanies Paulus Burgensis's extended *Additio* to Lyra's exposition of 4 Kings 20:9 (Figure 1.10). Likewise assimilated into printed editions of the *Glossa*, in particular the seven editions of the *Postilla super totam Bibliam* published by Anton Koberger in Nuremberg between 1479 and 1497, and frequently copied thereafter, this image was designed to display as clearly as possible how the clock would have operated, were it similar to an equatorial sundial organised around a perpendicular gnomon. The long supporting text begins: "In which figure the clock of Ahaz is composed of a certain column planted perpendicularly upon the equidistant base of the horizon."[28] The allusion to a meridian line in the left-hand disc of Van Liesveldt's illustration, and in both of Vorsterman's and Lempereur's disks, derives from this illustrative image first codified in print by Koberger, as also does the sequential marking of the hourly segments – lettered in Burgensis, numbered in the Bibles. Vorsterman's and Lempereur's solar face likewise stems from this prototype (Figures 1.1–1.3).

The exegetical origins of the *Dial of Ahaz* in the Van Liesveldt, Vorsterman, and Lempereur Bibles help to account for its interpretative complexity; they also explain how and why it works as a locus of allegorical entanglement, whose meaning attaches not only to Isaiah 38, but also to other scriptural loci closely associated with this place. In the Van Liesveldt Bible, the parallel passages to be read in tandem with Isaiah 38 are designated in marginal glosses as 4 Kings

First Printed Editions, ed. Kimberly van Kampen and Paul Henry Saenger (London: The British Library, 1999), 15–22; Froehlich, "The Fate of the *Glossa ordinaria in the Sixteenth Century*," in *Die Patristik in der Bibelexegese des 16. Jahrhunderts*, ed. David C. Steinmetz (Wiesbaden: Harrassowitz, 1999), 19–47; and Froehlich, "Makers and Takers: The Shaping of the Biblical *Glossa ordinaria*," in *Biblical Interpretation from the Church Fathers to the Reformation* (Farnham, Surrey: Ashgate, 2009), 1–19.

28 Koberger, ed., *Postilla super totam Bibliam* [Regum quartus. Capi. xxv.: "Vis ut ascendat umbra, id est, ut procedat ultra"]: "In qua figura horalogium achaz componitur ex quadam columna. Perpendicularit posita supeer basim orizontis eque distantemque quaedem columna sit."

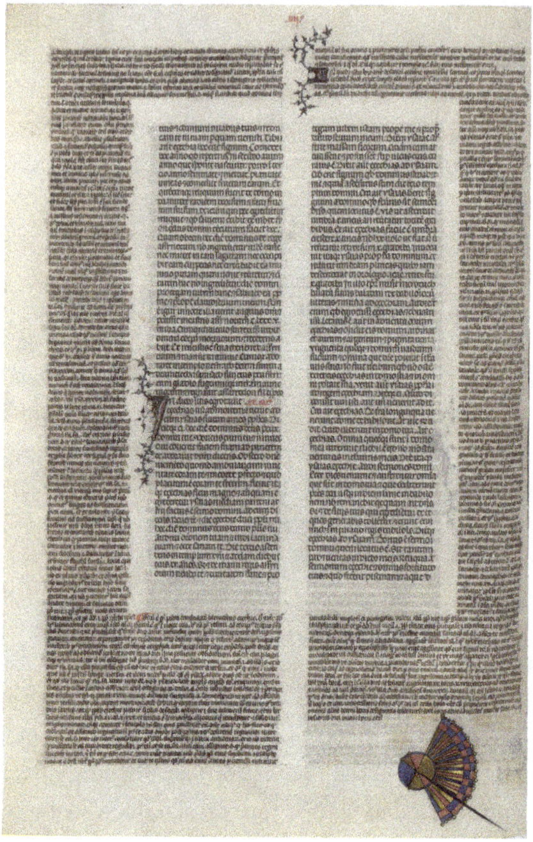

FIGURE 1.6
Dial of Ahaz, in the so-called *Bible of Jean, Duc de Berry*, ca. 1388–1390
PHOTO: BIBLIOTECA APOSTOLICA VATICANA, MS VAT. LAT. 50, FOL. 191V

20, 2 Paralipomenon 32, Joshua 10, and Wisdom of Solomon 11 (Figure 1.3). In the Vorsterman Bible, they are 4 Kings 20a [= verses 1–11], 2 Paralipomenon 32, 1 Kings 2g [= verse 34], Luke 2c [= verse 12], Ecclesiasticus 48d [= verse 26], and Job 4d [= verse 10] (Figure 1.2). In the Lempereur Bible, they are 4 Kings 20a [= verses 1–11], 2 Paralipomenon 32f [= verse 24], Psalm 50b [= verses 8–14], 1 Kings 2g [= verse 34], Luke 2b [= verse 12], Ecclesiasticus 48d [= verse 26], Joshua 10c [= verses 12–14], Job 4d [= verse 10], Psalm 6a [= verses 5–8], Psalm 113d [= verses 12–13], and Habacuc 3a [= verses 2–5] (Figure 1.1). Once identified as an allegorical diagram, the *Dial of Ahaz* can be seen to function as the hermeneutic fulcrum through which these various passages pivot in a mutually qualifying relation of parataxis, as if each were a marginal gloss to all. The choice of scriptural parallels, largely codified in the *Glossa ordinaria et interlinearis*, was further amplified by Nicholas de Lyra in the fourteenth century, and in the fifteenth by Paulus Burgensis in his *Additiones* and Matthias Torringus

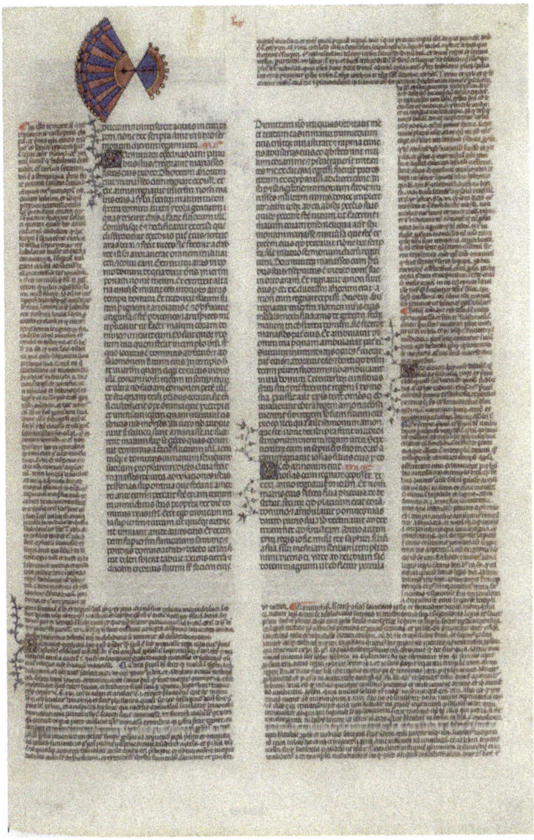

FIGURE 1.7
Dial of Ahaz, in the so-called *Bible of Jean, Duc de Berry*, ca. 1388–1390
PHOTO: BIBLIOTECA APOSTOLICA VATICANA, MS VAT. LAT. 50, FOL. 192R

in his *Defensorium postillae Nicolai Lyrani*. The *Dial of Ahaz*, parsed in light of these exegetical commentaries, distills the gist of their shared relation to the solar miracle narrated by Isaiah.

I want now to examine the connection between these pericopes, the exegetical texts underlying them, and the image of the miracle: that the texts can all be tied to a single image confederates them, underscoring their associated status, and identifying them as a kind of scriptural-exegetical intertext. Let us begin with the print's relation to Isaiah 38 and its attendant glosses. With reference to Isaiah 38:7, "And this *shall* be a sign unto thee," and to Lyra's postil thereon, Paulus Burgensis offers a reading of the miraculous sign that can equally be applied to the accompanying image, which is, after all, itself a *signum*; both sign and image are open to multiple interpretations, within the limits set by doctrinal and exegetical decorum:

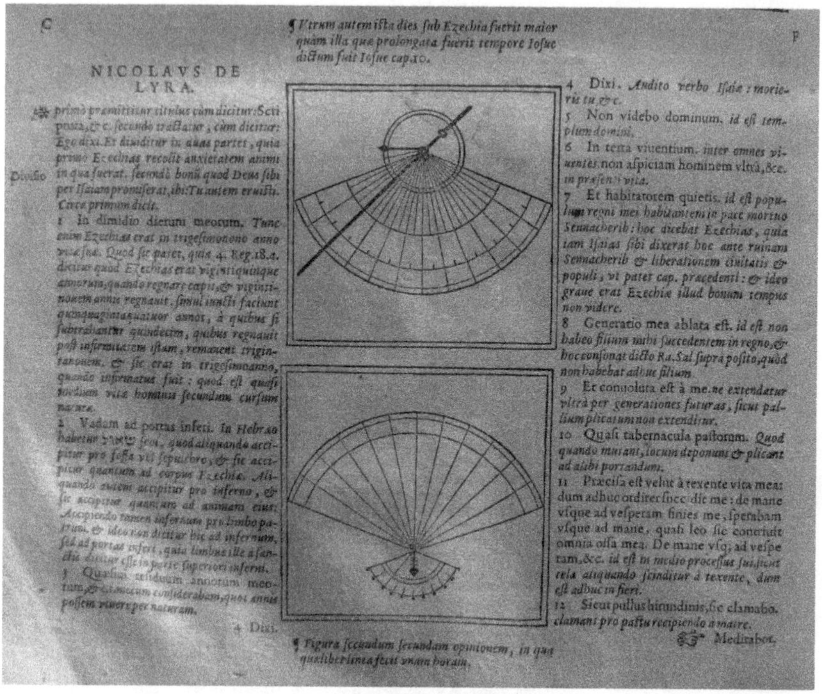

FIGURE 1.8 *Dial of Ahaz*, in Jacques de Cuilly, François Feuardent, and Jean de Dadré, eds., *Biblia Sacra, cum glossa ordinaria* [...] *et postilla Nicolai Lyrani.* 6 vols. (Paris: Jean-Baptiste Regnauld, 1590), 4: cols. 339–340
PHOTO: WALTER MELION, REPRODUCED WITH PERMISSION FROM UNIVERSITEITSBIBLIOTHEEK, BIJZONDERE COLLECTIES, LEIDEN

Although the common explanation by the holy doctors of this sign made to Hezekiah corresponds to that which the postil expounds, many things yet occur therein worthy of the highest admiration, as has been fully shown in the *additio* to 4 Kings 20, where this same sign has similarly been discussed. Wherefore by authority of the saints, and without any presumptive assertion, it appears right that this sign be variously explained, for the purpose of calling forth admiration, as mentioned above, even while cleaving to the literal truth.[29]

29 Cuilly, Feuardent, and Dadré, *Bibliorum Sacrorum cum Glossa ordinaria*, 4: col. 343: "Licet communis expositio doctorum sanctorum circa hoc signum factum Ezechiae sit conformis huic quod exponitur in postilla, multa tamen occurrunt in hoc maxima admiratione digna, prout in additione super 4. Reg. 20. cap. ubi de hoc eodem signo similiter tractatur fuit large ostensum. Unde ad praedictas admirationes tollendas, et ad veritatem

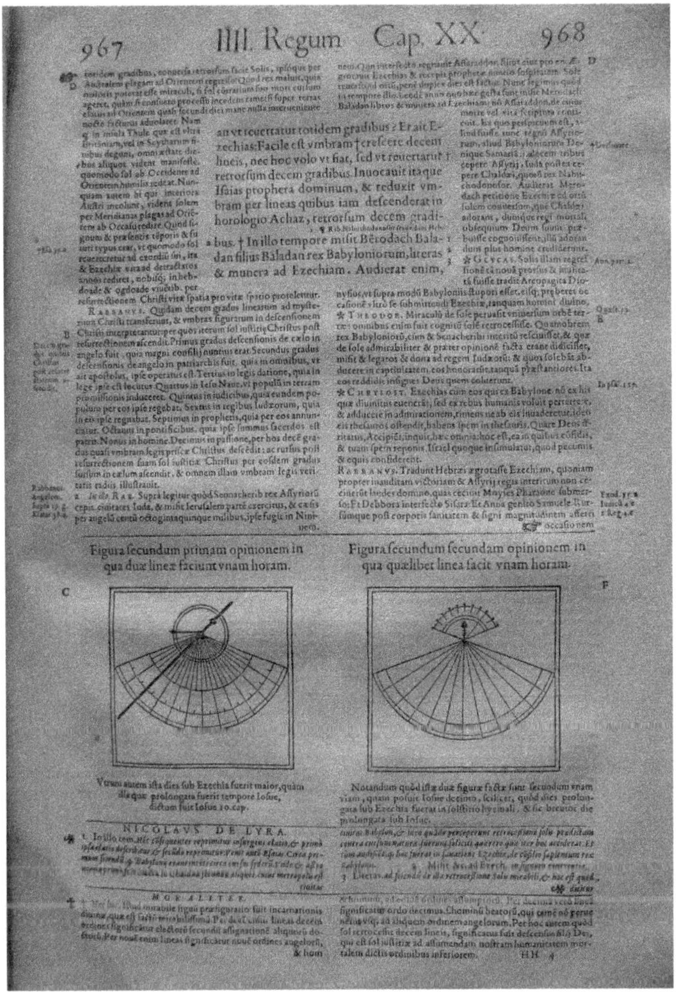

FIGURE 1.9 *Dial of Ahaz*, in Jacques de Cuilly, François Feuardent, and Jean de Dadré, eds., *Bibliorum Sacrorum cum Glossa ordinaria iam ante quidem a Strabo Fulgensi collecta: nunc autem novis, cum Graecorum, tum Latinorum Patrum expositionibus locupletata*. 6 vols. (Venice, 1603), 2: cols. 967–968
PHOTO: WALTER MELION, REPRODUCED WITH PERMISSION FROM UNIVERSITEITSBIBLIOTHEEK, BIJZONDERE COLLECTIES, LEIDEN

Thus seen, the image signifies its own semiotic multivalence: it can be read as an allegory of the richness of meaning to be discovered in the divine portents scripturally transmitted by the Lord through his prophets (Figures 1.1–1.3).

The Glossist, Lyra, and Paulus Burgensis elucidate the semiotic richness of the miraculous sign by dwelling on its bifold structure: in moving forward, then

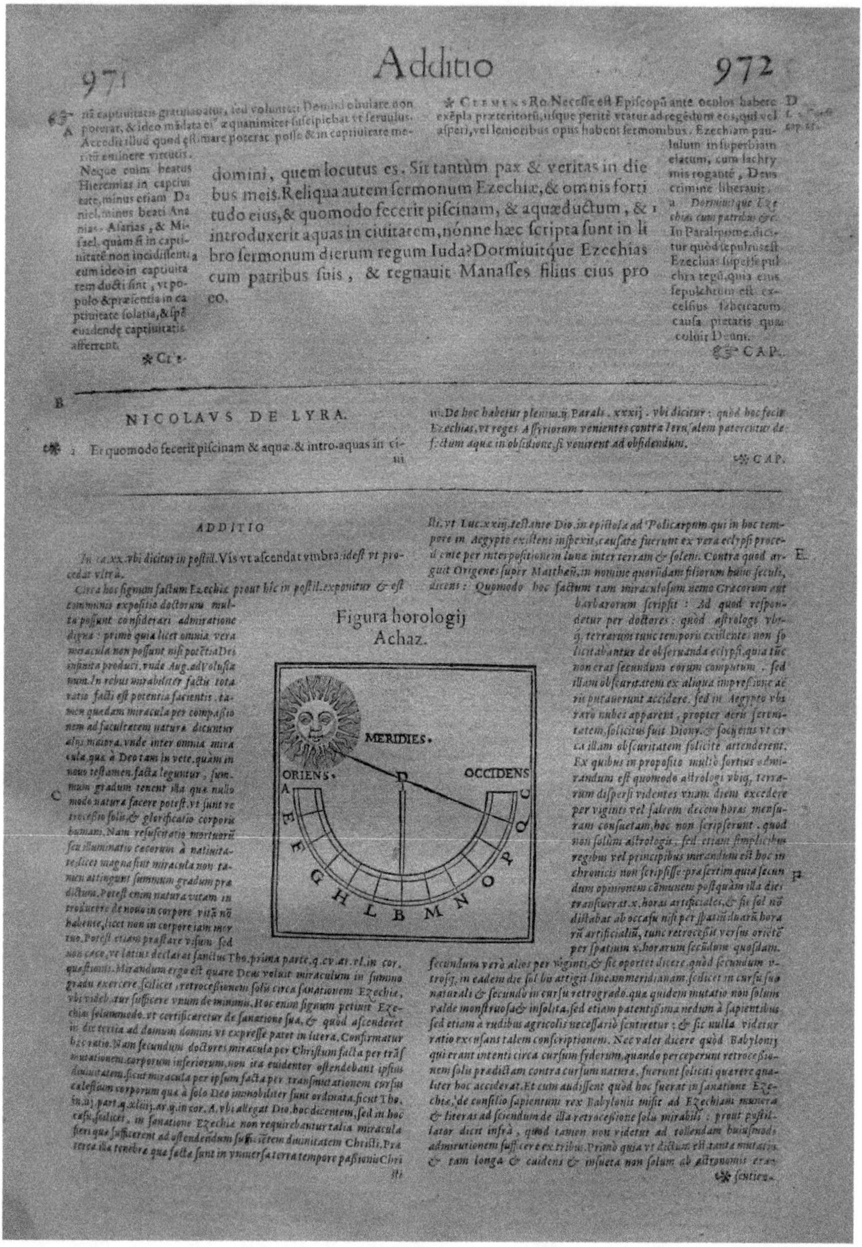

FIGURE 1.10 *Clock face*, in Jacques de Cuilly, François Feuardent, and Jean de Dadré, eds., *Bibliorum Sacrorum cum Glossa ordinaria iam ante quidem a Strabo Fulgensi collecta: nunc autem novis, cum Graecorum, tum Latinorum Patrum expositionibus locupletata.* 6 vols. (Venice, 1603), 2: cols. 971–972

PHOTO: WALTER MELION, REPRODUCED WITH PERMISSION FROM UNIVERSITEITSBIBLIOTHEEK, BIJZONDERE COLLECTIES, LEIDEN

backward, the sun stages a dual analogy to/for Hezekiah; his life was moving to its close, but now he recovers fifteen additional years, as if his life were being reset and his youth restored. The Glossist coins the term *congrua dispositio* (congruant disposition) to refer to this twofold analogy between the sun's two motions and the two motions – surcease and increase – of Hezekiah's reign. The dual form of the image – one disc standing for anterograde, the other for retrograde movement – accommodates this textual imagery of temporal progression and retrogression, functioning pictorially as its visual trope:

> Fitly conveyed to the king expectant of death, such a sign, which the Lord by congruant disposition caused the sun to fashion, befell most aptly, since just as the sun poised at the threshold of setting was returned to daybreak, so the king prepared for death was restored to the joy of incipient life.[30]

With reference to Isaiah 38:10, "I said in the cutting off of my days," the opening lines of the Canticle of Hezekiah (his prayer of thanks for the gift of newfound life), the two discs can also be read as an allusion to God's merciful doubling of a penitent sinner's days. The Glossist compares Hezekiah to Abraham who died only after having lived into ripe old age, and contrasts him with the unrepentant sinner of Psalm 54:24, whose days were halved: "Bloody and deceitful men shall not live out half their days."[31] The two discs, then, signify by congruant disposition both the renewal and the extension of Hezekiah's life, the course of which is first redrawn and then prolonged. Furthermore, when the image is coupled with the Canticle of Hezekiah (Isaiah 38:9–20), it operates phenomenologically as an allegory of the king's consciousness of the passage of time, its loss and repair, as Lyra states in his postil on Isaiah 38:10, "I am deprived of the residue of my years": Hezekiah, he argues, is bringing time to mind ("idest

literae prout plane sonat prosequendam, salva semper sanctorum autoritate et absque aliqua temeraria assertione, hoc signum aliter videtur exponendum, prout ibidem plane fuit declaratum."

30 Cuilly, Feuardent, and Dadré, *Bibliorum Sacrorum cum Glossa ordinaria*, 4: col. 339: "Convenienter autem tale signum regi in mortis expectatione posito, aptissimum evenit, quod dominus per solem fieri taliter, per solem congrua dispensatione disposuit, quoniam sicut sol in procinctu occasus sui positus, in diei initium reducitur, sic rex in mortis expectatione constitutus, quasi ad incipientis vitae gaudia revocatur."

31 Cuilly, Feuardent, and Dadré, *Bibliorum Sacrorum cum Glossa ordinaria*, 4: cols. 339–340: "Abraham, mortuus est plenus dierum senectute bona, peccatores in dimidio dierum moriuntur. [...] Viri impij et dolosi non dimidiabunt dies suos, quia non implent opera virtutum, nec poenitendo emendant delicta."

mecum considerabam"), pondering the flow of years in natural time ("quot annis possem vivere per naturam").[32] This reading is in line with the Glossist's account of Isaiah 38:12, "Mine age is departed, and is removed from me as a shepherd's tent," which he interprets as a prefiguration of the Pauline admonition, "the time *is* short," in 1 Corinthians 7:29. Like Paul, Hezekiah is urging us not to ignore time's transit or passing ("non ergo transitum vel decursum eius ignoramus").[33] Viewed in these terms, the allegory expands to encompass not only the movement of time, but also Hezekiah's acutely felt experience of time's coming and going.

Self-evidently, the *Dial of Ahaz*, in signifying the passage of time and God's alteration thereof, must also represent the Lord's power over nature and human affairs. The Glossist makes this point in his postil on Isaiah 38:15, "What shall I say? he hath both spoken unto me, and himself hath done *it*," by juxtaposing to it Romans 9:16, on the causative mercy of God who is the sole agent of all blessings: "So then it is not of him that willeth, nor of him that runneth, but of God that showeth mercy."[34] The Glossist's avowal of divine mercy complements his earlier reading of Isaiah 38:5, "behold, I will add unto thy days fifteen years," where he remarks that God's omnipotence is everywhere manifest in the books of Kings and Paralipomenon, to such an extent that history, the record of past events, and prophecy, the prediction of events to come, become virtually indistinguishable, as here in the case of Hezekiah. Calling it *praescientia* (presience, foreknowing), the Glossist defines this mixture of past and future as the distinguishing feature of divine omnipotence; whereas Hezekiah believed his life to have been extended by fifteen years, for God, who is all-knowing, and for whom nothing is contingent or adventitious, those fifteen years were preordained:

> 'Behold I will add to thy days fifteen years.' According to what was perceived [by Hezekiah]. For in the [divine] prescience within which these years existed, nothing was added. God foreknew that [Hezekiah] would be afflicted by infirmity of the spirit even unto death, but would convalesce through prayer and penitence. [...] History is mixed with prophecy in the book[s] of Kings and Paralipomenon, and prophecy is conjoined with history: [the prophecy/history] of the liberation of the

32 Cuilly, Feuardent, and Dadré, *Bibliorum Sacrorum cum Glossa ordinaria*, 4: col. 339.

33 Cuilly, Feuardent, and Dadré, *Bibliorum Sacrorum cum Glossa ordinaria*, 4: col. 340.

34 Cuilly, Feuardent, and Dadré, *Bibliorum Sacrorum cum Glossa ordinaria*, 4: col. 341: "Quod erravi emendabo, tu esto fide iussor et protector. Non est enim volentis neque currentis, sed Dei miserentis."

> city, of the fall of Assyria, of the reversion of the sun, of the years reckoned to Hezekiah.[35]

The liberation of Jerusalem and the fall of the Assyrian king Sennacherib take place in chapter 37, but are prophesied in chapter 38, because, as Lyra notes in his postil on Isaiah 38:22, "Hezekiah also had said, What *is* the sign," Scripture, like the solar miracle recorded by Isaiah, and like God himself, author of the miracle, is not bound by the natural order of time ('ordo temporis').[36] And the Glossist writes, with reference to Isaiah 38:1, "for thou shalt die," and 38:5, "I will add unto thy days," that God does not so much issue threats and then change his mind; rather, he strategically provokes us, as he did the negligent Hezekiah, to keep the Lord ever in one's thoughts.[37] It is therefore crucial, when thinking about God's treatment of Hezekiah, to follow Gregory the Great (*Moralia in Job* XVI:10) in distinguishing between divine *sententia* (judgment), which accommodates to the conditional operations of human speech ('sensum verborum'), and divine *consilium* (counsel), which issues from the *aeterna dispositio* (eternal order) of God.[38] In allusion to Psalms 51:6, "Behold, thou desirest truth in the inward parts: and in the hidden *part* thou shalt make me to know wisdom," the Glossist, still paraphrasing Gregory, declares:

> God had revealed his judgment ('sententiam') to his servants, but not his counsel ('consilium'): by threatening them, and inciting penitence, he was showing mercy. [...]. But it has to be understood that God considers certain things in two ways: according to [their] merits and the nature of the matter at hand (which are called inferior causes) and according to his counsel. Whence God foresaw that Hezekiah would die according to his

35 Cuilly, Feuardent, and Dadré, *Bibliorum Sacrorum cum Glossa ordinaria*, 4: cols. 337–338: "Ecce ego adijciam super dies. Secundum quod visum est. In praescientia enim in qua et hi anni erant, nihil additum est. Praesciebat enim Deus, eum quidem usque ad mortem infirmitate animi affligendum sed oratione et poenitentia convaliturum. [...] Historia libri Regum et Paralipom. prophetiae mixta est, quia et prophetia historiae copulata, de liberatione urbis, de interitu Assyrij, de reversione solis, et de annis Ezechiae additis."

36 Cuilly, Feuardent, and Dadré, *Bibliorum Sacrorum cum Glossa ordinaria*, 4: col. 342: "Scriptura enim sacra non semper observat ordinem temporis, sed frequenter prius dicta vel facta posterius scribuntur, et econverso."

37 Cuilly, Feuardent, and Dadré, *Bibliorum Sacrorum cum Glossa ordinaria*, 4: col. 335: "nec impleta non Deo mutante sententiam, sed provocante nos ad sui notitiam. Dominus enim patiens est super malicia."

38 Cuilly, Feuardent, and Dadré, *Bibliorum Sacrorum cum Glossa ordinaria*, 4: col. 336: "Sententiam debemus intelligere sensum verborum, consilium vero ab aeterna dispositione est."

> deserts, or if you will, his physical being, although this was not to be his future, nor would he die, according to [divine] counsel.[39]

For his part, Paulus Burgensis, when he argues in his second *additio* that Isaiah 38:22, properly translated from the Hebrew, should read "How marvelous the sign," rather than "What *is* the sign," is merely stating that Scripture, pace Lyra, is here following the *ordo temporis*.[40] In other respects, the opinions of the Glossist and of Lyra still apply: Hezekiah, having seen the sign, accepted and admired it, is party to the *sententia of* the Lord, but not to his *consilium*.

The *Dial of Ahaz* is evocative of or, better, attuned to these various considerations about the nature of time, and of divine judgment and counsel (Figures 1.1–1.3). For one thing, the status of the two discs oscillates between history and prophecy: they signify the passage of the sun forward and back, and as such, align with the prophecy in Isaiah 38:7, "And this *shall* be a sign," of what God intends to do; but they also tally with Isaiah 38:8, "So the sun returned ten degrees, by which degrees it was gone down," and in this sense correlate to history rather than prophecy. They can thus be viewed through a retrospective or anticipatory lens, either as recording what God has accomplished, or predicting what he purposes to accomplish. The discs are comprised by a clock designed to keep track of the *ordo temporis*, as well as any miraculous changes to it; directed to Hezekiah, the clock reveals how the Lord brings (or alternatively, promises to bring) his judgments into effect within the temporal order discernible by human eyes and minds. But because the discs are embedded within an atmospheric register of clouds nested within the planetary sphere of the sun, and viewed from above and outside the realm of human affairs, the image as a whole simultaneously alludes to the guiding presence of God – in short, to divine *praescientia*. (Although the clouds are absent from the Vorsterman and Lempereur Bible, the effect of a view seen from a very high vantage point is intensified by the greater interval of space between the sun and moon and the discs, which now appear to float within a spatial vacuum.)

39 Cuilly, Feuardent, and Dadré, *Bibliorum Sacrorum cum Glossa ordinaria*, 4: col. 336: "Sententiam, vero servis suis Deus revelaverat, non tamen consilium: set ut eorum comminationibus poenitentiam incuteret populis suis, et sic eorum miseretur. […] Sed sciendum est, quod Deus duobus modis aliquas res considerat secundum merita et secundum rei naturam (quae dicuntur causae inferiores esse) et secundum suum consilium. Unde Ezechiam moriturum Deus praeviderat secundum meritum, vel secundum physicam, licet non futurum et non moriturum, vel secundum consilium suum."

40 Cuilly, Feuardent, and Dadré, *Bibliorum Sacrorum cum Glossa ordinaria*, 4: col. 344: "sed potius significant acceptationem signi iam dati seu admirationem de tanto signo."

The paradoxical division of the *Dial of Ahaz* into two faces originating in mutually contradictory models of two types of clock, further implies that God, even as he exercises control over time, essentially transcends it. These and other curious features of the image, such as the dial's hugeness of scale and indeterminate location, suggest that the *ordo imaginis*, having been designed to contravene standards of legibility maintained amongst the bibles' other, more conventionally historical illustrations, can be interpreted as an allusion to the illegibility of divine *consilium*. Or, put in Lyra's terms, as applied in his reading of Isaiah 38:6, "And I will deliver thee and this city out of the hand," and 38:7, "that the LORD will do this thing that he hath spoken," the image alludes on the one hand to *prophetia praedestinationis*, lodged in the divine will and hence inscrutable, and on the other hand to the sort of prophecy that plays out *in causis inferioribus*, the causal sphere of the *ordo temporis*.[41]

Based on Jerome's *Commentary on Isaiah*, and supplemented by Rabanus Maurus's *Commentary on 4 Kings*, the ordinary glosses of 4 Kings 20:1–11 construe the solar miracle as a type for the mystery of the Incarnation and, more specifically, for the advent of Christ, the sun of justice. The component parts of the allegorical *Dial of Ahaz*, viewed through this lens, viz., as comprised by the sign of a great mystery whereby *prophetia praedestinationis* and *ordo temporis* cooperate, are to be understood as interactive and reconcilable rather than contradictory. The image becomes an allegory of the impact of the divine will, or alternatively, of divine *consilium*, *praescientia*, or *praedestinatio*, upon the *ordo temporis* that governs nature and humankind. The *signum* that prophesies a change in the present time of Hezekiah is simultaneously legible as a typological sign, relevant to all peoples, of the future coming of Christ. The *tertium comparationis* is the retrograde motion of the sundial's shadow, which signifies the rearward motion of Christ from heavenly perfection to terrestrial imperfection; conversely, the resumption of the sun's forward progress signifies the heavenward motion of Christ at the Resurrection. Referring to Hezekiah's request in 4 Kings 20:10 that the shadow of the sun go back ten degrees, in proof of his imminent recovery, the Glossist puts it thus:

41 Cuilly, Feuardent, and Dadré, *Bibliorum Sacrorum cum Glossa ordinaria*, 4: cols. 337–338: "Fuit prophetia comminationis quae est quando revelatur aliquid futurum, non prout est in mente divina, sed prout est in causis inferioribus: et sic erat de infirmitate Ezechiae, quae erat incurabilis per naturam, non tamen per hoc excludebatur divina virtute curaretur et ideo talis prophetia non semper impletur. [...] Hic consequenter datur signum mirabile ad confirmationem dictae prophetiae, quae erat praedestinationis, cum dicitur in persona domini."

> Which [sign] the king preferred, for what miracle could have been greater than that of the sun moving contrariwise to its usual course [...]. Never do they who inhabit the farthermost regions of the South see the sun return eastward, away from the place it sets, through southerly lands: which was both a sign of the present time and a type of the future, for even as the sun returned to the place of its rising, so the life of Hezekiah recovered the years subtracted from it; and so, too, for us who live in our seventies and eighties, the space of our lives is prolonged [eternally] by the space of the life of Christ, through his Resurrection.[42]

The Glossist adds, paraphrasing Rabanus, that the reversion of the sun by ten degrees, and its recursive advance through these same degrees, is a figurative image of Christ's descent from heaven at the Annunciation, and re-ascent into heaven at the Resurrection, with the degrees themselves respectively symbolising his predecessors, the patriarchs, judges, kings, and prophets, his followers, especially the popes, his humanity (degree nine), and his Passion (degree ten).

> Certain [exegetes] transfer the ten degrees of lines to the mystery of Christ, and interpret them as figurative shadows of his descent, through which [same degrees] Christ the sun of justice then rises after the Resurrection. The first degree of descent from heaven was in the angel who was a messenger of great counsel. [...] The ninth was in the man [i.e., Jesus]. The tenth was in the Passion: through which ten degrees Christ descended, as through the shadow of the old law: and again, after his Resurrection, Christ the sun of justice climbed upward through these same degrees, and illuminated every shadow of the law with the rays of truth.

Seeing the image of the *Dial of Ahaz* as a typological allegory of the mysteries of the Incarnation, Passion, and Resurrection of Christ entails reading the left-hand disc, numbered 1–12 (or, in the Vorsterman and Lempereur Bibles, I–XII), as a gauge of both the sun's regular passage east to west, and of its restored forward motion through the twelve hours of daylight, after its reversion by ten

---

42 Cuilly, Feuardent, and Dadré, *Bibliorum Sacrorum cum Glossa ordinaria*, 2: col. 973: "Quod rex maluit, quia maioris poterat esse miraculi, si sol contrarium suo motui cursum ageret [...]. Numquam autem hi qui interiora Austri incolunt, vident solem per Meridianas plagas ad Orientem ab occasu redire. Quod signum et praesentis temporis et futuri typus erat, ut quomodo sol reverteretur ad exordium sui, ita et Ezechiae vita ad detractatos annos rediret, nobisque in hebdoade et ogdoade viventibus per resurrectionem Christi vitae spatia pro vitae spatio protelentur."

degrees – viz., ten hours – of sunlight. It bears noting that while natural day, calculated from noon to noon, consists of twenty-four hours, whatever the latitude, artificial day consists of the time between sunrise and sunset, and varies with the latitude. The Glossist, Lyra, and Paulus Burgensis refer to the prolongation of artificial day, making little or no attempt to distinguish its hours from that of a natural day. Matthias Torringus alone refers to the prolongation of natural day. The *Dial of Ahaz* would seem to be an attempt to portray a sundial as it is described by the Glossist, Lyra, and Paulus Burgensis, on which the hours of artificial day exactly (and quite improbably) match the division of hours of natural day. Lyra, in his postil on Isaiah 38:8, "Behold, I will bring again the shadow of the degrees," explains that the *horologium* told the hours of the day either by two lines per degree, in which case the course of an artificial day was increased by ten hours (five degrees eastward from the point the sun had reached, then five degrees westward to that same point), or by one line per degree, in which case the course of an artificial day was increased by twenty hours (ten degrees eastward, ten degrees westward). Whether ten or twenty hours, since the shadow of the sun had already reached two hours before dusk, when the miracle transpired, these two hours would need to be added to the total number of extended hours: the miraculous artificial day was therefore either twenty-two hours long (10 + 5 + 5 + 2) or thirty-two hours long (10 + 10 + 10 + 2).[43] The image can of course be harmonised to either reading, since it is an allegorical not scientific instrument. The crucial point is that by reading the disc at left as an allusion to forward motion, both normative and recursive, and the disc at right as an allusion to backward motion, the image can be made perfectly to fit the typological allegory Lyra sets forth. This inflection of the allegory accords with the coupling of Isaiah 38:1–8 and Luke 2:12, "And this *shall be* a sign unto you," the angel's promulgation to the shepherds of the mystery of the Incarnation, in the Vorsterman and Lempereur Bibles.[44] It is also consistent with the moralising interpretation ('moraliter') of 4 Kings

43 Cuilly, Feuardent, and Dadré, *Bibliorum Sacrorum cum Glossa ordinaria*, 4: col. 338: "Et hoc patet, quod illud horologium erat ad solem ad cognoscendum horas diei per umbram, et dicunt aliqui, quod duae lineae faciebant unam horam: et secundum hoc illa dies artificialis fuit augmentata decem horis tantum, scilicet quinque horis in ingressu solis versus orientem et quinque horis redeundo ad locum ubi inceperat retrocessio. Alij vero dicunt, et conveniens videtur, quod quaelibet linea faciebat unam horam: et sic illa dies artificialis fuit triginta duarum horarum. Quia antequam sol retrocederet, habuerat decem horas, et postea per retrocessum et regressionem viginti: et tandem duas habuit sole tendente ad occasum."

44 The reference to Luke 2[:13] appears as a marginal gloss.

20:10, "nay, but let the shadow return backward ten degrees,"[45] appended to Lyra's exposition on the uni- and bi-linear versions of clock's degrees: "But by the sun's ten-line retrogression was signified the descent of the son of God, the sun of justice, for the purpose of assuming our mortal humanity inferior to the aforesaid order [of angels]."[46]

Paulus Burgensis and Matthias Torringus on Reading the *Dial of Ahaz*

This account of the *Dial of Ahaz* is predicated on the assumption, as noted above, that it evinces the omnipotence of the Lord. However, there was considerable variation in the way the nature of this wondrous showing of the Lord's power, and hence the significance of the allegory, were interpreted. For instance, in his *additio* to Lyra's postil on 4 Kings 20:9, "shall the shadow go forward,"[47] Paulus Burgensis argues that compared to the manifestation of divine power in Joshua 10:12–13 – "Then spake Joshua spoke to the LORD [...]: Sun, stand thou still [...], and thou, Moon [...]. And the sun stood still, and the moon stayed, until the people had avenged themselves upon their enemies" – the Isaian miracle was minified. It was tailored to the specific requirements and circumstances of Hezekiah. Paulus posits that the chief cause of all miracles is the potency of the miracle-worker ("tota ratio facti est potentia facientis"), and that the greater the miracle, the less its affinity with the natural disposition of the person or thing miraculously transformed ("summum gradum tenent illa quae nullo modo natura facere potuit").[48] An extreme case of such a major miracle is the stilling of the sun and moon in Joshua 10, whereas nothing in Isaiah points to the solar *corpus* as the locus of the miracle; rather, the locus was the sundial and, more particularly, the degrees traversed by the shadow. The artificial day was not lengthened, nor were the rays of the sun universally altered, for only those rays falling on the dial, and the shadow cast by the gnomon, were altered.[49] This is what the accompanying image of a sundial clock is intended

45 2 Kings 20:10 in the KJV.

46 Cuilly, Feuardent, and Dadré, *Bibliorum Sacrorum cum Glossa ordinaria*, 2: col. 974: "Per hoc autem quod sol retrocessit decem lineis, significatus fuit descensus filij Dei, qui est sol iustitiae ad assumendam nostram humanitatem mortalem dictis ordinibus inferiorem."

47 2 Kings 20:9 in the KJV.

48 Cuilly, Feuardent, and Dadré, *Bibliorum Sacrorum cum Glossa ordinaria*, 2: col. 977.

49 Cuilly, Feuardent, and Dadré, *Bibliorum Sacrorum cum Glossa ordinaria*, 2: col. 979: "Hoc signum secundum propriam significationem literae, necnon et rationis convenientiam non fuit factum in corpore solis, nec in ipsius cursu naturali, neque in omnibus radijs a

to indicate: namely, that unlike the earlier solar miracle, whereby God wished to manifest his glory and magnify the Israelites on campaign in the Promised Land, the later miracle was specifically attuned to the penitent Hezekiah. His sin, as 2 Deuteronomy 31:25 implies, and the Glossist asserts in his postil to Isaiah 38:1, "In those days was Hezekiah sick unto death," was pride and failure to render sufficient thanks to God.[50] The miraculous sign, according to Isaiah 38:5, "I have heard thy prayer," 4 Kings 20:5, "I have seen thy tears,"[51] and the Canticle of Hezekiah, especially Isaiah 38:14, "Like a crane *or* a swallow, so did I chatter: I did mourn as a dove," portends the power of the Lord retroactively to correct and forgive any transgression, restore and renew any transgressor, and consequently, the image of the *Dial of Ahaz* can be seen as a typological, diagrammatic allegory of the redemptive power of Christ, and of the power of penitence to secure divine mercy and benefaction. Lyra makes precisely this point in his postil on Isaiah 38:17, "thou hast cast all my sins behind thy back": "As if to say that [the Lord] is unwilling any longer to see them as worthy of punishment: and so, [Hezekiah says,] he has vivified me, not only in body but also in spirit."[52] The *Dial of Ahaz*, thus viewed, allegorically proclaims the power of penitence to secure divine mercy and benefaction.

Matthias Torringus offers a different perspective on the miracle in his scorching rebuttal of Paulus's *additio*, which he describes as a pointless exercise in *micrologia* – the study of minutiae. He applies the moniker because Paulus, as he sees it, has pointlessly endeavored to minify one of God's greatest miracles, easily equal in form and function to that recorded in Joshua 10. Citing Augustine's *De civitate Dei* XXI:7, he counterargues that this miracle, like all divine miracles, finds its final cause in the divine will, which is immeasurable

sole per orbem procedentibus, sed tantum in radijs solaribus, qui horologium Achaz tangebant, et in umbra ipsius horologij a radijs solaribus, et interpositionem corporis opaci eiusdem horologij causata."

50 Cuilly, Feuardent, and Dadré, *Bibliorum Sacrorum cum Glossa ordinaria*, 2. col. 979: "Disposuit tamen Deus miraculose, quod sole incedente cursu suo absque aliqua variatione seu retrocessione, radij solares in horologio Achaz tantum, et non alibi retrocesserunt per 10. gradus, et iterum reversi sunt per alios 10. gradus, in qua quidem retrocessione radiorum, et eorum reversione illa dies nec fuit maior vel minor solito: nam motus solis non fuit in aliquo variatus a cursu naturali, ut dictum est, nec etiam radij eius universales, sed tantum radij solares qui in horologio Achaz continebantur, prout finaliter imaginari potest in pertinenti figura."

51 2 Kings 20:5 in the KJV.

52 Cuilly, Feuardent, and Dadré, *Bibliorum Sacrorum cum Glossa ordinaria*, 4: col. 341: "Quasi nolens ea videre amplius ad puniendum: et sic vivificasti me, non solum corporaliter sed etiam spiritualiter."

and imponderable ("divina voluntas, cuius causa et ratio quaeri non debet").[53] Simply put, every miracle of God is great. He adds, however, that because the miraculous sign was a portent not simply of Hezekiah's recovery but of God's intention to save Judah and the Israelites from the Assyrians, as enacted soon thereafter in 2 Paralipomenon 32:21[54] – "And the LORD sent an angel, which cut off all the mighty men [...] of the king of Assyria" – its scope and consequences were very great indeed. Moreover, Ahaz, Hezekiah's father, was an apostate and idolator who diminished his people's faith in the Lord. The sign sent by God and, ironically, discernible in the *Dial of Ahaz*, was a necessary ameliorative to the Israelites' waning faith in the power of the Lord: in moving back the sun, God publicly authorises Hezekiah's project of restoring – bringing back – his people's piety and constancy:

> Whence it becomes apparent that in this matter, the healing of Hezekiah was of the least import, for it could have been effected by a power less than divine: but the other things only by [divine] virtue – namely, the incineration of [enemy] bodies without harm to the arms and vesture [of the Israelites]. Moreover, this sign was of greater necessity and utility than that one. For revenge upon their enemies could have been accomplished without the stopping of the sun, as was done by night in the time of Gideon, on the authority of Judges 7. But in this difficult enterprise, both confounding and unpredictable, it was needful that one of the greatest of all miracles be brought to pass, especially for the conversion of the people, and the restoration of their faith. For sinful Ahaz had led the whole people into idolatry, in 4 Kings 16. Hezekiah, the best of men, his son had led the people back in peace to the worship of the true God. Therefore, for the sanctioning of this difficult restoration its was useful, nay, essential that some great sign be produced.[55]

He adds, on the basis of Augustine Hibernicus's *De mirabilibus sacrae Scripturae* II.4, that just as in the case of Joshua the moon joined the sun in standing still, so now, for Hezekiah, the moon receded with the sun, thus ensuring the unified preservation of solar, lunar, and planetary order ("sive stando sive retrocedendo semper remansit idem aspectus corporum caelestium inter

53 Cuilly, Feuardent, and Dadré, *Bibliorum Sacrorum cum Glossa ordinaria*, 2: col. 980.

54 2 Chronicles 32:21 in the KJV.

55 Cuilly, Feuardent, and Dadré, *Bibliorum Sacrorum cum Glossa ordinaria*, 2: col. 980: "Unde apparet quod minimum in hac re fuit sanatio Ezechiae, nam illa fieri potuisset minore virtute quam divina: alia autem non nisi de virtu te fieri potuerunt, scilicet incineratio

se").[56] Finally, he chides Paulus for visualising the face of the dial as if it were inscribed with the hours of artificial rather than of natural day, and for treating the division of the former as if it were entirely coincident with the division of the latter. Torringus's rebuttal of Paulus, which amounts to a disquisition on the true nature of the miraculous sign, can just as well be applied to the image of the sign. For Torringus, the solar miracle signifies the imponderability of the divine will, its militancy and combativeness, and, perhaps most interestingly, its analogical power of instantiation – the backward motion of the sun is adduced as analog for the return of health and, more importantly, of faith. God fashions a sign, the allegorical meaning of which he expects the king and his people to parse. Torringus's allegorical reading of the sign provides a template for our reading of the image of the sign as an allegory of God's infinite potency: in converting an evil man's relic – the dial set up by the recalcitrant Ahaz – into a merciful sign of divine favor, he signals allegorically to every true penitent, on the model of Hezekiah, that the clock can be set back, grace can be restored, and God-given victory over one's foes, both bodily and spiritual, can be gained; at the same time, he licenses the prophylactic and reparative use of allegorical signs such as this one. Seen through Torringus's eyes, the *Dial of Ahaz* is describable as a meta-allegory of the restorative power of allegory to affirm the efficacy of divine and human goodness: "For God says, as is evident from the text, that he did what he did for the sake of his goodness, and that of his servant David."[57]

Viewing Isaiah 38 through the Lens of Wisdom of Solomon 11

Wisdom of Solomon 11:20, "but thou hast ordered all things in measure and number and weight," the second scriptural passage attached *in margine* to

---

corporum armis et vestibus illaesis, etc. Istud etiam signum fuit magis necessarium et utile quam illud. Nam aliter potuit fieri ultio de hostibus quam per stationem solis, potuit enim fieri in nocte sicut tempore Gedeonis, de quo Iudicum 7. In hoc autem inopinabili, et cunctis stupendo negotio necessarium fuit unum de maximis fieri signis, et potissimum pro conversione populi, et fidei reparatione. Achaz enim pessimus omnem populum induxit ad idololatriam 4. Reg. 16. Ezechias filius eius optimus populum reduxerat in pace ad cultum veri Dei. Ad autorizandum igitur hanc difficilem reductionem utile fuit, imo perneecessarium fieri aliquod magnum signum."

56 Cuilly, Feuardent, and Dadré, *Bibliorum Sacrorum cum Glossa ordinaria*, 2: col. 981.

57 Cuilly, Feuardent, and Dadré, *Bibliorum Sacrorum cum Glossa ordinaria*, 2: col. 981: "Quia dicit dominus Deus, ut patet in litera, quod fecit propter bonitatem suam, & propter David servum eius."

Isaiah 38 in the Van Liesveldt, Vorsterman, and Lempereur Bibles, accentuates the connection between the solar sign and the themes of divine power and mercy (Figures 1.1–1.3). The image of the *Dial of Ahaz*, since it displays a clock that numerically measures the passage of time, is of course perfectly suited to the task of allegorising the Lord's infinite capacity to instill and preserve universal order. Because Wisdom of Solomon 11:20–24 speaks of measure, number, and weight as indices of God's desire for spiritual order, which he mercifully restores by punishing sinners corporally, any allusion to verse 20 will perforce also evoke the relation between divine sway and grace, justice and compassion. The Glossist, in his canonical reading of Wisdom of Solomon 11:20, tropes measure, number, and weight, as follows:

> Thus does he fashion all things according to truth, and judgment, and justice. For he created all his works with regard to number, and measure, and weight. [Psalm 146:4:] 'Who telleth the number of the stars: and calleth them all by their names.' And elsewhere: [Job 28:25:] 'Who made a weight for the winds, and weighed the waters by measure.' For this was the principal pattern in the spirit of the Creator: in measure, quality; in number, quantity; in weight, order. In these did God constitute the world, and govern it, and in them he shall judge it. In these the just judge cannot be understood, nor held fast: all things lie hidden within the bounds of these three things, which are unlocked by the keys of the sun of judgment and divine mercy, when, having been [formerly] concealed, they shall be illuminated, and the counsels of hearts shall be unveiled.[58]

In the Van Liesveldt Bible, the paradoxical shrouding of the sunlit dial-faces by billowing clouds is perhaps an allusion to the Glossist's idea that number, measure, and weight lie hidden in plain sight, until God in his infinite mercy judges them fit to be revealed (Figure 1.3). The idea goes back to Augustine, who had defined number, measure, and weight as attributes of the created world comprised by but in no sense constitutive of the Creator, in *The Literal Meaning of Genesis*; as God limits, forms, and orders everything in such a manner that all things can potentially be grasped and represented by the heart and

58 Cuilly, Feuardent, and Dadré, *Bibliorum Sacrorum cum Glossa ordinaria*, 3: col. 1935: "Quia omnia secundum veritatem, et iudicium, et iustitiam facit. Omnia enim opera sua certo numero, et mensura, et pondere creavit. Unde: Qui numerat multitudinem stellarum et omnibus eis nomina vocans. Et alibi: Qui fecit ventis pondus, et aquas appendit mensura. Unde quidam ait: Omnia, quae a primaeva natura statuta sunt. Numerorum videtur ratione firmata. Hoc enim fuit principale in animo conditoris exemplar, etc. In mensura, qualitas, in numero, quantitas, in pondere, ratio. In his constituit deus

mind of man, so he can use them to unveil himself to whomever he chooses, for number, measure, and weight are in and of God.[59] For Augustine, the fact that the fiery sun has a place higher than air, and air higher than water and the vaporous clouds formed by air currents, was fundamentally expressive of God's benign governance over the world by the linked principles of number, measure, and weight.[60] The division of the Van Liesveldt image into three zones – the fiery sun above, the clouds below, and, mediating between them, the air permeated by the sun's rays – correlates to the Augustine system codified in *The Literal Meaning of Genesis*. Seen in this way, the dials emerging from banks of clouds, as coming to light, may function as an allegorical reference to the nature of Hezekiah's experience of God: for the Glossist, the process of divine unveiling in turn causes hearts to reveal themselves, as was the case for Hezekiah, whose Canticle makes known the secret thoughts that propelled him from despair to hopeful supplication, and finally, to joyful exaltation.

Furthermore, the juxtaposition of fiery sunlight and vaporous cloud in Van Liesveldt's *Dial of Ahaz* recalls the Glossist's paraphrase from Ambrose's *Hexaemeron* II.12, where the mutually tempering relation of fire and water – the one desiccating of moisture, the other dampening of flame – signifies the ordering of the world according to measure, number, and weight. Seen against the backdrop of Lyra's reading of Wisdom of Solomon 11:21, "For great power always belonged to thee alone,"[61] and 11:23, "But thou hast mercy upon all," both of which he subjoins to Wisdom of Solomon 11:20, the *Dial of Ahaz* becomes an allegory of the principle of mensuration underlying the divine order, and, more particularly, of its all-powerful and all-merciful effects upon God's chosen instruments: "You whose virtue is infinite from eternity. Therefore is added [the line]: […] 'Thou hast mercy upon all': for the preservation of their being."[62]

mundum, et gubernat, et iudicaturus est. In his iustus iudex comprehendi non potest, nec reprehendi, omnia in claustro horum trium abscondita latent, quae solis clavibus iudicij, et misericordiae dei reserabuntur, quando illuminabuntur abscondita tenebrarum, et revelabuntur consilia cordium."

59 John Hammond Taylor, trans., *St. Augustine, The Literal Meaning of Genesis*, vol. 2 (New York and Ramsey, NJ: Newman Press, 1982), col. 108.

60 Taylor, *St. Augustine*, 2: col. 49.

61 Wisdom 11:22 in the Douay-Rheims 1899 American Edition Bible is quoted here and corresponds to Wisdom of Solomon 11:21 in the KJV due to differing verse numbering. The KJV numbering has been retained in the body of the text to avoid contradiction with the other verses that are quoted from the KJV.

62 Taylor, *St. Augustine*, 2: col. 49. "Quae ab aeterno virtutis es infinitae. Ideo subditur. […] Et misereris omnium: conservando ea in esse."

Precisely because Wisdom of Solomon 11:20 was so closely attached to Isaiah 38:1–8 and 4 Kings 20:1–11, the Glossist, Lyra, Paulus Burgensis, and Torringus give weight to the issue of how long the extended day was, and whether this day was longer than the one recorded in Joshua 10:12–14.[63] The corollary issue of the exact meaning of Joshua 10:14, in relation to the miracle of Hezekiah, also greatly exercised exegetes. The fact that the image of the *Dial of Ahaz* describes the passage of time as exactly measurable, makes it readily adaptive to this exegetical problem. Lyra devotes his long postil on Joshua 10:14, "There was not before nor after so long a day," to this question. He reasons that if degrees are equated with hours, and if the sun receded suddenly, i.e., in an instant, then the artificial day recorded in Isaiah 38 and 4 Kings 20 must have lasted only 22 hours: the ten hours already elapsed, the ten additional hours traversed by the sun to reach the point it had previously reached, plus the two hours still to go before sunset. The day called forth by Joshua would thus have been longer by two hours, since Scripture says that that day, comprising 12 hours, was doubled.[64] But Lyra also considers the alternative position of Dionysius the Areopagite who hypothesised that Hezekiah's day was 32 hours long, the sun having moved gradually, not instantly, 10 hours back, before resuming its forward trajectory, whereas Joshua's was only 24.[65] However, Lyra demurs: he invokes the principle of decorum, averring that, if possible, the gist of Joshua 10:14 should be maintained. If Joshua's day occurred after the vernal equinox, toward the middle of June, when an artificial day lasts as long as 18 hours, then it must have been 36 hours in length. By contrast, Hezekiah's day lasted merely 32 hours.[66] Lyra then appraises another possibility: Jerusalem is in the third climatic zone ("in tertio vero climate"), where a day lasts about 14 hours around the time of the summer solstice, which is likely when Joshua captured Jericho

63 Curiously, this exegetical *catena* leaves out Daniel 5:24–28, which came to feature prominently in seventeenth-century exegesis of these related passages.

64 Taylor, *St. Augustine*, 2: col. 79: "et secundum hoc illa dies artificialis de qua est sermo, fuit tantum 22. horarum: dies autem sub Iosue fuit 24. horarum. Quia sol stetit immobilis spatio unius diei artificialis, et postea cursu suo naturali ivit ad occasum: et sic illa dies fuit duplicata. Sed quia Ezechias quaesivit signum notabile ab omnibus, ut videtur 4. Reg. 20. ca. et Esaiae 38."

65 Taylor, *St. Augustine*, 2: col. 80.

66 Taylor, *St. Augustine*, 2: col. 80: "In tot autem et tantis agendis bene et probabiliter potest dici quod fluxit tempus ab aequinoctio vernali usque ad medietatem Iunii vel circiter. Tunc vero dies artificialis habet 18. horas, et tunc fuit duplicata, ut dictum est, et per consequens fuit 46. [sic = 36] horarum: dies autem sub Ezechia fuit 32. horarum ad plus, ut patet ex supradictis: et sic videtur quod sub Iosue fuerit maior dies."

and the cities round about it: 2 x 14 equals a 28-hour day.[67] But if Hezekiah's day took place around the time of the winter solstice, which Lyra thinks likely, then it would have been shorter. This would be the case whether one calculates in terms of artificial or natural days.[68]

The image of the *Dial of Ahaz* in the Vorsterman and Lempereur Bibles, since it includes both the sun and the moon, is especially apt for such detailed corollary comparisons of the two solar miracles (Figures 1.1–1.2). Although Joshua 10:13 specifically refers to the moon, the miracle of Hezekiah mentions only the sun, and yet the Glossist, in his account of Isaiah 38:7, dares to presume that the moon, too, moved backward. With implicit reference to Wisdom of Solomon 11:20, the synchronous backward motion of the two heavenly bodies is taken by the exegetes for a sign of the principle of divine mensuration, which maintains order always and everywhere. Citing Augustinus Hibernicus (*Liber de mirabilibus sacrae scripturae* II.4), the Glossist extrapolates, with reference to Joshua 10:12, "Sun, stand thou still upon Gibeon; and thou, Moon, in the valley of Ajalon," that the sun and moon are typological images of Christ, for as they changed course in obedience to God, so he obeyed the Father in assuming the form and substance of a man.[69] Next, citing Rupert of Deutz, he describes the lengthened day as a prophetic allusion to Christ triumphant, who shall come as judge on the final day, elongating it while he sets about the

67 Taylor, *St. Augustine*, 2: col. 80: "Quantitas vero diei artificialis variatur secundum diversitatem climatum, et secundum appropinquationem maiorem ad solstitium aestivale et hyemale, et ideo licet in omni climate et quolibet tempore anni quaelibet dies artificialis habeat 12. horas artificiales, tamen horas naturales habet aliquando plures aliquando pauciores. In tertio vero climate dies brevissima, scilicet quando sol est in solstitio hyemali, habet 10. horas. Longissima vero scilicet quando sol est in solstitio aestivali habet 14. horas: quaestio autem nostra debet applicari ad tertium clima, quia in eo est Hierusalem ubi erat Ezechias, quando dies fuit sibi prolongata: et similiter Iosue erat satis prope Hierusalem quando prolongavit Deus diem ad eius petitionem."

68 Taylor, *St. Augustine*, 2: col. 81: "Et si propter oppositionem dicatur, quod dies prolongata sub Ezechia fuerit in solstitio hyemali, quod satis probabiliter potest dici: ut videbitur postea, sic dies ille habuit tantum 26. horas naturales et duas partes unius horae naturalis, quia quatuor horae artificiales […]. Probatio assumpti, scilicet, quod probabiliter potest dici diem sub Ezechia prolongatam fuisse in solstitio hymalie; et hoc ratione climatis; quia in tertio climate sol elevatur in meridie minoris diei per 36. gradus fere."

69 Taylor, *St. Augustine*, 2: col. 74: "Sol duos dies in unum conclusit, et luna diei spatio non occurrit. Non humani imperij authoritate luminaria requiescunt, sed Domini imperantis iussui obediunt: Dominus orationi servi sui obediens praecipit, et obediunt." Cited just before this passage, an excerpt from Jerome's *In Psalmum 65* anchors the analogy between Joshua and Jesus as obedient *servi Dei*; see ibid., 2: cols. 73–74: "Gratias agamus duci nostro Iesu: illo enim duce, illo pugnante, nos vincimus. Ille imperavit soli ut staret, et stetit, et

task of dividing the saved from the damned.[70] And finally, citing Adamantius (viz., Rufinus's translation of Origen's *Homilies* XI.3), he compares the longer day, during which Joshua avenged himself on his foes, not only to the final day of judgment, but also to the universal unfolding of the Church, chosen instrument of the sun of justice, at the end of time.[71] The *Dial of Ahaz*, read as an allegory implicitly comprising an allusion to Joshua 10, compounds its references to divine order, power, and counsel, with these further typological references to the advent and second coming of Christ.

Conclusion

In short, the allegorical *Dial of Ahaz* is an exegetically overdetermined image. It functions as a kind of allegorical palimpsest, onto which complementary exegetical inflections of Isaiah 38, 4 Kings 20, Wisdom of Solomon 11, and Joshua 10 are layered. Prompts for these intercalated readings of the image are supplied by the paratextual marginal references to relevant passages from these books, in the Van Liesveldt, Vorsterman, and Lempereur Bibles (Figures 1.1–1.3). This rich allegorical matrix is composite, at once historical, prophetic, and sapiential, in that it jointly responds to three genres of biblical book – history in the case of Kings, prophecy in the case of Isaiah, and wisdom in the case of the Wisdom of Solomon. By contrast, the illustration of Joshua 10:12–13 in Gerard de Jode's *Thesaurus sacrarum historiarum Veteris Testamenti* of 1579 and 1585 eschews allegory and marginal parataxis in favor of affective biblical narration of a singular event (Figure 1.11).[72] Engraved by Harmen Muller after Gerard van Groeningen, and inscribed "At Joshua's call, God prevents the sun and moon

istius typo, ille magnus erat: ille imperabat, sed Dominus efficiebat. Ille nos induxit in terram repromissionis."

70 Taylor, *St. Augustine*, 2: col. 74: "Futuri quoque universalis iudicii figura est. Quando enim Christus iudicaturus adveniet, tunc sole stante, longus dies erit, donec ulciscatur se de inimics suis."

71 Taylor, *St. Augustine*, 2: col. 74: "Sed retinuit et repressit diem consummationis. Videns enim Deus pater salutem gentium per ipsum solum posse constare, dicit ad eum: Pete a me, et dabo tibi gentes haereditatem tuam, etc. Dum ergo paterna pollicitatio impleatur, et ecclesia crescat, et plenitudo gentium introeat, ut tandem omnis Israel salvetur, dies producitur, et differtur occasus, nec unquam sol occumbit, sed semper oritur dum credentium cordibus sol iustitiae lumen veritatis infundit."

72 Eighth in a series of ten prints on Joshua, *The Stopping of the Sun and Moon* is inscribed with an excerpt from Joshua 10:12 (mislabeled 10:9); see Gerard de Jode, *Thesaurus sacrarum historiarum Veteris Testamenti* (Antwerp: Philip Galle, 1579). On De Jode's *Thesaurus*, see Hans Mielke, "Antwerpener Graphik in der 2. Hälfte des 16. Jahrhunderts: Der *Thesaurus*

from setting until Israel has revenged itself on his foes," the image forms part of a series of ten portraying scenes from chapters 2, 3, 5, 6, 7, 10, and 12 of the book of Joshua. Beneath the stationary disk of the sun, Joshua and the Israelites are shown slaying the soldiers of Adonisedec, Oham, Pharam, Japhia, and Dabir at Gibeon. At distant left, a slightly earlier event takes place – the killing rain of hailstones sent by God to defeat Israel's foes on the slopes of Beth-horon. Based in fact instead of allegory, this mode of visualising Scripture is narrative and illustrative rather than hermeneutic and diagrammatic. For the biblically literate viewer in sixteenth-century Antwerp, the *Dial of Ahaz*, almost uniquely amongst the images illustrating the Van Liesveldt, Vorsterman, and Lempereur Bibles, invited a sustained exploration of the exegetical meanings that nest within this complex visual allegory about the nature of divine agency. The challenge of addressing this virtually incommensurable subject resulted in the selection of allegory as the pictorial mode of choice.

Reading the Lempereur, Vorsterman, and Van Liesveldt versions of the *Dial of Ahaz*, discerning how they differ in what they show and how these differences register as inflections of the allegorical message they put forward, requires close and intensive viewing. In its closeness and intensity of engagement, such viewing resembles the kind of close reading exegetes were expected to apply to Scripture. The images, like the marginal glosses and the glossists' postils, adduce analogies across complementary scriptural texts, concatenating them. Together, these associated images and texts give us a fuller sense of how the solar miracles of Joshua and Hezekiah were understood by well-informed reader-viewers. These were textually but also visually literate persons, probably accustomed to hearing preachers sermonise on the stories of Joshua and Hezekiah and their moral and allegorical – or, to use exegetical parlance, their tropological and anagogical – meanings. After all, the ordinary glosses and Nicholas de Lyra's supplements to them were widely read by churchmen precisely because they provided a treasure house of topics for sermons and homilies, as well as spiritual exercises, applicable during

veteris et novi Testamenti des Gerard de Jode (1585) und seine Künstler," *Zeitschrift für Kunstgeschichte* 38 (1975): 29–83; and Peter van der Coelen, "Thesauri en Trezoren: Boeken en bundels met oudtestamentische prenten," in *Het Oude Testament in de schilderkunst van de Gouden Eeuw*, ed. Christian Tümpel, exh. cat., Joods Historisch Museum, Amsterdam (Zwolle: Waanders, 1991), 168–193. On the workshop circumstances that facilitated production of major undertakings such as the *Thesaurus*, see Peter van der Coelen and Marjolein Leesberg, "Depicting the Bible and Mapping the World: Gerard de Jode and his Legacy," in *The New Hollstein Dutch & Flemish Etchings, Engravings and Woodcuts 1450–1700: The De Jode Dynasty, Part 1*, ed. Peter van der Coelen and Huigen Leeflang, comp. Marjolein Leesberg (Ouderkerk aan den Ijssel: Sound & Vision, 2018), 28–93.

FIGURE 1.11 Harman Muller after Gerard van Groeningen, *Stopping of the Sun and Slaughter of the Amorrhites*, engraving, (plate) 21 x 28.7 cm, plate 8 from the *Joshua Series*, in Gerard de Jode, *Thesaurus sacrarum historiarum Veteris Testamenti* (Antwerp: Philip Galle, 1579)
PHOTO: WALTER MELION, REPRODUCED WITH PERMISSION FROM UNIVERSITEITSBIBLIOTHEEK, BIJZONDERE COLLECTIES, LEIDEN

the whole of the Church year. These allegories all turn on the motion of the sun, forward *ut solet*, retrograde or frozen in place, *per miraculum*. Just as the sun's geocentric orbit testifies to the Lord's governance over the laws of measure, number, and weight, so, too, its orbital eccentricities are seen explicitly to testify, *ab anomalia*, to his absolute power over human affairs, as epitomised by his dealings with Joshua, Isaiah, and Hezekiah. Visualised by reference to instruments of measurement that turn out, upon closer inspection, to be functionally impaired, the *Dial of Ahaz* in the Lempereur, Vorsterman, and Van Liesveldt Bibles evokes the immeasurable and incommensurable authority of Divine Wisdom. This curious image – part mimetic in that it purports to depict a sundial, part diagrammatic in that it consists mainly of geometrical forms, linear marks, and numerical signs, part exegetical in that it conjugates a string of biblical texts, and wholly allegorical in its compilatory method of construction and argument – constitutes one pregnant example of how preachers and their congregations, and the exegetes trained in the kinds of reading codified by the Glossist and Nicholas de Lyra, would have reckoned with the divine significance of the sun's motion, when viewing it through the lens of allegory. It offers us a glimpse of a pre-Copernican hermeneutics of the sun's relation to time.

CHAPTER 2

Mathematical Certainty and Biblical Inerrancy

Pedro Nunes and the Retrogradation of Shadows at the Dial of Ahaz

Henrique Leitão

Abstract

In 1537 – and again in 1566, in more detail – the Portuguese cosmographer Pedro Nunes (1502–1578) showed that, under certain conditions, the shadow of a vertical gnomon in a plane surface moves backwards for a brief period of time and then resumes normal progression. This happens twice during the course of the day. Nunes' prediction is remarkable for a number of reasons, but mostly because it is very counterintuitive and was made in stricly mathematical terms with no observation of the phenomenon itself. But what made this demonstration even more troubling was the fact that, by linking it to the well known biblical episode of the Horologium Ahaz (II Kings 20:8–11, Isaiah 38:8) Nunes launched a debate around this story and, by extension, about the possibility of a natural explanation of biblical miracles. The meeting of mathematical certainty and biblical inerrancy generated discussions that lasted until the late nineteenth century. Remarkably, the phenomenon remained unobserved: it was only by the late nineteenth century that Camille Flammarion reported having observed the phenomenon, together with a Mr Etienne Guillemin. The main objective of this presentation is to reconstruct the history of this surprising but nearly forgotten debate. This will add, in my view, to the understanding of the role and impact of the story of the dial of Ahaz in the history of science.

Pedro Nunes and the Dial of Ahaz (1537 and 1566)

Pedro Nunes's first printed book was published in 1537.[1] It is a collection of several works, of which the most important are two treatises on nautical matters.[2]

1 Pedro Nunes, *Tratado da sphera com a theorica do Sol e da Lua* […] (Lisbon: Germão Galharde, 1537).

2 The nautical treatises are *Tratado sobre certas dúvidas da navegação* [Treatise on certain doubts of navigation] and *Tratado em defensam da carta de marear* [Treatise in defence of the nautical chart]. Reference will be made to the modern edition: Pedro Nunes, "Tratado sobre certas dúvidas da navegação," in *Obras*, ed. Henrique Leitão (Lisbon: Academia das

 | DOI:10.1163/9789004677654_004

These two treatises are of exceptional importance in the history of navigation and mathematical cartography. They present the first discussion of the mathematical and geometrical properties of nautical charts, a complex topic that would give rise to a considerable debate, introduce the new concept of "rhumb line" (later called loxodromic curve), and discuss some fundamental ideas and concepts in mathematical navigation. In one of these texts, the *Tratado em defensam da carta de marear* (Treatise in defence of the nautical chart), Pedro Nunes tried to identify and correct some common errors made by sailors of his day. Portuguese pilots and sailors had been engaged in long distance oceanic voyages since the mid-fifteenth century and had devised a number of new navigational procedures and instruments and adapted existing ones to the new reality of deep ocean navigation. Nunes's aim was to carefully analyse these methods and instruments in order to extract their mathematical structure and, hopefully, to improve them.

One of the questions that Nunes analysed was the procedure used by sailors to estimate the hour of the day by looking at the direction (azimuth) of the sun. He showed that, contrary to what sailors believed, the knowledge of the latitude (height of the pole), the sun's declination, and the sun's azimuth, was not sufficient to determine the hour of the day.[3] While discussing this question he presented a diagram (Figure 2.1) and concluded:

> It follows, according to this figure and its demonstration, that in the regions that have their Zenith in point .d., with the Sun declining towards the North and raising continuously over the horizon until midday, *it changes the direction [azimuth] in which it was and returns again to it.* This happens before midday, but the same will happen after midday. *This is something surprising but cannot be denied because it is demonstrated with mathematical certainty and evidence.*[4]

Ciências de Lisboa and Fundação Calouste Gulbenkian, 2002), 1:105–119. Pedro Nunes, "Tratado em defensam da carta de marear," in *Obras*, 1:120–183.

3 Nunes, *Obras*, 1:155–157. A detailed analysis of the mathematical arguments is not the purpose of this essay; it will be presented elsewhere.

4 Emphasis mine. The original reads: "Segese segundo parece per esta figura com sua demonstração que na região que tem ho zenit no ponto .d. tendo ho sol declinação pera o norte: indo sobindo continuamente sobre ho horizonte at o meyo dia: muda o rumo em que antes estaua e torna outra vez a elle: e tudo antes do meo dia: e o mesmo fara também despois do meyo dia que parece cousa de admiração: e não se pode negar pois se demonstra com a certeza e a euidencia mathematica." Nunes, *Obras*, 1:156. All translations in this essay are mine, unless otherwise noted.

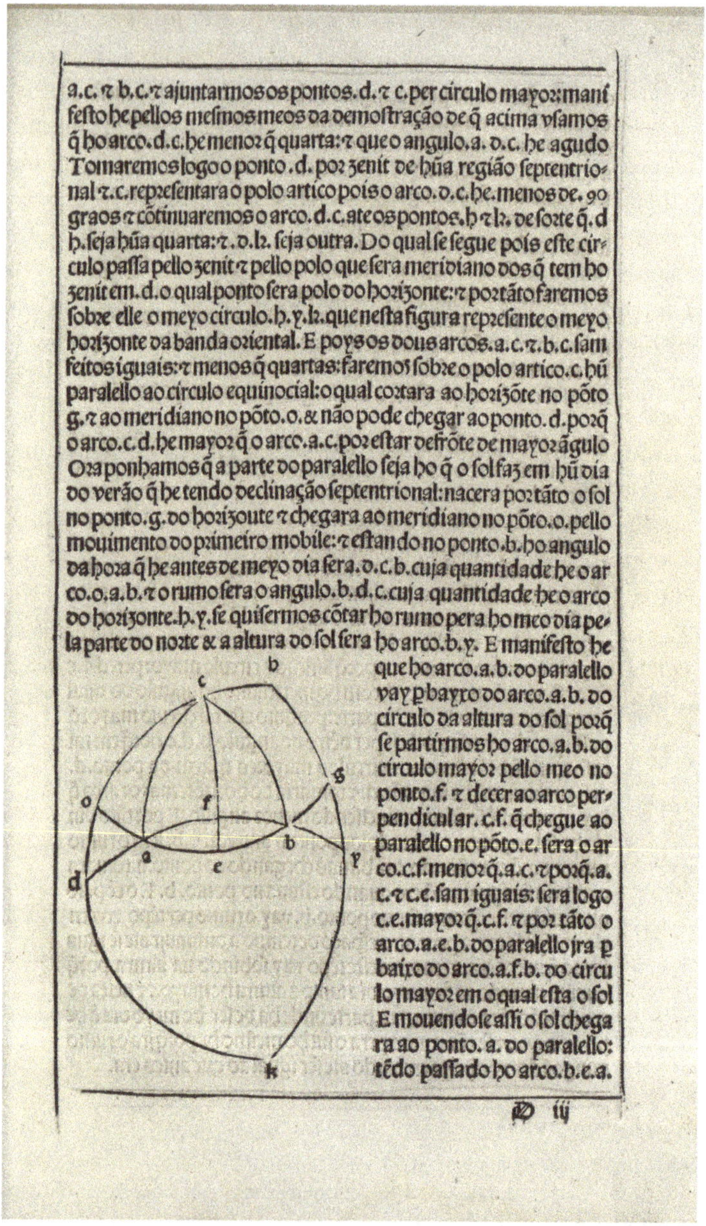

a.c. ⁊ b.c. ⁊ ajuntarmos os pontos. d. ⁊ c. per circulo mayor: manifesto he pellos mesmos meos da demostração de q̃ acima vsamos q̃ ho arco. d. c. he menor q̃ quarta: ⁊ que o angulo. a. d. c. he agudo Tomaremos logo o ponto. d. por zenit de hũa região septentrional ⁊. c. representara o polo artico pois o arco. d. c. he. menos de. 90 graos ⁊ cõtinuaremos o arco. d. c. ate os pontos. h ⁊ k. de sorte q̃. d h. seja hũa quarta: ⁊. d. k. seja outra. Do qual se segue pois este circulo passa pello zenit ⁊ pello polo que sera meridiano dos q̃ tem ho zenit em. d. o qual ponto sera polo do horizonte: ⁊ portãto faremos sobre elle o meyo circulo. h. y. k. que nesta figura represente o meyo horizonte da banda oriental. E poys os dous arcos. a. c. ⁊. b. c. sam feitos iguais: ⁊ menos q̃ quartas: faremos sobre o polo artico. c. hũ paralello ao circulo equinocial: o qual cortara ao horizõte no põto g. ⁊ ao meridiano no põto. o. & não pode chegar ao ponto. d. porq̃ o arco. c. d. he mayor q̃ o arco. a. c. por estar defrõte de mayor ãgulo Ora ponhamos q̃ a parte do paralello seja ho q̃ o sol faz em hũ dia do verão q̃ he tendo declinação septentrional: nacera portãto o sol no ponto. g. do horizoute ⁊ chegara ao meridiano no põto. o. pello mouimento do primeiro mobile: ⁊ estando no ponto. b. ho angulo da hora q̃ he antes de meyo dia sera. d. c. b. cuja quantidade he o arco. o. a. b. ⁊ o rumo sera o angulo. b. d. c. cuja quantidade he o arco do horizonte. h. y. se quisermos cõtar ho rumo pera ho meo dia pela parte do norte & a altura do sol sera ho arco. b. y. E manifesto he que ho arco. a. b. do paralello vay p baixo do arco. a. b. do circulo da altura do sol porq̃ se partirmos ho arco. a. b. do circulo mayor pello meo no ponto. f. ⁊ decer ao arco perpendicular. c. f. q̃ chegue ao paralello no põto. e. sera o arco. c. f. menor q̃. a. c. ⁊ porq̃. a. c. ⁊ c. e. sam iguais: sera logo c. e. mayor q̃. c. f. ⁊ por tãto o arco. a. e. b. do paralello jra p baixo do arco. a. f. b. do circulo mayor em o qual esta o sol E mouendose assi o sol chegara ao ponto. a. do paralello: tẽdo passado ho arco. b. e. a.

D iij

FIGURE 2.1 *Mathematical demonstration*, in Pedro Nunes, "Tratado em defensam da carta de marear," in Pedro Nunes, *Tratado da sphere com a theorica do Sol e da Lua* [...] (Lisbon: Germão Galharde, 1537), fol. 80r

PHOTO: THE LESSING J. ROSENWALD COLLECTION, RARE BOOK AND SPECIAL COLLECTIONS DIVISION, THE LIBRARY OF CONGRESS, WASHINGTON, DC

Surprising indeed. What Pedro Nunes showed with a diagram and a rigorous mathematical argument is that the shadow of a vertical gnomon moves back twice every day, and then resumes normal progression! Since this is the very first known instance of this result, it is advisable to pause for a moment and note how remarkable and extremely counterintuitive this prediction is. Consider the simplest of sundials: an horizontal flat surface upon which a vertical gnomon is placed. When illuminated by the sun, the gnomon projects a shadow onto the flat surface. During the course of the day this shadow moves, as a consequence of the sun's motion. Since the motion of the sun is always in the same direction, from East to West, the shadow projected by the gnomon in the flat surface should, one might think, obviously move continuously throughout the day, from West to East. However, as Nunes points out, given specific conditions, the shadow moves backwards (or retrogrades) for a brief period, and then resumes normal movement. This only happens in the torrid zone, that is, in places between the equator and any of the tropics, and where the latitude of the place is smaller than the sun's declination. This result is so surprising and counterintuitive that it is little-known today, even among experienced gnomonists.[5]

Yet, it is not only Nunes's result that is extraordinary; his confidence is also remarkable. After the mathematical demonstration he closes with a summary and a startling conclusion – a biblical conclusion:

> And from this demonstration it seems that the places where this will happen are those between the equator and the Cancer circle, when the Sun is more to the North than the zenith. In these places, in these conditions, the shadow of a vertical gnomon in the morning will be near to the noon shadow and then it will move backwards, and again it will move forwards, passing where it was, finally reaching the noon position. But in the regions between the parallel of Cancer and the North this will be

5 As written in a recent paper: "The phenomenon is very complex and difficult to understand and to explain, even for an experienced astronomer." Jaap van Dorp, "The prayer of Isaiah and the sundial of Ahaz (2 Kgs 20:11)," in *Psalms and Prayers*, ed. Bob Becking and Eric Peels (Leiden: Brill, 2007), 265. This is not exactly so; the phenomenon is indeed subtle, but not difficult to understand for a person with the basic knowledge of the mathematical techniques used in astronomy. It is true, however, that very few books on sundials mention or demonstrate it. For a modern treatment see, for example, Denis Savoie, *La Gnomonique* (Paris: Les Belles Lettres, 2001), 401–410, or Edmondo Marianeschi and Nicola Severino, "La Retrogradazione dell'Ombra," in *Atti del VI° Seminario di Gnomonica, 30 Sett. – 2 Ott.*, (San Benedetto del Tronto, 1994), 7–36.

> impossible, except by divine power, as one can read in *2 Kings*, where the shadow moved backwards ten degrees to manifest Hezekiah's health.[6]

Thus, Nunes makes a connection to the well-known biblical episode of the dial of Ahaz where shadows miraculously moved backwards (2 Kings 20:9–11). In a slightly different version (Isaiah 38:7–9) it is said that the retrograde motion of the shadows happened because the sun moved backwards.

The concept of a *natural* retrograde motion of the shadow of a sundial could clash with the biblical story's explicit reference to a miraculous event. However, since according to Nunes the retrogradation of shadows is only possible in the torrid zone, this allowed for a straightforward way to preserve the miraculous character of the event at the dial of Ahaz. Nunes established a simple distinction between *natural* shadow retrogradation (that occurs inside the torrid zone), and *miraculous* shadow retrogradation (that occurs outside the torrid zone).[7] A retrogradation observed in Jerusalem was thus a miraculous event.

Pedro Nunes's book of 1537 was written in Portuguese and likely had a somewhat limited circulation outside Iberia. In his *Opera,* written in Latin and published in 1566 in Basel, he presents greatly expanded versions of his nautical treatises from his 1537 text. Nunes's *Opera* had a wide dissemination, with later editions in Coimbra, in 1573, and again in Basel, in 1592.[8] In chapter 11 of Book II, Nunes returned to the demonstration of natural retrogradation:

> Furthermore it is worthy to note that in all places between the equator and the parallel of Cancer, when the Sun is nearer the arctic pole than

6 Nunes, *Obras*, 1:157.

7 Nunes's demonstration has been noted before, but only very briefly. See Francisco de Borja Garção Stockler, *Ensaio Historico sobre a Origem e Progressos das Mathematicas em Portugal* (Paris: Officina de P. N. Rougeron, 1819), 34 and 144–146; Francisco Gomes Teixeira, *História das Matemáticas em Portugal* (Lisbon: Academia das Ciências de Lisboa, 1934), 155–156. More recently: Henrique Leitão, "A brief note on the power of mathematics. Pedro Nunes and the retrogradation of shadows," in *Proceedings of Recreational Mathematics Colloquium V – G4G (Europe)*, ed. Jorge Nuno Silva (Lisbon: Ludus, 2017), 45–52; Peter Lynch, "Pedro Nunes and the Retrogression of the Sun," *Bulletin of the Irish Mathematical Society* 81 (Summer 2018): 23–30. See also my notes to the relevant passage in Pedro Nunes, *Obras*, ed. Henrique Leitão (Lisbon: Academia das Ciências de Lisboa and Fundação Calouste Gulbenkian, 2008), 4:724–727.

8 Heinrich Petri, *Petri Nonii Salaciensis Opera* (Basel, 1566). The book was republished under a different title as *De arte atque ratione navigandi* (Coimbra, 1573), and once again in *Petri Nonii Salaciensis Opera* (Basel, 1592). The relevant text is in Nunes, *Obras*, 4:136–144 (Latin); 4:395–403 (Portuguese translation); 4:724–727 (Notes, by H. Leitão).

> the zenith [is near the arctic pole], the Sun will pass the same vertical circle twice before noon and similarly, twice after noon [...] *Therefore, in those places, the gnomon's shadows will move backwards, with no miracle happening.*[9]

In this expanded Latin version Nunes presents a more detailed mathematical treatment, including a numerical calculation for a place in the torrid zone. He considers a place at latitude of 12° N – in "the Portuguese town of Cananor, in India," as he says – and an observation at the day of solstice (sun's declination: 23° 30′) and calculates that the retrograde arc will be 3° 41′.[10] The mathematical analysis concludes with a remarkable statement, where the issue of the observation of the phenomenon is raised for the first time:

> I confess, however, that I asked the men that travel frequently to these regions between the Equator and the parallel of Cancer if, in those regions, when the Sun is in Cancer, they saw, in the morning and in the afternoon, that the shadows move backwards for some time; but they told me they had never observed such thing. This is no cause for surprise, since the shadow's backward movement is very small. However, they should have observed that the shadow's length varies greatly (before noon and after) with no rotation around the gnomon.[11]

Nunes is quite right. Observing the phenomenon is not easy. He had never observed it and he knew of no person that ever did. As will be explained below, all evidence leads to the conclusion that the phenomenon of shadow retrogradation first predicted by Nunes in 1537 was only observed more than three

9 Emphasis mine. The original reads: "Praeterea annotatione dignum censemus, proprium esse omni loco posito inter aequinoctialem et circulum Cancri, cum Sol uicinior fuerit polo mundi arctico, quam uerticale punctum, ipsum Solem habere in uno atque eodem circulo ex uerticalibus bis ante meridiem, et bis similiter post meridiem, ita ut ex quo horizontis loco cum exoritur, leuatur, ex eodem rursus ante meridiem radios mittat. Quapropter gnomonum umbras in ipsis locis necesse est retrocedere, citra miraculum." Nunes, *Obras*, 4:136.

10 Nunes, *Obras*, 4:143.

11 "Fateor tamen me quaesiuisse ab iis hominibus qui ad ea orbis loca crebro adeunt, quae inter aequinoctialem et circulum Cancri posita sunt, utrum in ipsis locis quando sol in Cancro est, mane et sero umbras corporum rectorum supra horizontem aliquantisper regredi uidissent: at se hoc minime conspexisse responderunt: nec mirum, nam quia perexiguus est umbrarum regressus, idcirco non aduerterunt. At latere eos non debuit umbrarum longitudinem in spatio quatuor horarum nimium contrahi ante meridiem, post meridiem uero quam longissime produci, nulla interim circulari motione percepta circum gnomonis pedem." Nunes, *Obras*, 4:143–144.

centuries later, in the 1870s, and still with great difficulty. What is remarkable is the fact that Nunes was completely unperturbed by never having observed the phenomenon; for him, the mathematical demonstration was absolutely final. Whether or not anyone had already observed the phenomenon was of no importance for him. It is difficult to find a more solid trust in the power and certainty of mathematics in the sixteenth century.

Miracle of Natural Phenomenon?

The biblical commentary on the episode of the dial of Ahaz (2 Kings 20:9–11; Isaiah 38:7–9) is immense, in part because there are difficult textual problems with it, the most salient being that the Hebrew word *ma^calot* was variously translated in the Vulgate as *horologium*, *linea*, *gradus* etc. Thus, there was great discussion on what type of sundial was involved or, indeed, if it was a sundial at all. One other aspect that generated a considerable discussion was whether the shadows alone had moved backward, or the sun itself. As Volker Remmert has shown, the episode played a significant role in seventeenth-century debates about Copernicanism.[12] It is noted among historians of science as one of the biblical passages referenced by Galileo in his celebrated *Letter to the Grand Duchess Christina* (1615).[13]

Besides the efforts at clarification by biblical scholars and its use in the Copernican debate, it is known that attempts at a natural understanding of the phenomenon were made throughout the centuries. Already in the fourteenth century Levi ben Gerson (1288–1344), in his famous *Sefer Milhamot Ha-Shem* (The Wars of the Lord), written in the years 1317 to 1329, addressed the story of Hezekiah and suggested that the explanation was probably related to atmospheric phenomena.[14] Later, in a brief comment in his *Elucidatio fabricae*

12 Volker Remmert, "Picturing Jesuit Anti-Copernican Consensus: Astronomy and Biblical Exegesis in the Frontispiece of Clavius' *Opera mathematica* (1612)," in *The Jesuits II: Cultures, Sciences, and the Arts 1540–1773*, ed. John W. O'Malley et al. (Toronto: University of Toronto Press, 2006), 291–313 as well as Volker Remmert, *Picturing the Scientific Revolution: Title Engravings in Early Modern Scientific Publications* (Philadelphia: St. Joseph's University Press, 2011).

13 On this letter and the biblical sources used by Galileo see Bernard R. Goldstein, "Galileo's account of astronomical miracles in the Bible: a confusion of sources," *Nuncius* 5, no. 1 (January 1990): 3–16.

14 Goldstein, "Galileo's account of astronomical miracles," 7–8. On ben Gerson in general, see Bernard R. Goldstein, *The astronomy of Levi ben Gerson (1288–1344)* (New York: Springer, 1985).

ususque astrolabii (Oppenheim, 1524) Johannes Stöfler (1452–1531), noted that some "very learned mathematicians" think that the dial of Ahaz was made with unequal hours.[15] This tells us very little about the explanation, but shows that mathematicians had been thinking about it. In fact, instrument makers and gnomonists had proposed specific dials to reproduce the surprising shadow behaviour. If the report by a later mathematician is true, in the fifteenth century Johannes Regiomontanus had already constructed one such sundial.[16] What is known for sure is that by the sixteenth century, experts in the construction of sundials had figured out a way of reproducing the backward motion of the shadows using refractive sundials, that is, water-filled dials that explored the effect of refraction of light rays in water.[17] One of the most famous was made by the German instrument maker and engraver Georg Hartmann (1489–1564), who produced a so-called "Bowl of Ahaz," a refracting scaphe dial, in Nuremberg in 1548. Hartmann was perhaps one of the first to construct a refractive sundial, but he was not the only one.[18] All through the second half of the sixteenth century many more of these dials were built, by Christoph Schissler, Christopher Heiden, Guidobaldo del Monte, Ettore Ausonio, among

15 Johannes Stöfler, *Elucidatio fabricae ususque astrolabii* […], (Oppenheim, 1524), fol. 44r.

16 The information is conveyed by Mutio [Muzio] Oddi and refers to a refractive sundial: "& un giorno parlandone io col Padre Christoforo Clauio in Roma, mi disse, che Giouanni da Monteregio n' havea fatto uno ancor lui, per un Prencipe d' Alemagna." Mutio Oddi, *De gli Horologi Solari Trattato* (Venice: Per il Ginammi, 1638), 100. See Sven Dupré, "Galileo, the Telescope, and the Science of Optics in the Sixteenth Century" (PhD diss., University of Ghent, 2002), 97–98.

17 Julius F. Sachse, "Horologium Achaz," *Proceedings of the American Philosophical Society* 34, no. 147 (January 1895): 21–30; Harrold E. Gillingham, "The Horologium Achaz," *Connoisseur* 74 (1926): 220–222; Peter I. Drinkwater, "The miracle of Achaz," *Bulletin of the British Sundial Society* 92.1 (1992): 24; Allan Mills, "The 'dial of Ahaz', and Refractive Sundials in General. Part 1: Scaphe Dials." *Bulletin of the Scientific Instrument Society* 44 (1995): 21–24 and Allan Mills, "The 'dial of Ahaz', and Refractive Sundials in General. Part 2: Horizontal Planar Dials." *Bulletin of the Scientific Instrument Society* 45 (1995): 25–29; Philip M. Sadler, "An ancient time machine: The Dial of Ahaz," *American Journal of Physics* 63, no. 3 (1995): 211–216; Sven Dupré, "The Dioptrics of Refractive Dials in the Sixteenth Century," *Nuncius* 18 (2003): 39–67.

18 Jasmin Mersmann, "Moving Shadows, Moving Sun. Early Modern Sundials Restaging Miracles," *Nuncius* 32, no. 1 (January 2015): 96–123. For Hartmann's biography, see: Hans Gunther Klemm, *Georg Hartmann aus Eggolsheim (1489–1564): Leben und Wirken eines fränkischen Mathematikers* (Forchheim: Ehrenbürg-Gymnasium, 1990); see also Dupré, "Galileo, the Telescope, and the Science of Optics," 95–100, and Filippo Camerota, "Two new attributions: a refractive dial of Guidobaldo del Monte and the "Roverino compass" of Fabrizio Mordente," *Nuncius* 18 (2003): 25–38.

others.[19] Later, in the seventeenth century, another type of sundial, the analemmatic sundial, was introduced by Jean-Louis Vaulezard and used to reproduce the retrogradation of the shadows.[20]

It is not easy to determine what the status of these objects and the real intentions of those who made them were. It has been pointed out that Hartmann's bowl was not meant to "explain" the miracle (in the sense of providing a natural explanation of what was traditionally considered to be the result of God's action). According to available evidence, the bowl was intended only to *imitate* the phenomenon, as is apparent from the word "imitator" in its legend ("Hydraulicum quod mirabili artificio Horologium Achas in Quo Esaias Umbram Solis Retrorsum Duxit Decem Gradibus [...] *imitator*").[21]

But Nunes, it must be noted, had a completely different approach, and made a completely different claim. He did not propose a dial or the specific design of a dial in order to reproduce the phenomenon. He also did not suggest an atmospheric or optical phenomenon as a plausible explanation, as several others did and would continue to do in the sixteenth and seventeenth centuries. Nunes did not limit himself to showing that the phenomenon of shadow retrogression was *possible* in natural terms. What he did was much

19 Owen Gingerich and Philip Sadler, "Christopher Schissler's Wonderful 'Bowl of Ahaz' of 1578," *Bulletin of the American Astronomical Society* 21 (1989): 1218. On Schissler, see Peter Plaßmeyer, "Christoph Schissler: The Elector's Dealer," in *European Collections of Scientific Instruments, 1550–1750*, ed. Giorgio Strano, Stephen Johnston et al. (Leiden: Brill, 2009), 15–25, and Samuel Gessner, "'Geometricus et astronomicus faber': Chr. Schissler aus Augsburg," in *Weiter sehen/Seeing further*, ed. Jürgen Hamel and Michael Korey (Frankfurt am Main: Verlag Harry Deutsch, 2012), 123–154. On Schisslers refraction sundial cf. chapter 4 in this volume by Andrew Morrall, "Turning Back the Sun. Christoph Schissler's 'Horologium Ahaz' as 'Kunststück.'"

20 See Jean-Louis Vaulezard, *Traicté ou usage du Quadrant analematique* [...] (Paris: au Globe céleste, 1640). Later, more developed in: Jean-Louis Vaulezard, *Traitté de l'origine, demonstration, construction & usage du Quadrant analematique* (Paris: J. Le Brun; au Globe céleste, 1644). See Margarida Archinard, "Les Cadrans Solaires Analemmatiques," *Annals of Science* 62 (2005): 309–346. On the use of these sundials to reproduce the behaviour of the dial of Ahaz, see: Louis Chomard, "Le cadran analemmatique et la rétrogradation de l'ombre," *L'Astronomie* 20 (1906): 433–449; Jean-Paul Parisot, "La rétrogradation de l'ombre dans les cadrans solaires analemmatiques," *Journal for the History of Astronomy* 16 (1985): 43–48, and Gilbert Vincent, "Construction géométrique simplifiée d'un cadran solaire analemmatique et rappel de quelques propriétés," *Observations & Travaux* 63 (2006): 1–5.

21 Emphasis mine. See Allan Mills, "The "dial of Ahaz," and Refractive Sundials in General. Part I: Scaphe Dials," *Bulletin of the Scientific Instrument Society* 44 (March 1995): 23; Ivars Peterson, "Turning Back Time: An Antique Sundial Simulates a Biblical Miracle," *Ivars Science News* 137, no. 6 (February 1990): 91 and Mersmann, "Moving Shadows," 114–116.

more disturbing: he demonstrated with mathematical certainty that it was *necessary*. He presented a mathematical demonstration, that is, a strictly logical and universally valid argument that predicted the necessary occurrence of the phenomenon in nature.

Let us consider Nunes's result from two angles: firstly, simply as an astronomical and mathematical result, and secondly, in its relation to the biblical episode of the dial of Ahaz.

Nunes's demonstration is very interesting as a mathematical result. It is counterintuitive and was reached with no empirical basis: he never observed the phenomenon and, as far as he knew, nobody else had ever observed it either. From a technical point of view the demonstration is not difficult but it is subtle and requires considerable mathematical ingenuity. Nunes used the techniques of sixteenth-century cosmography and spherical trigonometry but throughout the centuries the demonstration was refashioned with more advanced mathematical techniques. As we will see, considered simply as a mathematical result, it was so startling that it acquired a life of its own. Mathematicians, astronomers and lovers of mathematical puzzles, discussed it until the end of the nineteenth century.

With respect to the biblical episode of the dial of Ahaz, Nunes's result (generalised, as we will see below, by Clavius) introduced a whole new, and rather disturbing issue in the discussion of the miracle. The result combines two striking aspects: it predicts a wholly natural and mundane event, and it is completely certain and necessary. It is mundane in the sense that it makes no appeal to an infrequent or rare event (as optical and meteorological phenomena, such as parhelia, that were offered as possible explanations),[22] and it does not depend on a dial intentionally designed for the purpose (refractive or analemmatic, etc). In fact, Nunes never commented on the sundial itself, he merely mentioned a vertical gnomon. He showed that (where the required conditions are fulfilled) retrogradation of shadows takes place every day, twice per day, regardless of the shape or form of sundials. More than questioning a supposed miracle, what he did was to reveal a natural phenomenon, completely unknown until then.

The issue of natural explanations of the Ahaz story requires a brief remark. As several commentators noted, natural occurrences (parhelia, etc) can easily

22 A parhelion (also known as a Sun dog or mock Sun) is a rare atmospheric phenomenon that consists of a bright spot to one or both sides of the Sun. These spots have the same altitude as the Sun, but different azimuth (that is, they stand in the same almicantarath as the Sun). Parhelia have been observed and described since Antiquity; for example by Aristotle (*Meteorology* III.2, 372a14).

be conceived of as miraculous, the miracle being the sudden appearance, at the Prophet's command, of an exceedingly rare natural event. This is not the case with Nunes's explanation. The fact that every shadow retrogrades twice every day challenges the supernatural nature of the Ahaz story at a much deeper level. It is quite simply the existence of a (new) natural phenomenon, whose very "naturalness" may render the Ahaz story problematic.

Yet, it bears noting that the natural aspect is peculiar because the phenomenon was never observed and was solely and uniquely supported by a mathematical demonstration. Pedro Nunes provided an unquestionable mathematical demonstration that was not dependent on any other conditions except mathematical ones. It is imposing precisely because it was absolutely certain. What clashed with the biblical story was not an observation, nor was it a historical fact, an archaeological find, or a philosophical argument. With Nunes's demonstration, what clashed with biblical inerrancy was mathematical certainty.

Nunes's Demonstration and the Jesuits

Christopher Clavius (1538–1612) was the first mathematician to comment on Nunes's demonstration and prediction. Clavius was a careful reader and an admirer of Nunes's works, which he referenced always in appreciative terms.[23] Indeed, via Clavius, Nunes became a greatly admired mathematical author among the Jesuits. The closing chapter of Clavius's *Fabrica et usus instrumenti ad horologiorum descriptionem peropportuni* (1586), bears the title, "Quod in omni loco terrae inter aequatorem, et tropicum [Cancer] uel [Capricorn] posito umbrae gnomorum, turrium atque arborem in planis horizonti aequidistantibus bis in die naturaliter retrocedant, cum Sol uertice loci borealior est."[24] The chapter opens with an explicit mention of Nunes's result:

---

23 See Eberhard Knobloch, "Nunes and Clavius," in *The Practice of Mathematics in Portugal. Proceedings of the International Conference*, ed. Luís Saraiva and Henrique Leitão (Coimbra: Imprensa da Universidade, 2004), 163–194. See also: Eberhard Knobloch, "Christoph Clavius. Ein Namen- und Schriftenverzeichnis zu seinen *Opera Mathematica*," *Bolletino di Storia delle Scienze Matematiche* 10 (1990): 135–189.

24 "In all places of the Earth situated between the Equator and the Tropic of Cancer or Capricorn the shadow of gnomons, towers, and trees [projected] in a plane parallel to the horizon retrogrades naturally twice every day when the sun is more to the South than the place's Zenith" (translation mine). Christopher Clavius, "Quod in omni loco terrae inter aequatorem" in *Fabrica et Usus Instrumenti ad Horologiorum Descriptionem Peropportuni* [...] (Rome: Bartolomeo Grassi, 1586), 105–112.

> It does not seem alien to our purpose as we treat the description of sundials in which the gnomon's shadow take different positions, to show, following Pedro Nunes, in chapter 11 of book two of his *De navigatione*, that it is proper of all the places in the torrid zone whose zenith is less distanced from the equator than the parallel in which the Sun lies, that is, whose latitude or pole's height, is less than the Sun's declination in any parallel, that the shadows of buildings, trees and gnomons in planes parallel to the horizon, retrograde twice every day with no miracle happening, that is, once before noon and once after noon.[25]

Clavius immediately adds his absolute trust in the mathematical demonstration:

> This may sound unbelievable to some less trained in astronomical matters, since the Sacred Writings state that in the dial of Ahaz, at the time of Hezekiah, king of Judea, the shadow moved backwards not naturally, but by divine action. *It is not only believable but absolutely necessary, and without miracle.*[26]

After these words, Clavius reproduces the mathematical demonstration given by Nunes. He has no hesitations in accepting Nunes's mathematical demonstration and is well aware that this is an extremely counterintuitive result that will puzzle non-mathematicians. But, following Nunes, he stresses that since it is proven mathematically, it cannot be doubted – "It is not only believable but absolutely necessary." It must be noted that Clavius, just as Nunes, attributes no relevance whatsoever to the fact that the phenomenon was never observed;

25 "Non alienum a nostro instituto uidetur, quando de horologiorum descriptione agimus, in quibus gnomonum umbrae uarias positiones uariis in horis fortiuntur, demonstrare cum Petro Nonio lib. 2. de Nauigatione, c. 11. proprium esse omni loco Zonae torridae, cuius uertex minus ab Aequatore recedat, quam parallelus, in quo Sol existit, hoc est, cuius latitudo, siue altitudo poli minor sit, quam declinatio Solis in quolibet parallelo constituti, ut umbrae aedificiorum, arborum, gnomonumque in planis Horizonti parallelis bis in die citra ullum miraculum retorcedant, semel quidem ante meridiem, et post meridiem iterum." Clavius, "Quod in omni loco terrae inter aequatorem," 105.

26 Emphasis mine. The original reads: "quod nonnullis parum in rebus Astronomicis exercitatis incredibile prorsus uideri possit, cum in horologio Achaz regis Iuda umbrarum tempore regis Ezechiae non naturaliter, sed uirtute diuina retrocessisse testentur sacrae literae. At uero esse hoc non solum credibile, sed omnino necessarium, nulloque modo miraculo illi, quod Deus in horologio Achaz in signum sanitatis recuperandae Ezechiae regi exhibuit, aduersari, aut repugnare, ex iis, quae sequuntur, liquido constabit." Clavius, "Quod in omni loco terrae inter aequatorem," 105.

for both the German Jesuit and Nunes, the mathematical demonstration provides absolute certainty, dispensing with the need for observation.

Clavius also mentions the biblical episode, and he is obviously forced to discuss the relation between the mathematical demonstration and the supernatural nature of the event at Ahaz:

> But if things are in this way, what shall we say about that famous miracle in the dial of Ahaz as a sign of recovery of King Hezekiah, on which not naturally, but by divine power, the shadow moved backwards ten degrees or lines, according to the Sacred Writings? Pedro Nunes in the place mentioned answers that the shadow's retrogradation we are talking about can only take place in the torrid zone, i.e. between the equator and one of the tropics, as can be deduced from the demonstration. Thus, since the city of Jerusalem, where Hezekiah was dying is placed on the temperate zone, the shadow cannot go backwards there without a miracle. For this reason, since the Sacred Writings testify that the shadow indeed went backwards, there is no doubt that a miracle happened.[27]

Clavius agrees with Nunes's demonstration and also refers to Nunes's distinction between *natural* and *miraculous* retrogradation based on the place's location; since Jerusalem is outside the torrid zone, retrogradation of shadows there can only happen miraculously. But here Clavius parts company with Nunes:

> What Nunes said – and must be contested – is in chap 11, when he writes: "Therefore, it is not absurd that in those places (between the Equator and the tropic of Cancer or of Capricorn) the shadows move forward and backward. However in our northerly regions, which are beyond the tropic of Cancer this cannot happen, unless miraculously, as one reads that it happened, by divine command, as a sign of health of King Hezekiah." These words seem to imply that the retrogradations in our regions are not different from the ones in the dial of Ahaz, except that

27 "Haec ita cum sint, quidnam de illo tam illustri miraculo in horologio Achaz in signum sanitatis regi Ezechiae ostenso dicemus, in quo non naturaliter, sed uirtute diuina umbram decem lineis, gradibusue retrocessisse narrant sacrae literae? Petrus Nonius loco citato respondet, regressum illum umbrae, de quo egimus, locum solum habere in Zona torrida, hoc est, inter Aequatorem et alterutrum tropicorum, ut ex demonstratis liquet. Cum ergo urbs Hierosolyma, ubi Ezechias aegrotabat, in Zona temperata citra tropicum [Cancer] sita fit, non potuit sine miraculo umbra retrocedere. Quare cum sacrae literae eam regressam fuisse testentur, miraculo procul dubio factum id est." Clavius, "Quod in omni loco terrae inter aequatorem," 109.

> ours are natural whereas the other was the result of a direct command by God. But in truth this cannot be accepted.[28]

Clavius does not accept Nunes's distinction between natural and miraculous shadow retrogradation depending on the place's location. In order to show that such a distinction cannot be accepted he makes a disturbing mathematical observation, recalling a simple fact from gnomonics: it was well known that a plane dial at a given latitude can be made to behave equivalently to a plane dial at a different latitude simply by inclining the sundial's plane; this is the concept of "equivalent latitude" or "equivalent dial."[29]

> In fact, it is possible to achieve the shadow's retrograde motion outside the torrid zone if we elevate a dial [...] Thus, if someone says that the plane of the dial of Ahaz was set in such a way – so that the shadow's backward motion happened naturally – who will not realise that by saying so one denies such a famous miracle? And this cannot be accepted in any way.[30]

By showing that shadows can retrograde naturally *at any latitude*, that is, in Jerusalem as well, Clavius destroyed Nunes's distinction between *miraculous* shadow retrogradation and *natural* shadow retrogradation. Thus, in order to maintain the truly miraculous nature of the event, Clavius is forced to conclude that the miracle at the dial of Ahaz was that the sun itself moved backwards.[31]

---

28 "Quod autem Nonius senserit, ita respondendum esse, manifestum est ex cap. 11 citato, ubi ita scribit [*Non est igitur absurdum, si in iis locis* (nempe inter Aequatorem, et tropicum ♋, vel ♑) *progrediantur umbrae, et retrocedant. In hac tamen plaga nostra Boreali, quae citra tropicum ♋, posita est, id citra miraculum fieri non posset, quemadmodum iussu Dei legitur accidisse in signum salutis regis Ezechiae.*] quibus uerbis significare uidetur, nostrum hunc regressum ab illo in horologio Achaz facto non differre, nisi quod noster naturalis est, ille uero iussu Dei effectus, propter varios locorum situs. Verum haec responsum non placet." Clavius, "Quod in omni loco terrae inter aequatorem," 109.

29 This is a standard result in gnomonics. See for example, Savoie, *La Gnomonique*, 37–39. An equivalent sundial can be obtained by inclining the plane of the sundial or by inclining the gnomon in a plane sundial.

30 "Nam licet regressus ille umbrae in Horizonte extra Zonam torridam fieri nequeat, si tamen eleuetur planum aliquod cum stylo, supra quod polus altitudinem habeat minorem, quam grad. 23 1/2 retrocedet in eo umbra non secus, atque in Horizonte aliquo zonae torridae, ut supra diximus. Si quis igitur diceret, planum horologii Achaz eiusmodi habuisse situm, quis non uidet, fieri potuisse uiribus naturae in eo regressum illum umbrae, atque adeo miraculum illud tam insigne, ac celebre euacuari? Quod tamen nulla ratione admittendum est." Clavius, "Quod in omni loco terrae inter aequatorem," 109–110.

31 Clavius, "Quod in omni loco terrae inter aequatorem," 110.

Clavius concludes by saying that the phenomenon at the dial of Ahaz had to be of a totally different type than normal (i.e. *natural*) shadow retrogradation. Indeed, how could a phenomenon be judged miraculous that took place twice per day, each and every day?[32] In sum, under the pressure of a mathematical demonstration, Clavius was forced to adopt a new criterium: shadow retrogradation is a natural phenomenon taking place daily (although never observed!). It can, however, be a miraculous occurrence, if it happens as a result of the retrograde motion of the sun, by divine command. As long as biblical commentators were willing to accept the possibility of a backward motion of the sun, Nunes's mathematical prediction could be domesticated.

After Clavius

The comment by Clavius increased enormously the impact of Nunes's demonstration, but it also changed somewhat that impact. Gone was Nunes's simple expedient of preserving the miracle by separation of earthly regions where shadows could retrograde naturally and regions where shadow retrogradation had to be miraculous.

The first echoes of Clavius's discussion of the dial of Ahaz among biblical exegetes can be traced to the beginning of the seventeenth century. The Jesuit Nicolaus Serarius (1555–1609) made a passing reference to it in his posthumously published commentary on the book of Kings, *In libros Regium et Paralipomenon*, of 1617.[33] But the first extended discussion was provided by the Jesuit Cornelius a Lapide [Cornelius Cornelissen van den Steen] (1567–1637), one of the most influential Catholic exegetes of the early seventeenth century, professor of Holy Scripture in Leuven and (after 1616) in Rome.[34] In 1622 his commentaries on Isaiah, Jeremiah, Ezekiel, and Daniel, *Commentaria*

32 Clavius, "Quod in omni loco terrae inter aequatorem," 110.

33 Nicolaus Serarius, *In Libros Regum et Paralipomenon Commentaria Posthuma* (Mainz: Balthasar Lipp, 1617), 352. Serarius underlines Clavius's conclusion that the miracle was the backward motion of the Sun. Juan de Mariana (1536–1624) does not cite Clavius explicitly in his commentary, but states the issue in terms that hint that he knew it: "Sane umbra reuerti non potuit, nisi sole retro conuerso." Juan de Mariana, *Scholia in Vetus et Novum Testamentum* (Paris, 1620), 180. I am indebted to Brent Purkaple for these indications.

34 See: John P. van Kasteren, "Cornelius Cornelii a Lapide," in *The Catholic Encyclopedia*, ed. Charles G. Herbermann, Edward A. Pace et al., vol. 4 (New York: Robert Appleton Company, 1908), 377; Luke Murray, *Jesuit Biblical Studies after Trent: Franciscus Toletus & Cornelius A Lapide* (Göttingen: Vandenhoeck & Ruprecht, 2019).

in Quatuor Prophetas Maiores, appeared, with a discussion, several pages-long, on "De horologio Achaz et egressu umbrae."[35] Like his other commentaries, this would be immensely popular, with many editions published.

When commenting on Isaiah 38:8 after a brief historical detour to establish that the Ahaz dial was indeed the first sundial on record, Cornelius posed six different questions. For our purposes, Quaestio 1 is the most important: "We inquire first whether it was only the shadow that moved backwards at the sundial of Ahaz, or if it was really the sun, and consequently its shadow all over the earth?"[36] As Cornelius explains, this question had been widely considered before and authors could be grouped around two positions: "Vatablus, Burgensis, Arias Montanus, et quidam alii" had asserted that only the shadow moved backwards, and they argued that it was possible for God to keep the sun moving in its normal direction while reverting the direction of motion of the shadow. In contrast, other authors stated that the sun itself had to move backwards and not only its shadow. This had been the position most commonly defended by biblical commentators in the past.[37] Cornelius presented three reasons in support of the argument that the sun itself moved backwards. The first is that the Babylonians also referred to having seen a "portent," hence it was not merely a shadow moving backwards; the second reason is the very words in Isaiah 38:8; the third reason is Clavius's argument:

> Thirdly, because such dials in which only the shadow retrogrades can be constructed naturally in any place, if they are made horizontally, where the height of the pole [= latitude] is less than 23 degrees. Thus, at those latitudes, that is, between the equator and any of the tropics of Cancer or Capricorn, this will happen naturally. The demonstration of this fact can be found in chapter 11 of *Fabrica instrumenti ad horologium* of our Clavius (where he also correctly refutes Pedro Nunes assertion that such was the case – and therefore naturally – of the retrogradation of the shadow in the dial of Ahaz, when it is stated that it was miraculous).[38]

35 Cornelius a Lapide, "De horologio Achaz et egressu umbrae," in *Commentaria in Quatuor prophetas Maiores* (Paris: Societas Minima, 1622), 339–342. Further editions were published in the seventeenth century (1625, 1634, 1654, 1664, 1679, 1689), the eighteenth century, as well as more recently.

36 "Quaeritur primo, an solum umbra retrocecesserit in horologio Achaz, an vero etiam sol, et consequenter eius umbra per totum orbem?" Lapide, "De horologio Achaz."

37 Lapide, "De horologio Achaz," 340.

38 "Tertio, quia talia horologia, in quibus sola umbra retrocedat, naturaliter erigi possunt ubiuis locorum, videlicet si ita erigantur plana, ut altitudinem poli habeant minorem quam graduum 23. Unde ubi ea est altitudo, scilicet inter Aequatorem et alterutrum

Cornelius makes no remark about the mathematics behind this argument; he accepts Clavius's analysis with no reservations, as exegetes usually did.[39] Somewhat suprisingly, however, he seems not to have understood the full power of Clavius's words. Cornelius actually affirms only Nunes's position, and seems not to have realised that Clavius had shown that shadow retrograde motion occurs naturally in all regions of the Earth.

Cornelius a Lapide was the first Jesuit exegete to engage seriously with Clavius's (and Nunes's) demonstration, but several other Jesuits, and not only biblical scholars, were also aware of it. In fact, the discussion presented by Clavius rapidly made the topic familiar among members of the Society of Jesus. Soon, Jesuit mathematicians took the issue up. In 1626, Jan Mikołaj Smogulecki (1610–1656), a young student at a Jesuit college, defended his mathematical thesis under the presidency of the mathematics teacher Georg Schönberger. Smogulecki's thesis, titled *Sol Illustratus ac Propugnatus*, is a long discussion of many phenomena involving the sun.[40] Among these, he treats extensively the question "regressio solis," in twelve points. Although in some parts he is clearly dependent on Cornelius a Lapide, his approach is much more mathematical. In Question 1 he addresses the conventional topic, "Utrum sol vere retrocesserit, an tantum umbra eius?" ("Whether the sun truly moves back or is it only its shadow?"), and it is in question 2 that he cites Nunes, via Clavius: "Utrum regressio umbrae in horologio naturaliter non possit contingere?" ("Whether the retrogradation of the shadow cannot happen naturally?"). To this he replies that within the torrid region this can happen naturally. Therefore, he continues in agreement with Clavius, those who live in "Goa, Calicut, Cananor, Golden Chersonesus, Mexico, Cape of Good Hope, in parts of Brasil, Peru, Java Major" will witness the phenomenon naturally without need for a miracle.[41]

Tropicorum, Cancri scilicet aut Capricorni, naturaliter id accidit. Huius rei demonstrationem habet noster P. Clauius in Fabrica instrumenti ad horolog. cap. 11. (ubi et recte refutat Petrum Nonium asserentem, talem, ac proinde naturalem fuissem regressum umbrae in hoc horologio Achaz, cum constet fuisse miraculosum)." Lapide, "De horologio Achaz," 340.

39 Volker Remmert, "'Our Mathematicians Have Learned and Verified This': Jesuits, Biblical Exegesis, and the Mathematical Sciences in the Late 16th and Early 17th Centuries," in *Nature and Scripture in the Abrahamic Religions: Up to 1700*, ed. Jitse M. van der Meer and Scott H. Mandelbrote (Leiden: Brill, 2008), 672 and 676f.

40 Jan Mikołaj Smogulecki, *Sol Illustratus ac Propugnatus ab Illustri ac Generoso Domino* [...] (Freiburg im Breisgau: Theodorus Meyer, 1626). The topic "Regressio solis" is addressed in pages 96–108. Cf. also chapter 3 in this volume by Brent Purkaple, "Appealing to Ahaz. The Jesuits and the Sundial of Ahaz in the Era of Copernican Astronomy."

41 Smogulecki, "Quaeritur 2. Utrum regressio umbrae in horologio naturaliter non possit contingere?," in *Sol Illustratus ac Propugnatus*, 97–98.

Then he approaches the more problematic issue: "Utrum non etiam in zona temperata, quam nos incolimus idipsum possit contingere?" ("Whether also in the temperate region where we live the same can happen?"), to which he replies that it will happen if the dial is adequately inclined.[42] Smogulecki is then obliged to explain how the miraculous character is preserved, which he does by noting that if the dial was horizontal, shadows would never retrograde in Jerusalem and, if the dial was inclined, surely the shadow movement had another cause. Smogulecki's text was well known among Jesuit mathematicians and he himself became a Jesuit with a distinguished career in China. It is not implausible to assume that these gnomonic puzzles were discussed in other occasions throughout his life.[43]

In the following years, the conclusions by Nunes and Clavius seem to have interested above all Jesuit mathematicians and astronomers. In 1651 the celebrated Giovanni Battista Riccioli (1598–1671), published his immensely influential *Almagestum novum*, a gigantic encyclopedia of astronomy and related matters, in two volumes.[44] In vol. 2, chapter 37, (*Quomodo Patres et Sacri Interpretes praedicta Sacrae Scripturae loca intellexerint. De motu Solis, ac Miraculis in Solis Motu diuinitus factis*), Riccioli discusses several issues related to the miracle of the sun in Joshua and then proceeds to the Hezekiah episode. He lists the commentators and exegetes that argued that only the shadow had moved, and those that argued that the sun itself had moved backwards. It is

42 Smogulecki, "Quaeritur 3. Utrum non etiam in zona temperata, quam nos incolimus idipsum possit contingere?," in *Sol Illustratus ac Propugnatus*, 98.

43 Smogulecki's tract *Sol Illustratus ac Propugnatus* was cited, for example, by Christoph Scheiner in his celebrated book *Rosa Ursina* (1626–1630) – but not on Ahaz. On Smogulecki, especially his career in China, see: Edward Kosibowicz, "Un missionnaire polonais oublié: Le Père Jean Nicolas Smogulecki S.J. missionnaire en Chine au XVIIe siècle," in *Revue d'Histoire des Missions* (Paris: Les Amis des missions, 1929), 6:335–360; Robert Danieluk, "Michal Boym, Andrzej Rudomina, and Jan Smogulecki: Three Seventeenth-Century Missionaries in China; A Selection of Documents from the Roman Jesuit Archives," in *Monumenta Serica* (Sankt Augustin: Monumenta Serica Insititute, 2011), 59:417–443.

44 Giovanni Riccioli, *Almagestum novum*, 2 vols. (Bologna: Haeredis Victorij Benatij, 1651). The discussion of the miracle of Ahaz is in vol. 2, pages 482–485. On Riccioli see: Christopher M. Graney, *Setting Aside All Authority: Giovanni Battista Riccioli and the Science against Copernicus in the Age of Galileo* (Notre Dame, IN: University of Notre Dame Press, 2015); Alfredo Dinis, *A Jesuit against Galileo? The Strange Case of Giovanni Battista Riccioli Cosmology*, ed. Ricardo Barroso Batista and Alvaro Balsas (Braga: Axioma-Publicações da Faculdade de Filosofia, 2017). Cf. also chapter 3 in this volume by Brent Purkaple, "Appealing to Ahaz. The Jesuits and the Sundial of Ahaz in the Era of Copernican Astronomy."

while presenting the justification for this last opinion that he introduces the observation:

> Furthermore, it is possible to have sundials in which only the shadow moves backward naturally, as Clavius demonstrates in chapter 21 of *Fabrica instrumentis ad horologia*, where he prudently refutes Nunes who had stated that it could be made with a sundial, and thus naturally, that which was miraculous and accomplished supernaturally.[45]

Riccioli does not give Nunes's argument a prominent place in the discussion, but uses it to substantiate the idea that the sun itself had moved backwards.

Among Jesuit mathematicians, Nunes's argument sometimes showed up in unexpected places. In 1660, in an advanced book on the geometrical properties of the cycloid, the distinguished Jesuit mathematician Antoine de Laloubère (1600–1664), felt compelled to address Nunes's result. In his *Veterum Geometria Promota in Septem de Cycloide Libris*, Laloubère studied the properties of the hyperbolic path of shadows projected by a vertical gnomon. In the fifth corollary to proposition XXXI he directed his attention to the phenomenon of shadow retrogradation shown by Nunes, discussing it in considerable detail.[46] This mention is interesting not only because Laloubère was an eminent mathematician, with outstanding contributions in geometry, but also because the analysis was strictly mathematical, with no biblical implications whatsoever.

As Laloubère's treatment hints, by the second half of the seventeenth century Nunes's demonstration had already acquired an importance of its own, purely as a mathematical result, independent of the story of the dial of Ahaz. It is in this manner that it appears in the very influential *Cursus seu Mundus Mathematicus* by the Jesuit Claude Dechales (1621–1678), where Nunes and Clavius are not even mentioned by name.[47] In the first edition (1674), Dechales addresses the topic in the part devoted to *Gnomonics*, in the proposition

---

45 "Accedit quod naturaliter fieri possunt horologia, in quibus sola umbra retrocedat, ut demonstrat Clauius in fabrica instrumentis ad horologia cap. 21. ubi prudenter Petrum Nonium refutat dicentem, illud ostentum factum fuisset artificio sciaterico, ac proinde naturaliter, cum fuerit miraculose ac supernaturaliter factum." Riccioli, *Almagestum Novum*, 483.

46 Antonio Lalovera, *Veterum Geometria Promota in Septem de Cycloide Libris* [...] (Toulouse: Arnaud Colomiez, 1660), 334–336.

47 Claude Dechales, *Cursus seu Mundus Mathematicus*, 3 vols. (Lyon: Anissoniana, 1674). On Dechales, see: Antonio Nardi, "Un galileiano eccentrico: il gesuita François Milliet Dechales tra Galileo e Newton," *Archives Internationales d'Histoire des Sciences* 49, no. 142 (1999): 32–74.

"Horologium Achaz describere." The title is somewhat misleading because what is presented is simply the mathematical demonstration of the shadow retrograde movement, with no discussion whatsoever of the type of dial or the miracle itself.[48] After Dechales's death a second extended edition was published (1690).[49] In this work, in addition to the discussion of retrogradation in the part on *Gnomonics*, as in the first edition, one finds also the result repeated in the book on *Geography*, now as the property of a gnomon's shadow in different parts of the earth.[50]

Shadow Retrogradation in the Seventeenth Century Outside the Society of Jesus

The dissemination among the Jesuits of Nunes's demonstration of the retrograde motion of shadows, and Clavius's extension of it, was only the first phase of the diffusion of these ideas. Immediately after, Nunes's argument started to be commented on by non-Jesuits as well. In 1640 John Wilkins (1614–1672), the famous bishop of Chester, one of the founders of the Royal Society, published his *Discourse concerning a new planet*, a discussion, and defence, of heliocentrism.[51] The text contained the proposition, "That there is not any place in scriptures from which (being rightly understood) we may infer the diurnal motion of the sun or heavens." It is while discussing the relevant Bible passages that Wilkins is led to comment on the story of Ahaz, about which he has the following to say:

48 Dechales, *Cursus seu Mundus Mathematicus,* vol. 3, 226–227.

49 Claude Dechales, *Cursus seu Mundus Mathematicus*, 2nd ed., 4 vols. (Lyon: Anissonios, Joan. Posuel & Claud. Rigaud, 1690).

50 On Gnomonics: Claude Dechales, "Propositio XIX. De Horologia Achaz," and "Propositio XX. Horologium Achaz exhibere," in *Cursus seu Mundus Mathematicus*, 4:262–263. On Geography: Claude Dechales, "Propositio LXII. Theorema. In zona torrida dum sol versatur inter tropicum, et loci parallelum umbra styli perpendiculariter erecti ad planum horizontale, mane et vesperi sua relegit vestigia," in *Cursus seu Mundus Mathematicus*, 2:432.

51 Originally published as: John Wilkins, *A discourse concerning a new planet, tending to prove, that 'tis probable our earth is one of the planets* (London: Printed by R.H for John Maynard, 1640), with many later editions. I will be referencing the following edition: John Wilkins, *The Mathematical and Philosophical Works of the Right Reverend John Wilkins* […], vol. 1 (London: Printed by C. Whittingham, for Vernor & Hood, Cuthell & Martin and J. Walker, 1802). See Felicity Henderson, "Taking the Moon Seriously: John Wilkins's *Discovery of a World in the Moone* (1638) and *Discourse Concerning a New World and Another Planet* (1640)," in *John Wilkins (1614–1672): New Essays*, ed. William Poole (Leiden: Brill, 2017), 129–157.

> As for this last place concerning the sun's returning ten degrees in the dial of Ahaz (2 Kings xx.11; Isa. xxxviii, 8): I think it may probably be affirmed, that it is to be understood only concerning the shadow: which though it do necessarily happen in all horizontal dials, for any latitude betwixt the tropics: and so consequently in all declining dials, the elevation of whose pole is less than the sun's greatest declination; as Clavius de Horol. cap. 21. observes: yet the circumstances of this relation in scripture, make the event to differ from that other which is common and natural: which against its nature did seem to go backwards, when the sun itself was not in the least manner altered from its usual course. Of this opinion were Abarbinel, Arius Montanus, Burgensis, Vatablus Sanctius, etc.[52]

Wilkins knows Clavius's text and understands well the consequences of the demonstration. He is thus aware that shadow retrogradation can have a wholly natural explanation: "necessarily happen in all horizontal dials, for any latitude betwixt the tropics [...] consequently in all declining dials." However, as a Copernican, he wishes to dissent from Clavius's opinion (most commonly held by Jesuits at the time), that the miracle had been a backward motion of the sun, and he therefore proceeds to discuss five reasons against the possibility of a retrograde motion of the sun.

Wilkins was a respected and very popular author, and after his book, the argument of Nunes was well known in scientific circles in England. But English theologians and biblical scholars were also aware of it. Rev. John Gregory (1607–1646) in his *Notes and Observations upon Passages of Scripture* (1646) makes a brief note that shows not only that he was aware of the demonstration but that the result was already widely known at the time. In the opening letter "To the Reader" he comments on the dial of Ahaz:

> For the kinde I'me sure 'twas like none of ours now in use, and if the Retrocession could be meant of the shadow (and some men look no further) the same thing may be made to fall out everyday upon an ordinary Dyall, and (notwithstanding what a good Mathematician hath said to the contrary) [on the margin: *Petr. Nonius*] in a Site and Position of Spheare

52 John Wilkins, "That the Earth may be a planet," in *The Mathematical and Philosophical Works of the Right Reverend John Wilkins* [...] (London: Printed by C. Whittingham, for Vernor & Hood, Cuthell & Martin and J. Walker, 1802), 1:155.

> without the Tropicks. Therefore the going back is to meant of the Sun it selfe.[53]

The matter-of-fact manner in which the natural retrogradation of shadows, happening every day, is referred to, and the marginal indication of Pedro Nunes, betray that John Gregory is talking about topics well known to his readers. Gregory wishes to defend that the miracle involved the sun directly, and there is perhaps a veiled critique to Wilkins as one of the men that think that it had to do with the shadow only. Given the great popularity of these *Notes and Observations* it is clear that already by the mid-seventeenth century learned authors in England – and via them, the not-so learned public – had been warned about the issues raised by Nunes's result.

Actually, by the mid-seventeenth century, it was especially the strictly "scientific" (astronomical, mathematical) aspect of the demonstration of the retrogradation of the shadows that was generating more interest. The famous *Geographia Generalis*, by Bernhard Varenius (1622–1650), was published for the first time in 1650.[54] It would enjoy great popularity, with several later editions successively improved; Isaac Newton himself contributed to the revised versions and was instrumental in the appearance of editions in 1672 and 1681. In Chapter XXVII, Varenius writes "Of the Shadows which upright bodies cast when enlightened by the Sun; and of the Divisions of the Earth that arise therefrom."[55] He discusses the topics in a series of mathematical propositions, reaching the important Proposition XIII:

53 John Gregory, *Gregorii Opuscula, or, Notes and Observations upon Passages of Scripture* [...], 2nd ed. (London: Printed for R. Royston, 1650). The first edition was in 1646 and there were many later editions. On Gregory see Gordon Goodwin, "GREGORY, JOHN (1607–1646)," in *Dictionary of National Biography*, ed. Leslie Stephen and Sidney Lee (New York: Macmillan and Co., 1890), 23:101–102.

54 Bernhard Varenius, *Geographia Generalis, in qua affectiones generales Telluris explicantur* [...] (Amsterdam: Lodewijk Elzevir, 1650). In addition to numerous Latin and English editions, in 1716 and in 1750 respectively an Italian and a Dutch translation were published. Other translations in French, Turkish and Russian were also published. I will be citing from the edition: Bernhard Varenius, *A Compleat System of General Geography* [...] *Improved and Illustrated by Sir Isaac Newton and Dr. Jurin* [...] (London: Stephen Austen, 1734). For an analysis of the many editions of the Geographia Generalis and the biography of Varenius, see: Margaret Schuchard, ed., *Bernhard Varenius (1622–1650)* (Leiden: Brill, 2007).

55 Bernhard Varenius, "Of the Shadows which upright bodies cast when enlightened by the Sun; and of the Divisions of the Earth that arise therefrom," in *A Compleat System of General Geography*, 619–632.

> *Proposition XIII*: In Places in the Torrid Zone, while the Sun is between the Parallel of the Place and the Tropic nearest to it, the shadow then goes back twice, repeating the directions of the same Points, once in Forenoon and once in the Afternoon; and the Sun will then seem to go back.[56]

A simple demonstration is produced. Varenius adds immediately a Corollary destined to clarify that, in addition to this natural behaviour, there is room for a different, miraculous, retrograde motion, the difference being perceived in the *suddenness* of the movement of the shadows:

> *Corollary*: 'Tis no Miracle that shadows go back on Dials, except they go back on a sudden; or if they point back to the same Hour-Lines; if the style be not perpendicular, but parallel to the Poles: and tho' it be perpendicular, yet the Line of Shadows doth not show the hour, except the Plane of the Dial be in the Equatorial Plane.[57]

Varenius's distinction would be used by later commentators. Later, he established the result that shadows could move backwards at any latitude. Varenius's presentation of these results, in a sequence of clear and terse propositions, was very influential and these propositions were cited or referred by many later authors.

Schools and Universities

Similar to what had already happened in Jesuit colleges, the prediction and demonstration by Nunes attracted considerable attention in the context of mathematical education. A number of mathematical dissertations and public theses dealt with it, in a confirmation of the interest that this result generated in academies and universities. In 1658, in the celebrated Amsterdam Athenaeum, one student, Alardus de Raedt, presented his mathematical thesis, *Disputatio Mathematica de Retrogressu Umbrae in Horologia Achaz*.[58] This *Disputatio* is an

56 Varenius, "Of the Shadows," 624–625.

57 Varenius, "Of the Shadows," 625.

58 Alardus de Raedt, *Disputatio Mathematica de Retrogressu Umbrae in Horologia Achaz* [...] (Amsterdam: Jan Banningh, 1658). On de Raedt, see: Doede Nauta, *Biografisch lexicon voor de geschiedenis van het Nederlands protestantisme*, vol. 6 (Kampen: Kok, 2006), 239–241. On the Amsterdam Athenaeum, see: Dirk van Miert, *Humanism in an Age of Science. The Amsterdam Athenaeum in the Golden Age, 1632–1704*, trans. Michiel Wielema and Anthony Ossa-Richardson (Leiden and Boston: Brill, 2009).

extended discussion, well ordered in thirty-one topics (thesis), of many different aspects related to the motion of shadows in the horologium Ahaz. Alardus de Raedt knew well the works that had treated the question before and cites directly from Nunes. After recalling, in thesis I and II, that Pedro Nunes had shown that shadows retrograde in the torrid zone, and that this truly happens in many places of the earth, Alardus explains (in thesis III) that the same will happen in the temperate region if the sundial is conveniently inclined. And then he directly challenges any doubter to check by himself, by inclining a sundial – in Amsterdam.[59] Having thus brought retrogradation of shadows to the daily experience of his audience, he addresses the question of the miraculous explanation of the Bible story, and after a long argument concludes that it has to be due to a miraculous backward motion of the sun.[60]

In 1673, in Wittenberg, one Gottfried Ulrich Mayer defended a mathematical thesis entitled *Disputatione Mathematica Solem tempore Hiskiae Retrogradum*. The learned discussion of the biblical story included a paragraph devoted to asserting the supernatural dimension of the event ("Causam phaenomeni huius non in natura quarendam, sed soli Deo pro liberrima volutate sua").[61] Mayer, however, felt the need to explain that the question presented some difficulties because Pedro Nunes had demonstrated mathematically that a *natural* explanation for the backward shadow motion was possible, for places within the torrid zone, and Clavius had afterwards extended the result to regions outside the torrid zone. Still, Mayer insisted, this natural behaviour could not be accepted as the explanation of the biblical story, and he proceeded to discuss, according to more traditional commentaries, whether the miracle had affected the shadows (thus, the light rays) or the sun itself.[62]

A similar discussion took place in June 1689, at Jena Academy, when student Joseph Heissius presented his thesis, *Sciatericum Achas et in eo miraculum*.[63] Heissius was familiar not only with the texts of Nunes and Clavius, but also with the *Geographia universalis* of Varenius, from which he also quotes. He was

59 Raedt, *Disputatio*, fol. A2v.

60 Raedt, *Disputatio*, fol. A4r.

61 Gottfried Ulrich Mayer, *Disputatione Mathematica Solem tempore Hiskiae Retrogradum Wittebergae* […] (Wittenberg: Johann Burckhard, 1673), fol. A3r.

62 This dissertation acquired some reputation and was published again in 1719 with a slightly different title: Gottfried Ulrich Mayer, *Disputatione Mathematica de Sole Tempore Hiskiae Retrogrado* […] (Wittenberg: ex officina Gerdesiana, 1719).

63 Joseph Heissius, *Sciatericum Achas et in eo miraculum* (Jena: Literis Krebsianis, 1689), 39–40. The thesis was reprinted two years later in a collection of works published under Johann Andreas Schmidt's name: Johann Andreas Schmidt, *Variorum Philosophicorum Decas* […] (Jena: Tobias Ohrling, Bibl., 1691).

also aware of the refractive sundials made by Georg Hartmann. After explaining the mathematical arguments put forward by Nunes, Clavius, Varenius, and the sundials made by Hartmann, he put foward the possible conclusion: "If, therefore, this can happen solely by natural means, then the retrogradation of the shadow was not against nature."[64] Heissius, of course, was not prepared to accept that. He therefore recalled the arguments put forward by Clavius and by Varenius to justify that the phenomenon at the dial of Ahaz was *different* from the natural backward displacement of the shadows.[65]

At the University of Wittenberg, in 1710, one Christian Hermann presented his dissertation *De Maʿalôt Aḥaz seu de horologio Achaz*. In this very learned discussion, fully dedicated to the episode of the Ahaz sundial, both Christian and Jewish authors are cited, and there is also a reference to Clavius and to a possible natural explanation (*ex naturali*) of the phenomenon.[66]

To many it was the counterintuitive, almost paradoxical, nature of the phenomenon that exerted the greatest appeal. In 1693 Patrick Gordon, FRS, published his *Geography Anatomiz'd or the Geographical Grammar* – a book that would have spectacular success, with some twenty editions published up until the mid-eighteenth century – in which a number of geographical and cosmographical paradoxes were proposed.[67] Paradox 12 proposed the following challenge:

> There is a considerable number of places, lying within the Torrid Zone, in any of which, if a certain kind of Sun-Dyal is duly erected, the shadow will go back several degrees upon the same at a certain time of the year, and that twice every day, for the space of divers weeks; yet no ways derogating from that miraculous returning of the shadow upon de Dyal of Ahaz, in the days of King Hezekiah.[68]

64 "Si ergo haec omnia naturae viribus praestari possunt, non fuit illa umbrae retrocessio contra naturam." Heissius, *Sciatericum Achas et in eo miraculum*, 40.

65 Heissius, *Sciatericum Achas et in eo miraculum*, 40–41. Citing Clavius, he writes: "regressum illum tempore Ezechia regis factum à notro valde diversum fuisse," and then, following Varenius, "tum demum miraculum est, si id subito fiat insigni spatio (cur non etiam successive, si id semel accidat)."

66 Christian Hermann, *De Maʿalôt Aḥaz Seu de horologio Achaz* [...] (Wittenberg: Martin Schultz, 1710), 41.

67 From 1693 to 1754 some twenty different editions of the *Geography Anatomiz'd: or, the Geographical Grammar* were published. I will be citing from the edition of 1719: Patrick Gordon, *Geography Anatomiz'd: or, the Geographical Grammar. Being a Short and Exact Analysis of the Whole Body of Modern Geography* [...] (London: J. and B. Sprint, 1719).

68 Gordon, "Containing some amazing Geographical Paradoxes," in *Geography Anatomiz'd*, 38.

It is evident that Gordon has in mind Nunes's result, because he mentions places within the torrid zone, but since he mentions also a "certain kind" of dial it is possible that he had not fully understood the result and had in mind perhaps an analemmatic or refractive sundial.

Gordon's challenge was answered some years later on purely mathematical grounds by an unidentified author who presented himself merely as "a lover of mathematics." In "An essay towards the probable solution of the forty-five surprizing Paradoxes in Gordon's Geography," published inside *A Mathematical Miscellany in Four Parts* (Dublin, 1730), the paradox is solved simply by stating the mathematical result of Pedro Nunes.[69]

Challenging the Miracle

By the second half of the seventeenth century many mathematicians, astronomers and gnomonists had come to terms with Nunes's mathematical prediction of a natural daily retrograde motion of the shadow of a sundial. Adjusting this result to the biblical episode of the dial of Ahaz had required some care, but had not proved to be exceedingly difficult or problematic. All scholars had agreed on the need to distinguish between a natural backward motion and a miraculous backward motion. The way to make this distinction was open to discussion and there were different proposals, but no one challenged that the Bible reported a truly miraculous event.

This conviction would slowly start to change and Nunes's demonstration was ready to be used in other ways. "The universal disposition of this Age is bent upon a rational religion," observed Thomas Sprat in 1667.[70] Not surprisingly, that demonstration and the fact that it opened the door for a natural explanation of a miracle in the Bible, had great appeal for those who wished to propose naturalistic readings of the Sacred Scripture. One of those men was Charles Blount (1654–1693), a social radical, a deist, and an avowed enemy of organised religion. Immediately after his death, a collection of his papers was published anonymously under the title *The Oracles of Reason* (1693). The book contained direct challenges against Scripture, which were responded to by Josiah King in *Mr Blount's Oracles of Reason, Examined and Answered* (Exeter,

69 Samuel Fuller, *A Mathematical Miscellany in Four Parts. By a Lover of Mathematics* (Dublin: S. Fuller, 1730), 10–11.

70 Thomas Sprat, *The History of the Royal Society*, ed. J. I. Cope and H. W. Jones (London: Routledge & Kegan Paul, 1959), 374. See also: John Spurr, ""Rational Religion" in Restoration England," *Journal of the History of Ideas*, 49, no. 4 (1988): 563–585.

1698).[71] One of these challenges arose because Blount had found impossible to accept, as many assumed, that the sun had moved backwards. King's response is interesting because it was based on the opinion of "all mathematicians:"

> All Mathematicians agree in this, that a Dial may be made between the Tropicks, on which the shadow may naturally go back. And *Clavius* hath demonstrated that the same may be done on a Dial made without the Tropicks. In our elevation here in Exeter, a Plane may be fitted for such a Polar Altitude, as will make a Retrocession of the Shadow Natural.[72]

King, of course, wished to insist on the fact that it was really the sun that moved backwards. The matter-of-fact manner in which he mentioned that the "shadow may naturally go back" is worth noticing, but rather more interesting is the suggestion that people had already observed the phenomenon:

> I Remember a good Mathematician told me, that he made a *South* Vertical Dial, for the Right Honourable the late Lord *Clifford* of *Chudleigh* in *Devon*; on which this *Phaenomenon* of the Shadows going back might be seen.[73]

There had been attempts at explaining the phenomenon as a result of natural causes, as a meteorological or an optical phenomenon. These had been proposed before and continued to be proposed. In his inaugural oration at Gresham college in 1657, Christopher Wren (1632–1723) argued that mathematics can help in explaining the Bible and provided a naturalistic explanation of the miracle of Ahaz by suggesting that it was due to a parhelion.[74] Years later, in 1670, in his famously polemical *Tractatus Theologico Politicus*, Baruch

71 Josiah King, *Mr. Blount's Oracles of Reason, Examined and Answered* [...] (Exeter: S. Darker, 1698).

72 King, *Mr. Blount's Oracles of Reason*, 19–20.

73 King, *Mr. Blount's Oracles of Reason*, 20.

74 Wren's inaugural lecture survives in two versions, an English draft and a Latin text. The English draft is in Christopher Wren Jr., *Parentalia: or, Memoirs of the Family of the Wrens* (London: T. Osborn, 1750), 200–206, and the Latin text in: John Ward, *The Lives of the Professors of Gresham College* (London: J. Moore, 1740), 199–206; Jim Bennett, "Christopher Wren's Greshamite history of astronomy and geometry," in *Sir Thomas Gresham and Gresham College. Studies in the intellectual history of London in the sixteenth and seventeenth centuries*, ed. Francis Ames-Lewis (Hampshire: Ashgate, 1999), 192. See also James A. Bennett, "Christopher Wren: Astronomy, Architecture and the Mathematical Sciences," *Journal for the History of Astronomy* 6 (1975): 149–184.

Spinoza (1632–1677) too suggested that the explanation was a parhelion of the sun.[75] Differently from these explanations, which implied rare and very infrequent events, Nunes's argument postulated a totally banal occurrence, a phenomenon that in fact took place every day, twice per day. Thus, implicit to Nunes's explanation was not the question how does it happen, but, quite the contrary, why hasn't it been observed before? It is no surprise therefore that, as his explanation became more known, reports on the observation of shadow retrogradation also appeared.

In 1703 witnesses reported that the shadow at the sundial in a convent in Metz had moved backwards. This episode generated great interest and many authors reported it.[76] The most common explanation was one that invoked the condensation of the atmosphere, as Thümming had first suggested.

Eighteenth-Century Bible Commentators and Mathematicians

During the eighteenth century, Bible commentators and exegetes, both Catholic and Protestant, seem to generally have been aware of Nunes's result, but they failed to engage deeply with it. Sometimes, so it seems, because they did not fully understand the mathematical argument or because they misjudged its implications.

Abraham Calovius [Calov, Kalau] (1612–1686), the Lutheran theologian and author of the famous "Calov Bible," did not expand much on the topic in his *Biblia Illustrata*, but did not hesitate to mention that the behaviour could be natural, as Nunes and Clavius had shown.[77] Propagating what was a common error, he assumed that Nunes had denied the miraculous nature of the phenomenon and that Clavius had corrected him. As we saw before, things were not that simple.

In 1709 the Dutch theologian Johannes Mark (1656–1731) published his immensely erudite *Scripturariae Exercitationes* in twenty five biblical places of

75 Baruch Spinoza, "De prophetis," in *Tractatus Theologico Politicus* (Hamburg: Henricus Kunraht, 1670), 15–30.

76 See: M. Parent, *Essais de Recherches de Mathematique et de Physique*, vol. 3 (Paris: Chez Jean de Nully, 1713), 34–43; Ludwig Philipp Thümmig, *Phaenomenon singulare solis coelo sereno pallescentis ad rationes revocatum* (Halle: Hiiliger, 1722), 19; Gabriel Christoph Benjamin Busch, *Versuch eines Handbuchs der Erfindungen*, vol. 6 (Eisenach: Wittekinde, 1795), 371.

77 Abraham Calovius, *Biblia Testam. Veteris et Novi Illustrata*, 2nd ed. (Dresden: Johann Christoph Zimmermann, 1719), 830. Originally published as: Abraham Calovius, *Biblia Testam. Veteris Illustrata*, 4 vols. (Frankfurt am Main: Balthasar Christoph Wust, 1672–1676).

the Old Testament.[78] In Exercitatio XV, "Ad 2.Regum. Cap. XX, vers. 8–11, Qua miraculosus umbrae Solaris Regressus per decem gradus tractatur," he discusses an impressive list of opinions, from authors ancient and modern. Citing directly from Cornelius a Lapide, he makes a reference to the Nunes/Clavius mathematical result, as supporting the idea that the miracle involved movement of the sun itself.[79] Mark does not elaborate much more on that result, but his discomfort becomes apparent a few pages later when he puts Clavius and the "impius" Spinoza side by side as proponents of natural explanations for the miracle.[80]

Rather more sensitive to the mathematical arguments was Johann Jacob Scheuchzer (1672–1733) in his famous and beautifully illustrated *Kupfer-Bibel* (1731–1733), originally published in German but released immediately afterwards in Latin and French.[81] Scheuchzer's discussion of the Ahaz episode follows a sequence traditional among exegetes: first, the dial itself, a topic that had generated a great quantity of commentaries; and, secondly, the miraculous retrogradation of the shadows. On this last aspect, Scheuchzer proceeded as usual, listing and analysing the opinion of those who said that the phenomenon had to do only with the shadows and not the sun, and afterwards discussing the reasons of those who held that it was really the sun that moved backwards. While discussing these arguments he notes that "one should not keep silent on what Varenius wrote"[82] and one should also consult about this topic "Petrus Nonius Salaciensis L. II. de arte et ratione nauigandi c. 11. Clauius de fabrica et usu Instrumenti ad Horologior. descript. c. 21. Daniel Schwenter Orat. de Opticae laudibus."[83] Scheuchzer then adds a cautionary reflection,

78 Johannes Mark, *Scripturariae Exercitationes ad Quinque et Viginti Selecta Loca Veteris Testamenti* (Amsterdam: Jacobus Borstius, 1709).

79 Mark, *Scripturariae Exercitationes*, 769–820. The argument by Nunes and Clavius is mentioned on page 800 in paragraph VIII, "Horum autorum argumenta pro ipsius Solis Regressu exhibentur."

80 "Sed et ex Parheliis in natura repertis, de quibus impius Spinoza loquitur, et ex Horologiis arte humana fabrefactis, de quibus ex Clavio narrantem à Lapide audivimus supra, par. VIII." Mark, *Scripturariae Exercitationes*, 819.

81 Johann Jakob Scheuchzer, *Kupfer-Bibel, in welcher die Physica Sacra, oder Geheiligte Natur-Wissenschafft derer in Heil. Schrifftvorkommenden Natürlichen Sachen*, 3 vols. (Augsburg: Christian Ulrich Wagner, 1731–1735). The comment to 2 Kings 20: 8–11 is in vol. 3, 268–274.

82 "Mithin darff man dasjenige nicht stillschweigend übergehen was uns Varenius lehret Geogr. Gener. L. II. c. 27. Prop.13." Scheuchzer, *Kupfer-Bibel*, 3:273.

83 After citing Nunes's mathematical result, Scheuchzer continues: "worüber auch zu lesen sind Petrus Nonius Salaciensis L. II. de arte et ratione nauigandi c. II. Clauius de fabrica et usu Instrumenti ad Horologior. descript. c. 21. Daniel Schwenter Orat. de Opticae laudibus, allwo er berichtet, daß Georg Hartmann, ein guter Mathematicus und Künstler zu

saying that Nunes's and Clavius's arguments "should in no way be interpreted as reducing the miracle that happened in Hezekiah times, or diminishing the glory of GOD that resulted from it."[84]

The mathematical *certainty* of Nunes's demonstration eliminated any doubts about the existence of a *natural* shadow regression. The problem that biblical scholars faced was how to relate this natural phenomenon to the story of Ahaz. Biblical exegetes wanting to guarantee the supernatural character of the Ahaz story had essentially three options: they could argue that the sun itself had moved backwards; they could explain that the retrogradation at Ahaz had been of another type, a non-natural one, resulting from a direct command by God; or they could simply keep silent about the mathematical result.

In fact, some Bible commentators chose to completely omit any mention of the mathematical demonstration about the retrograde motion of shadows. Such was, for example, the case of Antoine Augustin Calmet (1672–1757), the author of one of the most influential catholic Bible commentaries in the eighteenth century, first published between 1707 and 1716, and reprinted many times.[85] His *Commentaire littéral sur tous les livres de l'Ancien et du Nouveau Testaments*, which includes a long and learned "Dissertation sur la rétrogradation de l'ombre du Soleil sur l'horloge d'Achaz," makes not even a passing reference to the mathematical demonstration. This omission is so glaring that there is perhaps something revealing in it. Calmet was a very learned scholar who showed no hesitation in engaging even with the most detested adversaries (he treated Spinoza's natural explanation with sarcasm). Yet, he failed to comment on Nunes's argument that, by this period, was well known among exegetes. One may suspect that he was insensitive to the mathematical argument, or that he felt that it introduced nothing of relevance.

Nürnberg, eine Sonnen-Uhr ersonnen, welche er die Sonnen-Uhr Ahas genennet, worinnen der Schatten just 10. Stuffen, wie in der bishero erzehlten Wunder-Geschichte, zurück gegangen." Scheuchzer, *Kupfer-Bibel*, 3:273.

84 "Jedannoch solle dieses Anbringen keinesweges dahin gemeynet seyn, daß gegenwärtiges zu Ezechiä Zeiten geschehenes Wunder verringert, oder die daraus hervor leuchtende Ehre GOTTES geschmälert werde." Scheuchzer, *Kupfer-Bibel*, 3:273.

85 Calmet's *Commentaire littéral sur tous les livres de l'Ancien et du Nouveau Testament* was immensely popular. It was first published in 23 vols., between 1707 and 1716, with a second edition in 1714–1720 and a third in 1724–1726. Latin translations were published in Lucca (1730–1738), Venice (1730), Frankfurt (1730), Augsburg (1756) and Würzburg (1789). Calmet's commentary was also included in other Bible commentaries. On Calmet, see: Philippe Martin and Fabienne Henryot, *Dom Augustin Calmet; un itinéraire intellectuel* (Paris: Riveneuve éditions, 2008).

But the true reason was perhaps the opposite. In his "Dissertation," Calmet supports a form of scepticism not common among eighteenth-century Catholic discussions about miracles. According to him, natural explanations, that is, Philosophy, provide only hypotheses unable to attain true certainty ("la Philosophie n'a que des hypothéses, sans certitude réelle"); most secrets of nature are therefore forever inaccessible to our analysis ("la plûpart des secrets de la nature seront toujours impénétrables à toute nôtre attention"). On the contrary, Sacred Writings are clear. Thus, one should adjust Philosophy to Revelation and not the other way around ("N'est-il pas bien plus juste, et plus raisonnable de conformer la Philosophie à la révélation, que la révélation à la Philosophie?").[86] Needless to say, to someone defending this point of view, Nunes's mathematical demonstration was exceedingly problematic. As Nunes himself had stated clearly, and as many others after him concurred, including Clavius, his argument was a strict mathematical demonstration, hence absolutely certain and valid, impossible to be brushed away with a statement as "hypothéses, sans certitude réelle." For Calmet, silencing this explanation was surely the most advisable strategy.

Calmet was a Catholic exegete, but on the Protestant side the situation was not much different and commentators sometimes did not mention Nunes's demonstration. Charles Chais (1700–1785), the author of a famous *La Saincte Bible* [...] *Avec un Commentaire Littéral*, with a lengthy 26-page "Dissertation sur la retrogradation miraculeuse de l'ombre solaire au Cadran d'Achaz," made no reference to it.[87] Chais's silence is perhaps explainable by the fact that he defended the idea that it was utterly impossible for the sun to move backwards. To him the miracle of the biblical story happened only in the shadows. This position, of course, was one that had become considerably weaker after Nunes's demonstration. Perhaps because of this, Chais decided to not comment on it. According to Chais the phenomenon at Ahaz was a natural meteorological event ordered by God.[88]

86 Antoine Augustin Calmet, "Dissertation sur la Retrogradation du Soleil, à l'Horloge dAchaz," in *Commentaire littéral sur tous les livres de l'Ancien et du Nouveau Testament*, 2nd ed., vol. 2 (Paris: Emery, 1721), 398.

87 Charles Chais, "Dissertation sur la retrogradation miraculeuse de l'ombre solaire au Cadran d'Achaz," in *La Sainte Bible, avec un Commentaire Littéral, et des Notes choisies, tirée de divers Auteurs Anglois*, vol. 6 (Amsterdam: Marc Michel Rey, 1777).

88 Chais, "Dissertation," xxiii.

An Astronomical Paradox

If biblical commentators sometimes had difficulties in coming to terms with Nunes's argument, the reverse was happening with mathematicians and lovers of scientific questions. An important moment in the dissemination of the mathematical explanation of shadow retrograde movement was the publication of the immensely popular *Recréations Mathematiques et Physiques* by Jacques Ozanam (1640–1717). *Récréations Mathématiques* was originally published in 1694, but had many editions in the following years.[89] In the original version, among the various problems of gnomonics treated there is no mention of the dial of Ahaz. The book, however, was later revised, corrected, and much augmented by Jean-Étienne Montucla (1725–1799) and published in four volumes in 1778. In this new expanded version there are two problems where mathematical demonstrations of the retrogradation of the shadow are given.[90] These were surely a contribution of Montucla, perhaps based on Varenius. In Problem XXXII, "Comment l'ombre d'un style peut rétrograder sur un cadran solaire sans miracle," the mathematical argument by Nunes and a specific calculation are presented. In the following problem, "Sous une latitude quelconque, tracer un cadran où la rétrogradation de l'ombre ait lieu," the generalised situation, as explained by Clavius, is presented.

Ozanam's book, in the expanded version by Montucla, was translated into English by Charles Hutton (1737–1823) and published in London, in 1803, with several later editions.[91] Hutton's English version was very influential and

89 Jacques Ozanam, *Recréations Mathématiques*, 2 vols. (Paris: Jean Jombert, 1694). The book was immensely popular, with many editions afterwards. It was translated into English as *Recreations Mathematical and Physical* [...] (London, 1708), with many later editions and slightly different titles such as *Recreations for Gentlemen and Ladies, or Ingenious Amusements* [...] (Dublin, 1756). On Ozanam, see: William L. Schaaf, "Jacques Ozanam," in *Dictionary of Scientific Biography*, ed. Charles Coulston Gillispie (New York: Charles Scribner's Sons, 1974), 10:263–265; René R. J. Rohr, "Les cadrans solaires universels de Jacques Ozanam," *Centaurus. International Magazine of the History of Mathematics, Science, and Technology* 29 (1986): 165–177; René R. J. Rohr, "The universal sundials of Jacques Ozanam," *Antiquarian Horology* 18 (1989): 171–177.

90 Jacques Ozanam, *Récréations mathématiques et physiques* [...], 4 vols. (Paris: Chez Ant. Jombert, 1778). The problems on the retrogradation of shadow are in vol. 3, 270–273.

91 Jacques Ozanam, *Recreations in Mathematics and Natural Philosophy* [...], trans. Charles Hutton, 4 vols. (London: G. Kearsley, 1803). A second edition was published in 1814 and revised editions, by Edward Riddle, were published in London in 1840 and 1844. On Charles Hutton, see Benjamin Wardhaugh, introduction to *The Correspondence of Charles Hutton (1737–1823)* [...], (Oxford: Oxford University Press, 2017), xiv–xxxv; A. G. Howson, "Charles Hutton," in *A History of Mathematics Education in England* (Cambridge: Cambridge University Press, 1982), 59–74; Margaret E. Baron, "Hutton, Charles," in *Dictionary of*

magnified immensely the impact of Ozanam's book. The two relevant problems, in Hutton's translation, are as follows:

> Problem XXXII. *How the shadow of a style, on a Sun-dial, might go backwards, without a miracle.*
>
> This phenomenon, which on the first view may appear physically impossible, is however very natural, as we shall here show. It was first remarked by Nonius or Nugnez, a Portuguese mathematician, who lived about the end of the sixteenth century. It is founded on the following theorem.
>
> In all countries, the zenith of which is situated between the equator and the tropic, as long as the sun passes beyond the zenith, towards the apparent or elevated pole, he arrives twice before noon at the same azimuth, and the same thing takes place in the afternoon.[92]

Then follows the demonstration. The description by Ozanam (perhaps more correctly, by Montucla-Ozanam) is terse but correct. At the end he calculates actual values for the retrograde arc and for the time of retrogradation, at a specific location and time of year.[93]

Then follows the more general situation, for an inclined sundial at any latitude. The problem is made especially appealing by considering a latitude familiar to the reader – Paris in the French editions; London in the English ones – and showing that shadows retrograde naturally at that latitude:

> Problem XXXIII. *To construct a dial, for any latitude, on which the shadow shall retrograde or move backwards*
>
> For this purpose incline a plane, turned directly south, in such a manner, that its zenith shall fall between the tropic and the equator, and nearly about the middle of the distance between these two circles: in the latitude of London, for example, which is 51° 31′, the plane must make an angle of about 38°. In the middle of the plane, fix an upright style of such a length, that its shadow shall go beyond the plane; and if several angular lines be then drawn from the bottom of the style towards the south, about the time of the solstice the shadow will retrograde twice in the course of the day, as above mentioned.

Scientific Biography, ed. Charles Gillespie (New York: Charles Scribner's Sons, 1972), 6:576–577.

92 Ozanam, *Recreations in Mathematics*, 3:321.

93 Ozanam, *Recreations in Mathematics*, 3:322–323.

> This is evident, since the plane is parallel to the horizontal plane having its zenith under the same meridian, at the distance of 12 degrees from the equator towards the north: the shadows of the two styles must consequently move in the same manner in both.[94]

Lest any reader might think that these results and calculations contained an implicit attempt at explaining the biblical miracle, the text of Montucla-Ozanam closes with an important remark:

> Some may here say that this is a natural explanation of the miracle, which, as we are told in the sacred Scriptures, was performed in favour of Hezekiah king of Jerusalem; but God forbids that we should entertain any idea of lessening the credibility of this miracle. Besides, it is very improbable, if the retrogradation which took place on the dial of that prince had been a natural effect, that it should not have been observed till the prophet announced it to him, as a sign of his cure; for in that case it must have always occured when the sun was between the tropic and the zenith: the miracle therefore, recorded in the Scriptures, remains unimpeached.[95]

The miracle thus remained unimpeached, according to Montucla-Ozanam. It would not take very long for other authors to disagree.

As said, the text of Montucla-Ozanam (also in its English version, by Hutton) was very popular, and many other publications copied the demonstration from it. One such case was the famous *Dictionnaire Encyclopedique des Amusemens des Sciences Mathématiques et Physiques* (1792), a collection of curiosa on matters scientific and technological. In its long note on "Gnomonique," the authors copy Ozanam's text and discuss the two situations, namely "Comment l'ombre d'un style peut retrograder sur un cadran solaire sans miracle" and also "Sous une latitude quelconque, tracer un cadran où la rétrogradation de l'ombre ait lieu."[96]

Despite its circulation in works of popular science, the topic preserved its respectability, to the point of being discussed in the academies of the learned.

---

94 Ozanam, *Recreations in Mathematics*, 3:323. Naturally, the original French version gave Paris as the example: "sous la latitude de Paris, qui est de 49° 50'."

95 Ozanam, *Recreations in Mathematics*, 3:323–324. For unknown reasons, this remark was not included in the 1814 edition.

96 Jacques Lacombe, *Dictionnaire Encyclopedique des amusemens des Sciences Mathématiques et Physiques* (Paris: Chez Panckoucke, 1792), 605–606.

On 26 August 1772, Pietro Borrani presented a *Dissertatione sopra il retrocedimento dell'ombra solare sull'orologio di Acaz* at the Accademia degli Infecondi, a literary and artistic academy established in 1715 in Prato, Tuscany.[97] Borrani's discussion of the Ahaz dial is erudite and deeply influenced by mathematical ideas; not suprisingly, the results of Nunes and Clavius are used.

More importantly, perhaps, the result was presented and discussed in scholarly works. Jean-Étienne Montucla, who had already noted Nunes's demonstration when working on the 1778 augmented edition of Ozanam's *Récréations Mathématiques*, returned to the question in his famous *Histoire des mathématiques*.[98] In a long supplement devoted to the history of gnomonics, "Supplement au quatrième livre, contenant l'histoire de la Gnomonique ancienne et moderne,"[99] Montucla stresses the importance of Nunes's demonstration in the history of the discipline but, cautiously, avoids entering any biblical ground:

> Although the Portuguese geometer and astronomer Nonius has not written any treatise on Gnomonics, he deserves here a distinguished place due to the note and explanation of very unusual gnomonic phenomenon; that is, the retrogradation of the shadow on a sundial in certain latitudes. This phenomenon has seemed to some persons as adequate to explain naturally the occurrence of shadow retrogradation in the dial of Hezekiah. It is not our intent, however, to enter such discussion; we will confine ourselves to the phenomenon described by Nonius.[100]

97 Pietro Borrani, *Dissertatione sopra il retrocedimento dell'ombra solare sull'orologio di Acaz* [...] (Pistoia: Stampatore del Pubblico, 1773). On the Accademia degli Infecondi, see M. Maylender, *Storia delle Accademie d'Italia* (Bologna, 1926–1930).

98 First published in 1758. An expanded second edition was released in 1798–1799 – this is the edition I will be citing from. On Montucla, see: George Sarton, "Montucla (1725–1799): His Life and Works," *Osiris* 1 (1936): 519–567; Kurt Vogel, "Montucla, Jean-Etienne" in *Dictionary of Scientific Biography*, ed. Charles Gillespie (New York: Charles Scribner's Sons, 1974), 9:500–501; Noel M. Swerdlow, "Montucla's Legacy: The History of Exact Sciences," *Journal of the History of Ideas* 54 (1993): 299–328.

99 Jean-Étienne Montucla, *Histoire des mathématiques* [...], 2nd ed., vol. 1 (Paris: Henri Agasse, 1798), 715–736.

100 "Quoique le géomètre et astronome portugais Nonius n'ait pás écrit de traité de Gnomonique, il mérite ici une place distinguée, par la remarque et l'explication d'un phénomène gnomonique fort singulier; c'est celui de la rétrogradation de l'ombre sur un cadran sous certaines latitudes. Ce phénomène a paru á quelques personnes propre à expliquer naturellement celui de l'ombre rétrogradant sur le cadran d'Ezéchias; mais il n'est pas de notre objet d'entrer dans cette discussion; nous nous bornerons au Phénomène remarqué par Nonius. Comme néanmoins, cette explication, nécessairement un peu longue et compliquée jusqu'à un certain point, couperoit trop notre narration,

The detailed explanation of Nunes's argument is presented in a final note, "Note du Supplément au Quatrième livre. Sur le phénomène de la rétrogradation de l'ombre dans un cadran solaire."[101] Montucla presents a detailed analysis, with calculations, of Nunes's demonstration. He then explains that the same will happen at any latitude, as for example, at the latitude of Paris, provided one incline the plane of the sundial in such a way that its zenith lies between the equator and the nearest tropic, and concludes: "Donc ce qui se passera sur l'un se passera aussi sur l'autre." Montucla closes his careful analysis with a simple and prudent observation about the issue of the miracle:

> I leave to others the task of discussing to what extend one can explain with this the famous miraculous phenomenon performed on behalf of Hezekiah. I will limit myself to observe that Jerusalem is outside the torrid zone and that the Holy Scripture explains these facts in such a way that makes it extremely doubtful that that has been called the dial of Ahaz was in reality one.[102]

In the anglophone world Hutton's translation of Ozanam's *Récréations Mathématiques* assured that Nunes's mathematical demonstration became widely known in the first decades of the nineteenth century. A confirmation of this fact is the frequent appearance of information about the phenomenon in some influential encyclopaedias published in the United Kingdom.[103] The *Encyclopaedia Londinensis; or, universal dictionary of arts, sciences, and literature*, published in London between 1801 and 1828, in 24 volumes, under the direction of John Wilkes (1750–1810), presents a very erudite and extensive note on "Horology." There one finds a discussion on "How the Shadow of a Style, on

on la trouvera dans une note placée à la fin de ce supplément." Montucla, *Histoire des Mathématiques*, 1:730; see also 1:60–61.

101 Montucla, *Histoire des Mathématiques*, 1:737–739.

102 "Je laisse au surplus à d'autres le soin de discuter jusqu'à quel point on peut expliquer par-lá le fameux phénomène miraculeux operé pour Ezéchias. Je me bornerai à observer que Jerusalem est hors de la zone torride et que l'Ecriture Sainte s'explique sur ce fait d'une manière qui rend plus que douteux, que ce qu'on a appellé le cadran d'Achaz en fut un." Montucla, *Histoire des Mathématiques*, 1:739.

103 For broader context and a discussion of encyclopedias in the history of culture, see: Robert Collison, *Encyclopedias: their history throughout the ages*, 2nd ed. (New York: Haffner Publishing Company, 1966).

a Sun-Dial, might go backwards without a miracle is treated." The discussion is simply copied from Hutton.[104]

Between 1808 and 1830 the 18 volumes of the celebrated *Edinburgh Encyclopædia* appeared, under the editorial supervision of David Brewster (1781–1868). In the excellent note on "Dialling," possibly written by Brewster himself, it is stated:

> The Portuguese astronomer Nonius, deserves to be mentioned here, on account of his having noticed and explained the phenomenon of the retrogradation of the shadow on a dial in certain latitudes. Some have supposed that it was in this way the shadow went back on the sun dial of Ahaz, but this explanation seems quite inadmissible.[105]

The paragraph is not wholly correct, of course, because Nunes did not notice the phenomenon; his approach was strictly theoretical. Somewhat further, the issue of the retrogradation of the shadows is discussed in modern terms. The text explains what the retrogradation of the shadow is and concludes with the observation:

> The phenomenon we have been describing will appear very simple to the mathematicians, and perhaps hardly worthy of particular notice. It seems, however, to have excited attention, probably because of the apparent solution it affords to the return of the shadow on the sun-dial of Ahaz. But this explanation is not admissible, because in that case, the return of the shadow must have been a thing altogether miraculous, otherwise it would not have excited attention.[106]

At the beginning of the nineteenth century, fully naturalistic readings of the Bible with frontal refusals of some (or all) miracles were not uncommon. Biblical scholars were keenly aware of these ideas and these radical points of view. In the famous biblical commentary by Adam Clarke (1762–1832), *The Holy Bible: containing the Old and New Testaments* (1817), when commenting on 2 Kings 20, and the dial of Ahaz, the editor decided to include also a mention

104 "How the Shadow of a Style, on a Sun-Dial, might go backwards without a miracle is treated," in *Encyclopedia Londinensis*, ed. John Wilkes (London: J. Adlard, 1811), 10:300. A carefully drawn mathematical diagram is added.

105 "Dialling," in *The Edinburgh Encyclopædia*, ed. David Brewster (Edinburgh: William Blackwood, 1830), 7:693.

106 "Dialling," 7:706.

of the existence of a natural retrograde motion of the shadow in a sundial. He resorted to Hutton's English version of Ozanam's text.[107] After a very brief explanation of the physical phenomenon he closed his comments on the topic by referring the "unholy" attempts that had been made to use this result to deny the miracle:

> Of these principles some have endeavoured to make an unholy use, contending that what the Holy Scriptures consider to be a miracle, in the case of the retrogradation of the shadow on the dial of Ahaz, was the effect of a mere natural cause, without any thing miraculous in it. On this subject Dr. Hutton very properly remarks: "It is very improbable, if the retrogradation which took place on the dial of that prince had been a natural effect, that it should not have been observed till the prophet announced it to him as the sign of his cure; for in that case it must have always occurred when the sun was between the tropic and the zenith."[108]

Clarke does not elaborate much more in his response, but his words of caution are sufficient to confirm that some were proposing a completely natural reading, an "unholy" reading, of the miracle at the dial of Ahaz, and that the mathematical demonstration played an important part in the discussion. To Clarke, of course, this was completely unacceptable and he closed his explanations on the topic with strong words denouncing the infidelity of those that proposed such naturalistic interpretations:

> But the miraculous interposition is so obvious, that infidelity must be driven to pitiful shifts when it is obliged to have recourse to the insinuation of imposture, in a case where the miraculous interference of God is so strikingly evident.[109]

---

107 In the commentary to 2 Kings 20 in Adam Clarke, *The Holy Bible: containing the Old and New Testaments* [...], 8 vols. (New York: Daniel Hitt and Abraham Paul, 1817). On Adam Clarke, see: Phillip Schaff and Jakob Herzog, "Clarke, Adam," in *The New Schaff-Herzog Encyclopedia of Religious Knowledge* (Grand Rapids: Baker, 1953), 3:126; Ian Sellers, *Adam Clarke, controversialist: Wesleyanism and the historic faith in the age of Bunting* (St Columb Major: Wesley Historical Society, 1976); and Ian Sellers, "Clarke, Adam," in *Oxford Dictionary of National Biography*, ed. H. C. G. Matthew and Brian Harrison (Oxford: Oxford University Press, 2004).

108 Clarke, *The Holy Bible*, 1:878.

109 He adds: "Besides, such a dial could not be constructed for the latitude of Jerusalem without having the north end elevated twenty degrees seven minutes; which could not be used for the purpose which is indicated in the text. See No. 3 of the preceding observations." Clarke, *The Holy Bible*, 1:878.

Retrogradation of Shadows is Finally Observed

Events were about to take an exciting turn. In August 1877, a military engineer, Etienne Guillemin, sent a note on the retrogradation of the shadow in a sundial to the Helvetic Scientific Society.[110] In a succinct manner Guillemin described the mathematical aspects and also the sundial he had built for the purpose of observing the phenomenon.[111] Guillemin then proceeded to explain his observations and measurements. These are the first confirmed observations of shadow retrogression: it had taken 340 years for Nunes's prediction from 1537 to finally be confirmed by observation:

> One can see, from the moment the Sun appears above the plane of the dial, the shadow moving away from the meridian, and then approaching it until noon and continuing to progress in the same direction during a certain amount of time that depends on the relation of the declinations of the gnomon and the sun. After this, the shadow retrogrades again, approaching the meridian; very slowly first and afterwards faster.[112]

Having described both the mathematics of the phenomenon and his own observations, Guillemin proceeds to a series of final remarks that reveal his true intent. Assuming that the phenomenon of shadow retrogradation was known for more than 2.500 years (and always considered to be a miracle), he then asks, somewhat theatrically: "How is it possible that a phenomenon so natural and simple was ignored by mathematicians and astronomers and that no scientific work, that we know of, mentions it?"[113] Given that the phenomenon had been studied and described by authors whose works were easily accessible, such as Varenius, Ozanam, Montucla, Hutton, not to mention

---

110 Etienne Guillemin, "De la rétrogradation de l'ombre sur le cadran solaire," *Verhandlungen der Schweizerischen Naturforschenden Gesellschaft = Actes de la Société Helvétique des Sciences Naturelles = Atti della Società Elvetica di Scienze Naturali* 60 (1878): 201–207.

111 Guillemin, "De la rétrogradation," 203–204.

112 "On voit, dès le moment où le soleil apparaît audessus du plan du cadran, l'ombre s'écarter du méridien, puis s'en rapprocher jusqu'à midi et continue à progresser dans le même sens, pendant un temps qui dépend du rapport des déclinaisons du style et du soleil. Après quoi, l'ombre rétrograde de nouveau en se rapprochant du méridien; très lentement d'abord, plus rapidement ensuite." Guillemin, "De la rétrogradation," 204. These words are the first recorded observation of shadow retrograde motion, but see below for the claim by a former French Navy officer of having noted the phenomenon already in 1872.

113 "Comment se fait-il qu'un phénomène naturel si simple, soit ignoré des mathématiciens et des astronomes, et qu'aucun ouvrage scientifique, à notre connaissance du moins, n'en fasse mention?" Guillemin, "De la rétrogradation," 205.

the more ancient ones, whose works were perhaps not so accessible, the only thing that is surprising here is Guillemin's vast ignorance. All that Guillemin knew about the history of the question was what he could gather from the *Dictionnaire encyclopedique des amusements des sciences mathématiques et physiques* (1792), which is a mere copy of Ozanam's text. There he learned that Nunes had first explained mathematically the phenomenon, but, making a very hasty reading, concludes that Nunes had performed no calculations and had committed a serious error.[114] This sloppy and inattentive reading is enough for Guillemin to make his very predictable judgement: it was due to Nunes's errors and the influence of an "intolerant clergy" that the explanation of the phenomenon was forgotten.[115]

Completely unperturbed by the fact that his information came solely from an unreliable, second-hand source, Guillemin then started a fantastic tirade that would have delighted the anti-Catholic sensibilities of many in late nineteenth-century France. He postulated that Nunes's (putative) error had been a "voluntary error" by a man who had not dared to speak his mind. As usual, this type of speculation always ends up in invocations of persecutions and the Holy Inquisition, and Guillemin does not disappoint. Having mentioned the Inquisition, the name Galileo should soon appear, and again, Guillemin is wholly predictable. Actually, in his deeply confused state of mind, he treats Galileo as a contemporary (!) of Nunes. And he ends, spectacularly, by saying that, "if he had been permitted to show clearly that a fact described in the Bible as miraculous is nothing but a natural phenomenon, dependent of the sun's declination and the will of the operator, Nunes, we say, would have probably been burned as a heretic."[116]

There is something moving in Guillemin's narrow-mindedness and great ignorance. His intention, of course, was quite far from scientific. For him, the mathematical explanation of the phenomenon provided a clear proof of the errors and lies in the Bible: mathematical certainty exposed biblical falsity. It was, in his own words, a triumphal affirmation that "la science va de l'avant,

114 Guillemin writes: "un géomètre portugais, nommé Nonius ou Nugnez, qui vivait au XVIe siècle, avait déjà, à l'aide d'une figure, donné l'explication du fait dont il s'agit. Nous en trouvons la description dans un traité d'escamotage, publié a Paris en 1792 et intitulé: *Dictionnaire des amusements des sciences mathématiques et physiques*, […]. Malheureusement, dans sa démonstration, Nugnez ne donne aucun calcul et commet une forte erreur." Guillemin, "De la rétrogradation," 205. The reference is to Jacques Lacombe, *Dictionnaire Encyclopedique des amusements des Sciences Mathématiques et Physiques* (Paris: Chez Panckoucke, 1792), 605–606.

115 Guillemin, "De la rétrogradation," 206.

116 Guillemin, "De la rétrogradation," 206–207.

la vérité se fait jour." Guillemin's text was much too good to be left unused by rationalist thinkers. In 1879, Émile Littré (1801–1881), a noted freemason strongly influenced by Auguste Comte, used Guillemin's argument and his mention of Nunes in his corrosive denounciation of the miracles in the Bible.[117]

But despite Guillemin's propagandistic animus, his contribution was not without value. He was the first person ever to have observed the phenomenon of shadow retrogradation. In addition, he was instrumental in bringing into this topic a major figure of late nineteenth-century French science, the well-known populariser Camille Flammarion (1842–1925). Flammarion relates that he was visited in Paris in the year 1880 by Etienne Guillemin, "who came to explain that the miracle of Isaiah could be reproduced, and that it had been since the sixteenth century by Nonius (Nunes)."[118] The following year, in 1881, Flammarion himself went to Lausanne, to analyse the phenomenon with Guillemin in detail and after a few months began explaining that the miracle of Ahaz could be reproduced. This was not enough to quench his interest and in 1884 he built a sundial especially designed to observe shadow retrogradation in the Observatory of Juvisy, and began making systematic observations. Finally, in 1885 Flammarion published a detailed report of his investigations and observations in *L'Astronomie*, a journal devoted to "astronomie populaire, météorologie et physique du globe."[119] Flammarion's paper was followed by an article by Guillemin with a detailed mathematical analysis.[120] To Flammarion, Nunes's mathematical result showed that there was no miracle. It was simply a

117 É. Littré, "De quelques phénoménes naturels donnés ou pris dans la Bible comme miraculeux," *La Philosophie positive* 22 (1879): 147–149. Not suprisingly, Littré's (and Guillemin's) arguments were challenged by apologists such as Fulcran Vigouroux (1837–1915), in Fulcran Vigouroux, *Les Livres Saints et la Critique Rationaliste*, vol. 5 (Paris: Roger & Chernoviz, 1891), 126–133.

118 Camille Flammarion, "Le cadran solaire a rétrogradation de l'Observatoire de Juvisy, et le miracle d'Isaie," *L'Astronomie: Revue d'astronomie populaire, de météorologie et de physique du globe* 4 (1885): 321–329. On Flammarion, see: Danielle Chaperon, *Camille Flammarion: Entre astronomie et litterature* (Paris: Imago, 1998); Bernardette Bensaude-Vincent, "Camille Flammarion: Prestige de la science populaire," *Romantisme: Revue Du 19e Siecle* 65 (1989): 93–104; and Philippe de la Cotardiere and Patrick Fuentes, *Camille Flammarion* (Paris: Flammarion, 1994).

119 Flammarion, "Le cadran solaire a rétrogradation," 321–329.

120 Étienne Guillemin, "Théorie, formules et épures de la rétrogradation de l'ombre," *L'Astronomie: Revue d'astronomie populaire, de météorologie et de physique du globe* 4 (1885): 329–340. For more information, see: D. Roguet, "Le cadran solaire de l'Observatoire Flammarion de Juvisy et les Cadrans Solaires en general," *L'Astronomie: Revue d'astronomie populaire, de météorologie et de physique du globe* 26 (1912): 441–464; and G. Delmotte, "Cadran solaire Flammarion à Rétrogradation," *L'Astronomie: Revue d'astronomie populaire, de météorologie et de physique du globe* 37 (1923): 408–409.

natural phenomenon that had been known since antiquity and had been used to deceive the king. The explanation by Nunes was a superb opportunity to launch a naturalistic attack on the miracles in the Bible.

The papers by Flammarion and Guillemin gave rise to two immediate responses that appeared in the following issue of *L'Astronomie*. A priest of Poitiers, one *abbé* A. Blain, questioned the interpretation arguing that the biblical text required that the sundial had to be vertical and not horizontal, as Flammarion's (and Nunes's) mathematical treatment assumed.[121] Interestingly, Blain shows himself to be quite well informed about the history and discussions on shadow retrograde motion, mentioning Clavius and Cornelius a Lapide. The director of the Journal also intervened.[122] Of especial interest was a letter by a former French navy officer, claiming to have observed the phenomenon during one of his past voyages.[123]

The question by now had acquired a renewed interest and, in 1887, an engineer called Jules Gaudard published a 130-page long study of the topic: *Le cadran d'Achaz et les miracles*.[124] The essay started with a mention of the result obtained by Nunes and the work of Flammarion and Guillemin, but soon diverted from a technical discussion to become a lengthy study on the nature of miracles and, in particular, a discussion about the miracles in the Bible. Gaudard's sympathies were clear: he intended to follow where Flammarion had left and launch a critique of traditional reading of the Bible. The point that he wanted to stress was that the *possibility* of a natural retrogradation of the shadows eliminated what once had been considered a prodigy or a miracle.[125] The mathematical *certainty* of Nunes's demonstration provided the solid ground from which to launch the attack.

But not all were anti-biblical diatribes. By the end of the nineteenth century the question was still capable of generating some surprise among astronomers and mathematicians. In the December 1898 issue of the journal *Popular*

121 A. Blain, "La retrogradation de l'ombre," *L'Astronomie: Revue d'astronomie populaire, de météorologie et de physique du globe* 4 (1885): 387–389.

122 "Réponse," *L'Astronomie Revue d'astronomie populaire, de météorologie et de physique du globe* 4 (1885): 389–390.

123 Paul Jaffé, *L'Astronomie: Revue d'astronomie populaire, de météorologie et de physique du globe* 4 (1885): 390–391. The letter writer claims to have noticed shadow retrograde motion during a voyage on a ship in 1872. If this is so, this would have been the first observation of the phenomenon, but there is no way to confirm this statement.

124 Jules Gaudard, *Le cadran d'Achaz et les miracles* (Lausanne: G. Bridel, 1887).

125 "le fait capital de la possibilité d'une rétrogradation naturelle de l'ombre est dûment constaté, en sorte que ce qui était prodige alors ne le serait plus auhjourd'hui, sous réserve de certains arrangements." Gaudard, *Le cadran d'Achaz*, 7.

Astronomy one J. Morrison presented a paper stating that "we purpose to discuss the subject mathematically and to show by the rigorous formulae of mathematics, that this apparently extraordinary conduct of the sun dial of Ahaz, is not only possible from an astronomical point of view but that the statement of the prophet is, without the shadow of a doubt, literally and absolutely true."[126] Morrison presented a modern and correct mathematical explanation of the retrogradation of the shadows. He was, however, totally unaware of the long history related to this phenomenon. In the following issues of *Popular Astronomy* a small exchange of comments and replies took place.[127] One Herman S. Davis questioned: "I only wish to know to what extent the author [J. Morrison] is willing to acknowledge priority for the suggestion to the "method of demonstrating how the shadow of a style on a Sun-dial may go backwards without a miracle" first given, I believe, by Nonius the Portuguese astronomer who lived from 1492 to 1577. His demonstration is given on Page 321 of Vol. III of Hutton's edition of Montucla's Ozanam's *Recreations in Mathematics and Natural Philosophy* [...]. Surely there is "nothing new under the Sun," not even in the shadow thereof!"[128]

Conclusion

Pedro Nunes's surprising mathematical prediction, first presented in 1537 and again, in more detail, in 1566, introduced a new and unexpected topic in the immense body of commentaries on the dial of Ahaz (2 Kings 20:9–11; Isaiah 38:7–9). The story of the impact of this mathematical result in the interpretation of that famous biblical tale has never been told, and this is what I tried to do in this essay.

Nunes demonstrated that shadows naturally retrograde in places in the torrid region wherever the place's latitude is smaller than the sun's declination. In those places, every day of the year, twice a day, the shadow of a sundial's gnomon stops in its normal progression from West to East, moves backwards for a period, and then resumes its normal path. This is a very surprising result: it is extremely counterintuitive and it was achieved solely with mathematical arguments, in the complete absence of any observation.

---

126 Joseph Morrison, "The Sun Dial of Ahaz," *Popular Astronomy* 6 (1898): 540.

127 A lively discussion and exchange of letters and small articles took place throughout 1899 in *Popular Astronomy* 7 (1899): 51, 102–104, 108, 166–167, 206–208, 242–243, 277, 368–371, 381.

128 Herman S. Davis, "General Notes – The Dial of Ahaz," *Popular Astronomy* 7 (1899): 102–103.

A few years afterwards this result was shown to be even more general by Christopher Clavius, who noted that shadow retrogradation was not confined to places in the torrid region and actually could happen in all parts of the Earth. Despite being very counterintuitive, and also despite the fact that it is difficult to observe, this result is impossible to dispute because it is grounded on a rigorous mathematical demonstration. As Nunes and several of his commentators stated: "it cannot be doubted." For more than three centuries, from 1537 when it was first predicted, to 1877 when it was unquestionably observed for the first time, the natural occurrence of shadow retrograde motion was discussed, analysed, and studied solely based on the fact that it had been mathematically demonstrated.

This mathematical result had a long story and a complex trajectory with at least three lines of development. First, "men of science," be they gnomonists, mathematicians or astronomers, took it as an important result that required careful examination. Although nobody ever doubted the correctness of the result, its extreme counterintuitiveness introduced an element of surprise that had to be conquered by a close examination of the mathematical arguments. Through the years mathematicians and astronomers demonstrated it using different techniques and it was a frequent topic for mathematical examinations and dissertations. Secondly, popularisers of science and lovers of paradoxes were fascinated by it and used it as a puzzle. To them, it was precisely the result's total unexpectedness that made it interesting. And, finally, biblical exegetes had somehow to come to terms with its importance and consequences. The crucial issue was how to coordinate and harmonise mathematical certainty and biblical inerrancy. Nunes's result, by being an unquestionable mathematical demonstration and by showing that shadow retrograde motion was a natural, mundane, daily occurrence, clearly posed a challenge to biblical exegetes. In different manners, and to varying degrees of relevance, they incorporated it into their interpretations of the episode of the dial of Ahaz, but this required establishing delicate distinctions between natural and miraculous events in nature. Not surprisingly, the demonstration ended up being powerful ammunition in the hands of nineteenth-century freethinkers and all those who wanted to question the possibility of miracles and of a supernatural reading of the Bible.

Montucla said that the result of Nunes was so interesting that he deserved a distinguished place of honour in the history of gnomonics. This judgement, however, falls somewhat short of what Nunes really accomplished. Nunes's demonstration was a *prediction*. By mathematically predicting a natural phenomenon never observed, by initiating a debate that lasted for centuries before any observation was made, Nunes gave a startling demonstration of the power of mathematics. This prediction should perhaps be put in line with much later but similar instances in the history of science. That this prediction also affected the interpretation of a famous biblical episode only added to the surprise; mathematicians and astronomers were no less startled by the prediction than exegetes and biblical scholars.

CHAPTER 3

Appealing to Ahaz

The Jesuits and the Sundial of Ahaz in the Era of Copernican Astronomy

Brent Purkaple

Abstract

The biblical story of the Sundial of Ahaz was important in the early modern period for a variety of reasons, one of which was its relationship to Copernican astronomy. A miracle of a moving shadow seemed to clearly imply a moving sun, hence disproving Copernican astronomy. Yet, what is intriguing is that when one traces this biblical miracle among many members of the Society of Jesus, for whom the rejection of Copernican astronomy was important, it is not altogether apparent that this text was important because of its bearing upon Copernicanism. Rather, the fundamental interest of the Jesuits with regard to this biblical event was defending its miraculousness. The bearing of the miracle upon Copernicanism was secondary, and even then such a connection was not altogether clear. This chapter covers these aspects and others as a reminder that Copernicanism came about at a time when the Sundial of Ahaz was an important topic, even apart from its cosmological implications.

The biblical story of the Sundial of Ahaz as found in 2 Kings 20, Isaiah 38, and Ecclesiasticus 48 was important in the early modern period. For example, both Martin Luther and the Jesuit theologian, Robert Bellarmine, made reference to the Sundial of Ahaz – though with different interpretations – in the midst of disputes concerning the nature of religious signs and symbols. In a book about the Christian faith, Robert Bellarmine defends the significance of 'signs' according to Catholicism in contrast to that of the Lutherans and the Calvinists, mentioning the sundial as a type of 'sacrament,' along with the Battle of Gibeon, and Noah's rainbow. In this context, Bellarmine was directly engaging with Martin Luther who had previously appealed to the Sundial of Ahaz in his *Babylonian Captivity of the Church* as an explanation of what a sign entailed. In addition to these overt confessional lines drawn around the exegesis of this biblical miracle, it is also evident that it was not uncommon for interpretations of the Sundial of Ahaz to be circulated alongside those of the Battle of Gibeon for symbolic purposes alone. For instance, the Jesuit theologian Jacob Masens

 | DOI:10.1163/9789004677654_005

lists the Battle of Gibeon and the Sundial of Ahaz side by side as emblematic of the Church (Joshua) and of the Wisdom of God (Ahaz).[1]

In this chapter, however, I am setting aside these overt religious appeals, to focus only on those relating to sixteenth- and seventeenth-century astronomy. After 1543 and the publication of Nicolaus Copernicus's *De revolutionibus*, many early modern intellectuals turned toward biblical texts, such as the Sundial of Ahaz, as an arbiter as to whether or not the sun was stationary. As a consequence, for those advocating Copernicanism it became not only important to explain the astronomy, but equally so the coherence of the new astronomical system with certain biblical texts, such as with the story of the Sundial of Ahaz. Despite various attempts to do so, such as by Galileo Galilei who, in his 1615 *Letter to the Grand Duchess Christina*, claimed the miracle occurred not in the sun but in the shadow, by the second decade of the seventeenth century the Catholic theological consensus was clear: Copernicanism stood contrary to Scripture.

One curious aspect of this exegetical consensus against Copernicanism pertains to whether the biblical story of the Sundial of Ahaz was obviously utilised as an anti-Copernican argument. While certain early moderns such as Galileo, Foscarini, Campanella, and Caccini clearly drew the passage into a relationship to Copernicanism, and while the historiography generally asserts it as an important passage in this regard, the actual identification of this biblical event as significant for Copernicanism is much more ambivalent in the early modern period than might be expected.[2] This chapter establishes this aspect by considering the interpretive meaning of the Sundial of Ahaz among the Society of Jesus. As I demonstrate, while select Jesuit authors did interpret this particular biblical text as an anti-Copernican argument, the more pervasive

1 Robert Bellarmine, *Disputationum Roberti Bellarmini Politiani, Societatis Iesu, de controversiis christianae fidei, adversus huius temporis Haereticos*, vol. 2 (Ingolstadt: David Sartorius, 1589), 242; Martin Luther, *The Babylonian Captivity of the Church, 1520*, edited by Erik H. Herrmann and Paul W. Robinson (Minneapolis, MN: Fortress Press, 2016), 67; Jacob Masens, *Speculum imaginum veritatis occultae, exhibens symbola, emblemata, hieroglyphica, aenigmata* [...] *editio nova* (Cologne: Johannes Kinckius, 1664), 1102.

2 On Foscarini see Paolo Antonio Foscarini, "A Letter to Fr. Sebastiano Fantone, General of the Order, Concerning the Opinion of the Pythagoreans and Copernicus About the Mobility of the Earth and the Stability of the Sun and the New Pythagorean System of the World," in Richard J. Blackwell, *Galileo, Bellarmine, and the Bible: Including a Translation of Foscarini's Letter on the Motion of the Earth* (Notre Dame: University of Notre Dame Press, 1991), 224–236. On Caccini's sermon against Copernicanism, see Blackwell, *Galileo, Bellarmine, and the Bible*, 112. For a typical identification of the Sundial of Ahaz as clearly implying the anti-Copernican message, see Pietro Omodeo, *Copernicus in the Cultural Debates of the Renaissance: Reception, Legacy, Transformation* (Leiden: Brill, 2014), 57.

interpretation focused on the literal textual meaning, that the event was a miracle and an important sign of God's activity within the world.[3] As shown, the implications of this biblical event for Copernican astronomy are often noticeably absent. I contend that appreciating this complexity is an important aspect to recover when reconstructing the broader history of the way the books of nature and of Scripture were read together in the period after Copernicus's publication.

To explain this, I focus upon key mathematicians and astronomers among the Jesuit Order who all provide noteworthy discussion of the Sundial of Ahaz as it relates to early modern astronomy. The main authors I address are Christopher Clavius (1538–1612), Johannes Smogulecki (1610–1656), Giovanni Battista Riccioli (1598–1671), Mario Bettini (1582–1657), and Athanasius Kircher (1602–1680), on account of the extended treatment that they provide for this biblical event within their texts and the way in which all except Riccioli provide a physical recreation of the sundial. What becomes evident is that each of these authors advocated the miraculous nature of the biblical event, even though the precise lines of explanation and identification of the sundial varied.[4] In order to further reinforce the complex identity of this biblical event in the period after Copernicanism, I also proffer important considerations from Jesuit theologians in the final section, raising questions about the exegesis of these biblical texts and the way certain exegetical problems influenced the appeal of this biblical miracle to astronomers. In the end, this analysis shows that while Jesuit authors are uniform in their understanding of this biblical event as a miracle, the theological and scientific considerations that are brought to the passage are not merely – and perhaps not even fundamentally – applied to Copernicanism.

---

3 While this article takes the Jesuits as an interesting and important case study, it should be noted that the interest in the Sundial of Ahaz within the context of Copernicanism extended also to the Protestants. See Rienk Vermij, *The Calvinist Copernicans: The Reception of the New Astronomy in the Dutch Republic, 1575–1750* (Amsterdam: Koninklijke Nederlandse Akademie van Wetenschappen, 2002), 242, 328.

4 This goes in contrast to Sven Dupré who states that in "the sixteenth century, but maybe earlier, in the fifteenth century, as will become evident, the dial of Ahaz was taken to have been a refractive sundial. A scaphe dial, filled unto the rim with water, shows the shadow of the sun retreating as in the biblical passage." Sven Dupré, "Galileo, the Telescope, and the Science of Optics in the Sixteenth Century: A Case Study of Instrumental Practice in Art and Science" (PhD diss., University of Ghent, 2002), 86.

Christopher Clavius and the Early Opinion on Ahaz's Sundial

The most important Jesuit who dealt with the Sundial of Ahaz during the first decades of the Society's existence was Christopher Clavius. His works are seminal in this period and were utilised in forming the anti-Copernican consensus, among Jesuit mathematicians, astronomers, and theologians, and was routinely cited through at least the end of the seventeenth century. Recent interpretations, largely founded upon Clavius's published comments on two biblical miracles, the Sundial of Ahaz and the Battle of Gibeon, in his 1570 *Commentary on the Sphere of Sacrobosco* and their later figuration on the engraved title page of Clavius's collective works in 1612, have suggested that Clavius guided the Jesuit consensus with regard to Copernicanism and that the Sundial of Ahaz was among the important biblical texts that demonstrated the invalidity of Copernicanism.[5] While it is true that Clavius does include the Sundial of Ahaz among the biblical texts contrary to Copernicanism in 1570, in his subsequent and much more expansive writings on the Sundial of Ahaz this correlation is noticeably absent, an observation which challenges the belief in widespread utilisation of the Sundial of Ahaz as an anti-Copernican text.

In his 1570 *Commentary on the Sphere of Sacrobosco* Clavius laid out reasons for rejecting Copernicanism based upon biblical passages. After citing passages from the Psalms and Ecclesiastes, all of which supported the stationary earth and mobile sun, Clavius concluded by saying, "And again, it is recounted among the miracles that God sometimes causes the sun to go back or to stand still together."[6] As has been pointed out, in this particular instance what Clavius was referring to was the miracle of the Sundial of Ahaz (2 Kings

5 Volker Remmert, "Catholic Biblical Exegesis and the Origins of the Galileo Affair: The Engraved Title Page of Clavius's Opera Mathematica (1612)," in *Picturing the Scientific Revolution: Title Engravings in Early Modern Scientific Publications* (Philadelphia: St. Joseph's University Press, 2011), 17–45.

6 The full Clavius quotation: "The meaning of the Scriptures affirm in many places that the Earth is immobile and that the Sun and the rest of the stars move. Thus we read in Psalm 104:5 Who laid the foundations of the Earth, that it should not be removed forever. Similarly in Ecclesiastes 1:4–6, The Earth abideth for ever. The Sun also ariseth, and the Sun goeth down, and hasteth to his place where he arose: and there rising againe compasseth by the South, and bendeth to the North. What could be clearer? Also the testimony presented to us in Psalm 19:4–6 states very clearly that the Sun moves. There we read: In them hath he set a tabernacle for the Sun, which is a bridegroom coming out of his chamber, and rejoiceth as a strong man to run a race. His going forth is from the end of the heaven, and his circuit unto the ends of it: and there is nothing hid from the heat thereof. And again, it is recounted among the miracles that God sometimes causes the sun to go back or to stand still together." Quoted in Remmert, "Catholic Biblical Exegesis," 20.

20, Isaiah 38, Ecclesiasticus 48) and the miracle at the Battle of Gibeon (Joshua 10). Setting aside the other important miracle, that of the Battle of Gibeon, which is beyond the scope of this chapter, in this particular context it seems clear that Clavius assumes the interpretive utility of the Sundial of Ahaz as indicating the falseness of Copernicanism. Given this aspect one might think that this indicates the overall influence of this biblical text in shaping the anti-Copernican consensus, particularly in a text that was reprinted throughout the latter part of the sixteenth century and first half of the seventeenth century.

What is striking, however, is that aside from his brief comments here in his *Commentary on the Sphere of Sacrobosco*, Clavius does not provide any further elaboration on the Sundial of Ahaz. In fact the only mention of sundials Clavius makes in the 1570 edition, as well as in subsequent editions, is with regard to the way a sundial makes sensible the celestial movements.[7] Clavius's fullest explanations for the Sundial of Ahaz are seen in his printed works on sundials produced in the 1580s: the *Gnomonices libri octo* (1581) and the *Fabrica et Usus Instrumenti* (1586). Both proved important in his intellectual efforts as they were published in the same decade that Clavius finalised his calendrical reforms.[8]

In the *Gnomonices libri octo* Clavius addresses what will be reiterated by many later Jesuits: the Sundial of Ahaz is significant, in part, as an earlier historical attestation to sundials than that of Anaximenes of Miletus, as recorded by Pliny.[9] Clavius expounds a more detailed explanation of the Sundial of Ahaz five years later in 1586, in the *Gnomonices*'s companion, the *Fabrica et Usus Instrumenti*.[10] It is here that he presents his most significant interpretation of the Sundial of Ahaz, which is utilised by later Jesuit authors,

7 Christopher Clavius, *In sphaeram Ioannis de Sacro Bosco Commentarius* (Rome: Victor Helianus, 1570), 481. The movements of the celestial sky are of course important for Clavius's point. It remains unchanged in editions up to the time of the publication of Christopher Clavius, *Opera Mathematica*, 1612.

8 For a background on this period see James M. Lattis, *Between Copernicus and Galileo: Christoph Clavius and the Collapse of Ptolemaic Cosmology* (Chicago: University of Chicago Press, 1994).

9 Christopher Clavius, *Gnominices libri octo* (Rome: Francesco Zanetti, 1581), 7. He also adds in general that sundials began with the Egyptians, moved to the Greeks, then to the Latins. He also includes a lot of information on how ancient sundials were produced, information which will be cited by later Jesuit authors, particularly Book 11, proposition 60, which clarifies how an ancient equatorial sundial would work. Such a pattern of interpretation is interesting because Book 11, proposition 60 makes no mention of the Sundial of Ahaz. Pliny records Anaximenes as the inventor of sundials in Book 2, chapter 76 of Pliny, *Historia mundi naturalis* [...] (Frankfurt am Main: Martin Lechler, 1587), 21.

10 Christopher Clavius, *Fabrica et Usus Instrumenti ad Horologiorum Descriptionem Peropportuni* [...] (Rome: Bartolomeo Grassi, 1586).

whether mathematician, astronomer, or even theologian. In this text, Clavius is responding to a recent challenge to the miraculous nature of the event, wherein a natural regression of a shadow was proven to be a mathematical possibility by Pedro Nunes. Interestingly enough, the issue for Clavius was not Copernicanism explicitly, but whether or not the shadow's regression was miraculous and thus supernatural, as the text indicates, or whether it could be natural as Nunes's mathematics described.[11]

Nunes first introduced the mathematical argument in 1537 in Portuguese and again in 1566 in Latin. According to Nunes, the shadow produced by a gnomon would naturally recede under specific conditions, such as the viewer being situated between the equator and the Tropic of Cancer. As Nunes contended, mathematically, if one were located south of the Tropic of Cancer, the shadow on a gnomon should retrograde twice naturally, once before the meridian and once afterwards. The natural implications for the interpretation of the Sundial of Ahaz, as Henrique Leitão shows in this volume, were not lost on Nunes.[12] Yet, as Nunes and later Clavius would argue, the geographical location of Jerusalem precluded the influence of Nunes's argument for its bearing upon the interpretation of the biblical text. Following both Nunes's mathematical argument and his explanation as to why this did not affect the interpretation of the biblical passage, Clavius explained that Jerusalem had a higher latitude than that necessary for Nunes's mathematical regression, thereby implying that the event of Ahaz's Sundial "cannot happen by the forces of nature."[13]

Notably, Clavius does introduce a natural regression for the shadow, in which one could mimic from any location a latitude south of the Tropic of Cancer. Thus, even if one were north of the Tropic of Cancer, it stands to reason that one could reproduce a different latitude through the proper tilt of the sundial, thereby reproducing Nunes's claim. As Clavius reasons, though, the regression of the shadow either prior to the meridian or after is so minor, that it would not correspond to the ten degree regression as specified in the biblical text (Isaiah 38:8).[14] In addition to this, since the biblical text identifies the regression of the shadow as being 'new' and 'uncommon,' Clavius contends that it could not have occurred through a known trick, thus Nunes's claim as well as any reproduction of it would not fit with the wording of the text.[15]

---

11 Clavius, *Fabrica et Usus Instrumenti*, 110.

12 See chapter 2 in this volume: Henrique Leitão, "Mathematical Certainty and Biblical Inerrancy. Pedro Nunes and the Retrogradation of Shadows at the Dial of Ahaz."

13 "viribus naturae fieri non potest." Clavius, *Fabrica et Usus Instrumenti*, 110.

14 Clavius, *Fabrica et Usus Instrumenti*, 110.

15 "res illa fuit in horologio Achaz nova atque insolita, noster autem regressus in eodem plano singulis annis per plures dies apparet." Clavius, *Fabrica et Usus Instrumenti*, 110.

In the end Clavius's explanation of the sundial in the *Fabrica* developed as a response to Nunes. As such, the exegesis that Clavius applies to the text pertains to its miraculous nature, in contrast to a natural explanation, thereby maintaining the literal interpretation of the biblical text that the event was a miracle. Yet it is important to note that Clavius makes no statement with regard to Copernicanism in the context of the Sundial of Ahaz. While it is true that Clavius assumes the geocentric universe in the broader context of his explanation, as he identifies in the context the "diurnal motion of the Sun," one of only a few such references in the *Fabrica*, he nevertheless does not draw any clear connection between the biblical event and its bearing upon the Copernican question.[16] Instead, Clavius utilises mathematics to maintain the literal, and thus miraculous, interpretation of Scripture.

Within Clavius's explanation of the Sundial, he introduces an important aspect with regard to the Sundial of Ahaz and Nunes's rejection, which is particular to the Jesuit Order: their global mission. Due to their geographical locations, certain Jesuit missions could actually experience the natural regression of the sundial as Nunes had claimed. As a consequence, Clavius lists several locations within the Torrid Zone, where individuals could experience this, such as in Goa, Calecut, East Indies, Malacca, Mexico, Cape of Good Hope, Brazil, Peru, and Sumatra.[17] This reappears in works by later Jesuit authors and provides a unique connection between the Society of Jesus and this miracle. The importance of this particular biblical event and its mathematical explanation is given added weight in light of the global dimension of the Jesuit mission, not only affirming the significance of the mathematical response to Nunes, but also the important role that sundials played within the global mission of the Jesuit Order.

Due to his influence Clavius's mathematical explanation of the Sundial of Ahaz was picked up by later Jesuit authors, among them mathematicians, astronomers, and even theologians. For instance, in Serarius's *In libros Regem et Paralipomenon commentaria posthuma*, Serarius states that the event was not natural, citing chapter 21 of Clavius's *Horologiis*. The text in question is not the *Horologiis*, but rather Clavius's *Fabrica* chapter 21, which argued for the miraculous nature of the event in contrast to Nunes without reference to

16 "Sol motu diurno." Clavius, *Fabrica et Usus Instrumenti*, 107.

17 The locations listed are as follows: Goa, Calecut, Cananor, Aureae Chersonesi, Mexicum Hispaniae nova, Capitis bonae spei, Brasil, Peru, Jova Maior. Clavius, *Fabrica et Usus Instrumenti*, 109.

Copernicanism.[18] Additionally the commentary by Cornelius a Lapide cites both Clavius's *Gnomonices* and his *Fabrica*, neither of which, as noted above, pertains to issues of Copernicanism.[19] Similarly, in 1674 and then again in 1690, in Claude Dechales's *Cursus seu Mundus Mathematicus*, Dechales utilises mathematical explanations that are quite similar to Clavius's mathematical explanations, although neither Nunes or Clavius himself are mentioned.[20] In contrast to Clavius's explanation, as well as those of the theologians, Dechales does not engage the wording of the biblical text, but instead only reproduces the assumed mathematical question which Nunes had introduced. Differences aside, the broader interest on the part of Dechales was to explain the mathematics of the natural regression, "not to diminish faith by the miracle, but as an example to disclose their industry."[21] Even though Dechales does not mention Clavius his conclusions are similar: mathematical explanation reinforces an understanding of the miraculous nature of the event as entirely supernatural.

Interpreting Clavius's interest in the Sundial of Ahaz within this broader context aids in the analysis of two prominent depictions of the event in the decade surrounding Galileo's *Letter to the Grand Duchess Christina* of 1615 and the Catholic Church's condemnation of Copernicanism in 1616. The two publications, Jean Voel's *De horologiis sciothericis libri tres* (Figure 3.1), printed in 1608, and Christopher Clavius's *Opera Mathematica* (Figure 3.2) printed in 1612, both prominently picture the two miracles on their engraved title pages and have been used as evidence that the Sundial of Ahaz played a prominent role in the anti-Copernican consensus.[22]

---

18 Nicolaus Serarius, *In libros Regem et Paralipomenon commentaria posthuma* (Mainz: Balthasar Lipp, 1617), 351. It is also worth noting that Serarius makes no mention of Copernicanism in the context, an aspect which I will draw attention to in the later section "An Exegetical Conundrum."

19 Cornelius a Lapide, *Commentaria in Quatuor prophetas Maiores* (Antwerp: Martin Nuyts, 1625), 312.

20 Claude Dechales, *Cursus seu Mundus Mathematicus*, vol. 3 (Lyon: Anissoniana, 1674), 226–227; Claude Dechales, *Cursus seu Mundus Mathematicus*, vol. 4 (Lyon: Anissonios, Joan. Posuel & Claud. Rigaud, 1690), 262–263. The edition published in 1690 has a slight expansion on its basic mathematical explanation as compared to that of the 1674 edition.

21 "Aliquid non dico simile, sed tantum affine tentat Gnomonica, non ut fidem miraculo deroget, sed ut specimen edat suae industriae." Dechales, *Cursus seu Mundus Mathematicus*, 3:227. Dechales's explanation of the sundial follows an identification of the three main historical sundials from antiquity, the astronomical, Italian, and Babylonian. Such a connection between the history of sundials and that of Ahaz's Sundial conceptually follows Clavius's own interest in the event.

22 Jean Voel, *De horologiis sciothericis libri tres* (Tournon: Claude Michel & Thomas Soubron, 1608); Christopher Clavius, *Opera Mathematica*, 5 vols. (Mainz: Johann Volmar, 1612).

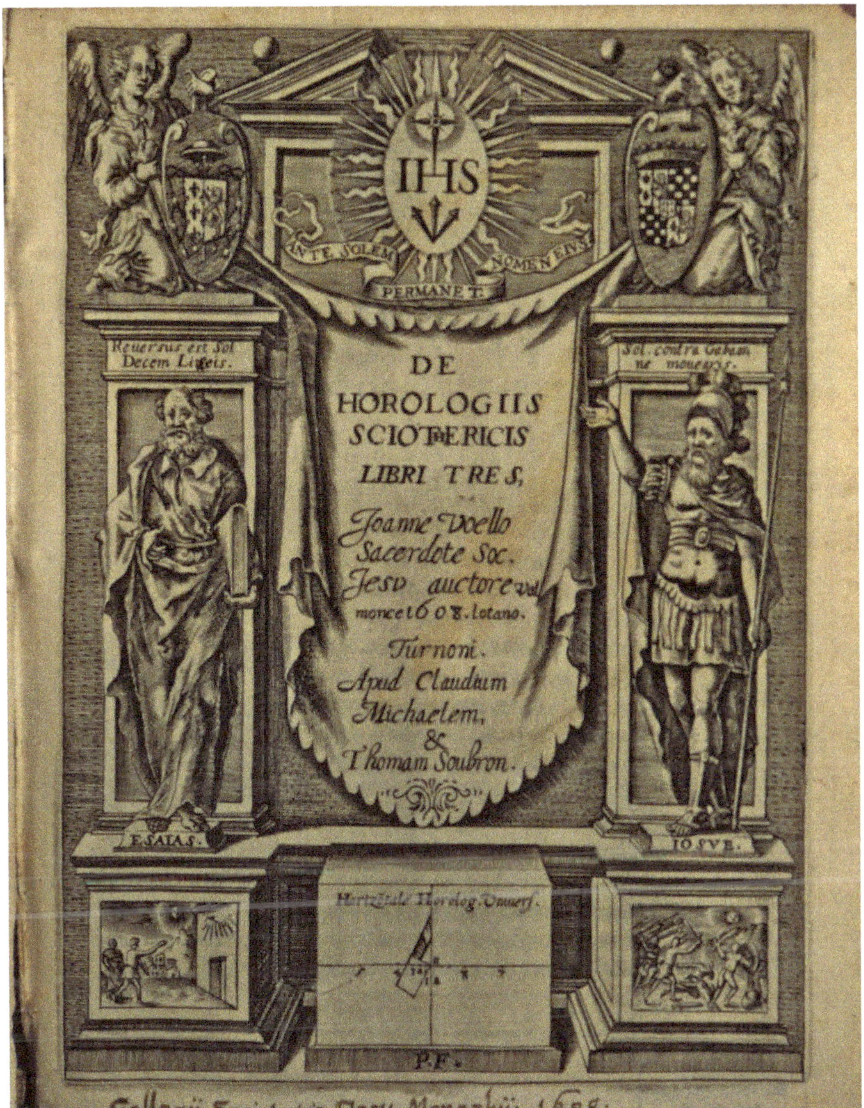

FIGURE 3.1 *Engraved title page* of Jean Voel, *De horologiis sciothericis libri tres* (Tournon: Claude Michel & Thomas Soubron, 1608)
PHOTO: BAYERISCHE STAATSBIBLIOTHEK, MUNICH, 4 MATH.A.315, TITLE PAGE

When looking at Voel's Preface in his *De horologiis sciothericis libri tres*, it is evident he is situating himself in line with Clavius's mathematical work on the sundial, particularly Clavius's *Gnomonices* from 1581. Similar to Clavius, in

FIGURE 3.2 *Engraved title page* of Christopher Clavius, *Opera Mathematica*, 5 vols. (Mainz: Johann Volmar, 1612)

PHOTO: LINDA HALL LIBRARY OF SCIENCE, ENGINEERING & TECHNOLOGY, KANSAS CITY

the dedication Voel states the important historical aspect of Ahaz's Sundial, namely that Ahaz was within the first Olympiad and so was the earliest attestation of a sundial.[23] While questions still remain as to the relationship between Voel's text and the Battle of Gibeon – which is not mentioned anywhere in the entire book – Voel's presentation of the Sundial of Ahaz suggests that it is less probable that the publication of the book was intended to situate it in the context of Copernicanism in 1608. Given these important aspects, what is much more likely is that the image provided an apt engraved title page, heightening the text's value through the figuration of such an important miraculous event.

Yet Voel's engraved title page also implies more about the value of sundials, particularly the universal sundial at the base of Voel's title page (Figure 3.1), a construction which he borrows from Clavius's *Gnomonices*. For Voel, as well as other Jesuits, Clavius's universal sundial came to be a sign of important mathematical and theological value, as it was applied to the construction of a cross, creating a tangible occasion in which mathematics was utilised for theology (Figure 3.3). In Clavius's universal sundial, which Voel reproduces in his 1608 book, the edges of the cross worked as gnomons and the moveable arm beneath the cross meant that it could be adjusted for any latitude.[24] Such a device draws attention to the practical theology of sundial crosses and the way in which the movement of light and shadow across the arms of the cross could serve as a source of theological experience.[25]

An additional aspect worth addressing is the partial quotation from the Psalms at the top of Voel's title page: "His Name Shall be continued as long as the sun" (Psalm 72). There is some suggestion that such an identification came to be used on sundials at churches, a further attempt to draw a correlation between sundials and aspects of its theological interpretation.[26] Upon looking at the title page, though, there is a more subtle figuration. With the IHS monogram of the Society situated above the biblical text the title page

23 Voel, *De horologiis*, a2b–a3a.

24 The sundial cross is included in both Clavius's *Gnomonices* and Voel's *De horologiis*. Clavius, *Gnomonices*, 636; Voel, *De horologiis*, 302. The protestant Georg Hartman also had constructed a cross that allowed for use in diverse latitudes. See Jasmin Mersmann, "Moving Shadows, Moving Sun. Early Modern Sundials Restaging Miracles," *Nuncius* 32, no. 1 (January 2015): 104.

25 It is clear that Clavius's universal sundial cross proved a source of theological reflection for the Jesuit mission as the uniformity of the sundial cross could be experienced in diverse geographical contexts. See Hadriano Lyraeo, *De Imitatione Iesu Patientis sive de Morte et Vita* (Antwerp: Jacob van Meurs, 1655), 344.

26 Mrs. Alfred Gatty, *The Book of Sun-Dials*, ed. H. K. F. Eden and Eleanor Lloyd, 4th ed. (London: George Bell & Sons, 1900), 211.

636 GNOMONICES

in centro infixum moueri poſſit, tranſeatque per centrum, vt in Aſtrolabio fieri ſolet, in cuius extremitatibus duæ tabellæ C E, D F, ad angulos rectos erigantur, ita vt deprimi poſſint, & erigi: quarum altitudines ita explorabimus. Sumpta in recta G H, diametro quadrati A B C D, æquali, ducatur H I, ad G H, perpendicularis; deſcriptoque ex G, vt centro, arcu circuli, ſupputentur in eo declinationes ſignorum, & aliorum graduum Eclipticæ, (Nos integrorum tantum ſignorum declinationes in exemplum adduximus) per quas rectæ ex G, emiſſę ſecent rectam H I, in K, L, M. Deinde in dioptra fiant in recta C E, quæ ipſam dioptram bifariam ſecet, duo foramina parua, vt per illa radius Solis tranſire poſſit, tanto interuallo inter ſe diſtantia, quanta eſt recta H K, quibus reſpondentia puncta in recta D F, quæ alteram tabellam bifariam ſecet, notentur, tanto ſpatio à puncto D, remota, quanto foramina à puncto C, abſunt. Si enim puncta K, L, M, transferantur in rectam D F, inter dicta duo puncta interiectam, initio facto à puncto prope D, & iterum initio facto ab altero puncto prope F, imprimendo puncta in recta D F, per quæ parallelæ lineę ducantur, inter quas 12. ſigna Zodiaci deſcribantur, vt in figura apparet, perfecta erit dioptra, cuius hic vſus eſt. Poſito quadrato horologii A B C D, in propria altitudine, ita vt Aequatori æquidiſtet, vertatur ipſum horologium, & dioptra hinc inde, donec radius Solis per foramen prope C, Sole in ſignis auſtralibus exiſtente, vel per foramen prope E, Sole in borealibus ſignis commorante, incidens cadat in locum Solis in recta D F, hoc ſolo notato, quòd antemeridiano tempore (quod cognoſcetur ex vmbra cuiuſcunque rei in plano Horizonti æquidiſtante. Hæc enim ſi decreſcit, Sol nondum ad Meridianum peruenit, ſi vero creſcit, ipſum iam pertranſijt) dioptra dirigenda eſt verſus ortum, pomeridiano vero verſus occaſum. Hac etenim ratione horologium proprium ſitum obtinebit, & linea fiduciæ vltra centrum verſus tabellam dioptræ oppoſitam indicabit horam præſentem, ſiue Sol exiſtat in borealibus ſignis, ſiue in auſtralibus.

COLLOCATO horologio ſiue per acum Magnete illitam, ſiue per lineam meridianam inuentam, in proprio ſitu, inuenietur hora multo facilius ex dioptra. Tunc enim dioptra ſolum circumducenda eſt, donec radius Solis cadat in rectam D F, etiamſi in tabella non ſint deſcripta ſigna. Quòd ſi deſcripta ſint ſigna, indicabit idem radius, in quonam ſigno Sol exiſtat.

IAM vero ſi horologium in plano Horizonti æquidiſtante verſus Solem dirigatur, ita vt vmbra axis cadat in lineam meridianam, vel radius Solis per dioptram incidens in rectam D F, & quadratum A B C D, eleuetur, donec à Sole non amplius illuminetur, ſed productum per Solem tranſeat, vel donec radius Solis in duo puncta rectæ D F, quæ foraminibus dioptræ oppoſita ſunt, cadat, indicabit quadratum ipſum in quadrante altitudinem Solis ſupra Horizontem. Ex qua, ſi meridiana fuerit, id eſt, maxima illo die, & ex declinatione Solis altitudo poli elicietur, vt in commentariis in ſphæram docuimus.

DE HOROLOGIO VNIVERSALI IN FORMAM crucis constructo. CAP. II.

Compoſitio horologii vniuerſalis in formam crucis.

FIAT ex eadem materia crux plana A B C D, & alia ſolida, concaua tamen, A E F G, quæ cum priori ita coniungatur, vt libere poſſit erigi & deprimi; ſintque tria brachia E, F, G, inter ſe prorſus æqualia. Deinde ad longitudinem ſtyli vni brachiorum æqualis deſcribatur tam horologium Meridianum, quàm Polare: Et Meridianum quidem orientale in latere crucis A F, depingatur, ductis lineis ad rectam E K, perpendicularibus; occidentale autem in oppoſito latere, vt numeri indicant: Polare vero in ſuperiore parte brachiorum E, H, hoc eſt, in plano crucis ad planum E G, recto, horę quidem antemeridianæ ad partes H, pomeridianæ vero ad partes E, vt numeri oſtendunt, ductis lineis ad rectam E H, perpendicularibus. In cruce denique A B C D, acus Magnete illita ſtatuatur, vt beneficio ipſius horologium ita poſſit collocari in plano, quod Horizonti ęquidiſtet, vt A, ad Boream, & C, ad Auſtrum vergat, latuſque A C, meridianam lineam referat. Poſito enim horologio in tali ſitu, ſi beneficio quadrantis crux A E F G, ita eleuetur, vt angulus C A L, ſit angulus con[p]lementi altitudinis poli, iacebit planum F M, in Aequatore, & brachia E, H, I, G, L, F, &c. horas

FIGURE 3.3 *Universal sundial cross*, in Christopher Clavius, *Gnominices libri octo* (Rome: Francesco Zanetti, 1581), 636

PHOTO: MAX PLANCK INSTITUTE FOR THE HISTORY OF SCIENCE, BERLIN

reads rather ambiguously as to whether the intended referent of 'His Name' is the intended reference of the Psalms – the LORD – or whether it is the Society itself. Such ambiguity becomes even more interesting when considering the Society's relationship to sundials throughout the seventeenth century, wherein the global reach of the Society came to be identified through the relative location of sundials, as illustrated in Figure 3.4.[27] As noted above in Clavius's own text, the global dimension of the Society and its close connection to sundials was a significant aspect that Clavius drew attention to when considering the Sundial of Ahaz. Of interest, however, is the way in which it fits within the broader rhetorical interests of the Society as they sought to use visual means to communicate the worldwide influence of their efforts.[28]

Whether or not Voel's book on sundials is an early visual attestation for the close identification between the Jesuit mission and sundials is a point yet to be determined. However, the questions it raises as well as the broader history of Clavius's role in the Sundial of Ahaz suggests a new perspective with which to consider Clavius's *Opera Mathematica*, published in 1612. Within this framework, the presentation of the Sundial of Ahaz, alongside other prominent biblical miracles, argues for Clavius's own activity in using the mathematical sciences to uphold the 'queen of the sciences,' theology itself. This would fit with the position of Clavius at the bottom of his own *Opera Mathematica*, wherein a portion of the biblical text from Wisdom 7 is placed on Clavius's lips "God gave me to know the cycles of the years and positions of the stars."[29] Understanding the title page from this perspective, wherein mathematics upholds the miraculous, fits the tradition and reception of Clavius's work on the Sundial of Ahaz and also upholds the existence of the other miracles on

27 Other examples are in Georg Schönberger, *Exegeses fundamentorum gnomonicorum* (Ingolstadt: Elisabeth Angermaier, 1615), 66, as well as Athanasius Kircher, *Ars magna lucis et umbrae* (Rome: Hermann Scheus, 1646), fol. 553. The only image that changes between the first edition of Kircher's *Ars magna* (1646) and the second edition (Amsterdam: Johannes Janssonius van Waesberge & Elizaeus Weyerstraet, 1671) is the sundial foldout. In addition to this, at the Jesuit Trinity College in Lyons, there was a universal sundial that would show the time in 37 colleges of the Society of Jesus. See Denis Ribouillault, "Sundials on the Quirinal: Astronomy and the Early Modern Garden," in *Gardens, Knowledge and the Sciences in the Early Modern Period*, ed. Hubertus Fischer, Volker Remmert, and Joachim Wolschke-Bulmahn (Cham: Spring International Publishing, 2016), 119.

28 See also Volker Remmert, "Visuelle Strategien zur Konturierung eines jesuitischen Wissensreiches," in *Le monde est une peinture. Jesuitische Identität und die Rolle der Bilder*, ed. Elisabeth Oy-Marra and Volker Remmert (Berlin: Akademie Verlag, 2011), 85–108.

29 For a helpful exposition of this text in the context of Clavius's frontispiece see Volker Remmert, "Catholic Biblical Exegesis," 33–34.

FIGURE 3.4 *A sundial map depicting Jesuit colleges relatively aligned based on the sundial time*, in Athanasius Kircher, *Ars magna lucis et umbrae* (Rome: Hermann Scheus, 1646)

PHOTO: HISTORY OF SCIENCE COLLECTIONS, UNIVERSITY OF OKLAHOMA LIBRARIES, NORMAN

the title page: the Sun at the Battle of Gibeon, the Rainbow at Noah's flood, and the Star of Bethlehem.[30]

While it is tempting to interpret these two images as clearly involved in the Copernican question, they are best situated within the broader Jesuit interest in locating the mathematical sciences as useful for theological interpretation, an aspect important for Clavius at the time when the Jesuit educational curriculum, the *Ratio studiorum,* was being established.[31] Suggestions that these images actively engage in the Copernican debate, while appealing, are not supported by the broader textual context.[32] Rather than a subtle sign of an anti-Copernican rhetoric, they advocate the usefulness of mathematics for biblical interpretation.[33] In the midst of questions about whether the biblical text depicted a miracle or not, Clavius's mathematical tradition provides an explanation to support both the miracle as well as the veracity of the Bible. While one might anticipate that a regressing sun would provide useful commentary on Copernicanism, it is important to recover the complex relationship between Copernicanism and Scripture, such as the absent connection between an otherwise obvious application. For some reason Ahaz's Sundial is not drawn into relationship to Copernicanism.

At the time when Clavius's *Opera* were published it was not mathematicians and astronomers alone who were interested in integrating the mathematical sciences with theology. For instance, the Jesuit theologian Robert Bellarmine articulates a similar point in his *De Ascensione* from 1615, a treatise in the spiritual tradition of Bonaventure. Bellarmine writes "all things are distributed, and each in measure, number, and weight, as Wisdom states," a reference which comes from the same biblical book cited at the bottom of

30 As Volker Remmert has noted, both the Rainbow and the Star of Bethlehem were topics of discussion among early modern Jesuits, particularly as it related to whether or not God could use natural means to act in supernatural manners. See Remmert, "Catholic Biblical Exegesis," 34–35.

31 For background on this see Dennis C. Smolarski, "The Jesuit Ratio Studiorum, Christopher Clavius, and the Study of Mathematical Sciences in Universities," *Science in context* 15, no. 3 (2002): 447–457.

32 That these images do not engage in anti-Copernican rhetoric goes in contrast to previous work by Volker Remmert. For an alternative to what is presented in this chapter, see Volker Remmert, "Catholic Biblical Exegesis," 17–45.

33 For more examples of the open exchange between Jesuit theologians and mathematicians, see Volker Remmert, "'Our Mathematicians Have Learned and Verified This': Jesuits, Biblical Exegesis, and the Mathematical Sciences in the Late Sixteenth and Early Seventeenth Centuries," in *Nature and Scripture in the Abrahamic Religions: Up to 1700*, ed. Jitse M. van der Meer and Scott Mandelbrote (Leiden: Brill, 2008), 665–690.

Clavius's title page.[34] In 1651, Bellarmine's biblical reference (Wisdom 11:20), reoccurs on the frontispiece of the astronomer Giovanni Battista Riccioli, in his *Almagestum novum*. The importance of this is that the interest in utilising the mathematical sciences for theology did not stem merely from the mathematicians themselves, but also from the theologians. If there was a consensus emerging in the early decades of the seventeenth century, what it pertained to was the utility of the mathematical sciences for theology.[35] From this perspective, then, Clavius's work on the Sundial of Ahaz is situated within this broader rhetorical context where the mathematical sciences were deemed of theological importance. Thus, the figuration of the Sundial of Ahaz on the title page depicts Clavius's work as a rhetorical appeal for the way in which the practice of mathematics unified the books of nature and Scripture.

This is not to deny that Clavius's works contributed to the anti-Copernican consensus. For instance, when looking at two early seventeenth-century Jesuit theologians, Juan Pineda and Jean Lorin, it is clear that they utilise the authority of Clavius to establish their anti-Copernican arguments. However, in their particular contexts they refer not to Clavius's explanation of the Sundial of Ahaz but more generally to the first chapter of Clavius's *Commentary on the Sphere of Sacrobosco* wherein Clavius articulates the basic understanding with regard to the geocentric universe, that the celestial sphere circles the earth.[36] So while it is evident that Clavius possessed a noticeable appeal in establishing the anti-Copernican opinion, this came not from his theological insights, such as in the case of the Sundial of Ahaz, but instead came from his mathematical

34 "Disposuit enim omnia, & singula in mensura, & numero, & pondere, ut dicit Sapiens." Robert Bellarmine, *De Ascensione mentis in Deum* (Rome: Giacomo Mascardi, 1615), 296. Paul G. Kuntz, "The Hierarchical Vision of St. Robert Bellarmino," in *Acta Conventus Neo-Latini Turonensis*, ed. Jean-Claude Margolin (Paris: J. Vrin, 1980), 959–977.

35 Alongside this we might also place the disputation of Georg Schönberger, a student of Christoph Scheiner. In his disputation he puts forward a similar explanation to Clavius, namely that the Sundial of Ahaz was important on account of its demonstration of the historical priority of sundials within Judaism. He also includes a brief mention of seven biblical events from the Old Testament, all of which demonstrate the importance of time and sundial especially in the way in which God interacts within the world (Exodus 9, Deuteronomy 28, 3 Esdras 1, 4 Esdras 5 and 6, Tobit 11 and 12, Ecclesiastes 12). If the Sundial of Ahaz should be read in the context of anti-Copernican rhetoric, one would expect a stronger anti-Copernican message here, especially given the fact that this was produced in 1615, the year before Bellarmine's stipulation to Galileo. Schönberger, *Exegeses fundamentorum gnomonicorum*, 3.

36 Juan Pineda, *Commentariorum in Iob*, vol. 1 (Cologne: Anton Hierat, 1600), 340; Jean Lorin, *In Acta Apostolorum commentaria* (Leiden: Horace Cardon, 1605), 215. It is quite possible that Lorin was following the text of Pineda.

and astronomical aptitude. Therefore, while the Sundial of Ahaz implies the geocentric movement of the celestial sphere, it is nevertheless important to maintain the historical complexity to note that in Clavius's own work and in the reception of his work, his analysis of the Sundial was not involved in the anti-Copernican rhetoric.

Smogulecki, Riccioli, and the anti-Copernican Interpretation

In the period after the 1616 condemnation of Copernicanism the explicit connection between the Sundial of Ahaz and Copernicanism changed. This is most notably evident in a disputation in Freiburg im Breisgau by a student of Georg Schönberger, Johannes Smogulecki, in 1626, and in the mid-century *Almagestum novum* by Riccioli.[37] A brief analysis of these texts reinforces the assertion that there were many physical representations of the Sundial of Ahaz in common circulation. It also provides a stark contrast to the picture presented with regard to Clavius and Voel, as both Smogulecki and Riccioli do present the Sundial of Ahaz as an anti-Copernican message.

Smogulecki's disputation, *Sol illustratus ac propugnatus*, contains one of the fullest treatments of Ahaz's Sundial produced by a Jesuit in the seventeenth century. The book itself is organized into six sections with the first five covering various aspects about the nature of the sun, its size, its motion, and its relationship to the Ptolemaic and Tychonic systems. Smogulecki considers the Tychonic the superior system, not only on account of its astronomical explanations, but also for the way in which it accords with biblical interpretation and patristic sources.[38] The explanation of the Sundial of Ahaz is in the sixth section of the book, which discusses the abilities of the sun, and is the second of seven analyses within this section.[39]

Before analysing Smogulecki's presentation of this event, it is worth noting that the Battle of Gibeon, which is often featured alongside the Sundial of Ahaz, receives particular treatment in the book, grabbing the attention of the reader from the beginning with a full-page engraving comparing Charles V to Joshua, as seen in Figure 3.5.

37 Although it is recognized it is difficult to determine authorship of disputations, particularly whether the author was the teacher or student, for ease of communication the author of the disputation is identified as Smogulecki.

38 Nicolas Smogulecki, *Sol illustratus ac propugnatus ab illustri ac generoso domino* (Freiburg im Breisgau: Theodorus Meyer, 1626), 22–85.

39 Smogulecki, *Sol illustratus ac propugnatus*, 86–125.

FIGURE 3.5 *Frontispiece: Charles v in Battle*, in Nicolas Smogulecki, *Sol illustratus ac propugnatus ab illustri ac generoso domino* (Freiburg im Breisgau: Theodorus Meyer, 1626)
PHOTO: ETH BIBLIOTHEK, ZURICH / HTTPS://DOI.ORG/10.3931/E-RARA-3002 / PUBLIC DOMAIN MARK

The importance of this, as becomes clear throughout the text, pertains to an account wherein Charles V, similar to Joshua, slowed down the sun.[40] The seemingly miraculous abilities of Charles V are discussed in the third part of the sixth section, in which Smogulecki analyses an event from April 24, 1547, where Charles V both won a decisive battle and slowed down the sun. When considering this miracle, Smogulecki has no reason to doubt its occurrence on account of multiple confirmed reports.[41] The book takes on a clear counter-reformational tone, recasting the history of the Habsburgs with the 'little Joshua,' Charles V, and his battle with the 'iconoclasts.'[42] The comparative militaristic and prophetic significance of both Joshua and Charles V provided an apt dedication to Leopold V, who stood in the lineage of Charles V, and who was involved in militaristic struggles for the sake of religion in the middle of the seventeenth century.

Within the text, it is prior to the sun's slowing in the Battle of Charles V, and yet after the explanation of the Battle of Gibeon, that the analysis of the Sundial of Ahaz occurs. At this point in the book, Smogulecki had already identified various biblical passages, including those pertaining to the Sundial of Ahaz, to uphold a stationary earth and mobile sun.[43] Thus, it is no surprise that in introducing the sun standing still, Smogulecki references Pope Paul V's declaration in 1616 against Copernicanism. As Smogulecki states, "the astronomer Copernicus [...] attributes motion to the earth contrary to Sacred Scripture."[44] On account of the dedication and the way the Battle of Gibeon is analysed just after the mentioning of the decree, textually there is a much stronger rhetorical connection to Joshua 10 than any of the other biblical texts.[45]

The first part of the analysis of the sundial miracle follows similarly to that of the previous discussion of Nunes and Clavius. Smogulecki's explanation of the Sundial event assumes the background knowledge of the works by Nunes and Clavius, although he does not utilise a detailed mathematical analysis in order to explain the occurrence of the event. Rather, relying on their testimony, he states that the event would not have been natural, but miraculous, due to the geographical location of Jerusalem, which was eight degrees and

40 Smogulecki, *Sol illustratus ac propugnatus*, 108.

41 Smogulecki, *Sol illustratus ac propugnatus*, 110. Note that Riccioli also records this event in his *Almagestum novum*. Giovanni Battista Riccioli, *Almagestum novum*, vol. 2 (Bologna: Vittorio Benacci, 1651), 482.

42 Smogulecki, *Sol illustratus ac propugnatu*, "Dedicatoria," fol. D.

43 Smogulecki, *Sol illustratus ac propugnatus*, 52–53. Of note, he mentions the Battle of Gibeon, the Sundial of Ahaz, as well as several other scriptural texts.

44 "Copernicus Astronomus [...] contra Scripturam Sacram terrae motum attribuit." Smogulecki, *Sol illustratus ac propugnatus*, 85.

45 Smogulecki, *Sol illustratus ac propugnatus*, 85–96.

ten minutes north of the Tropic of Cancer.[46] Moreover, similar to Clavius, he notes that understanding this event is of significance for members of the Jesuit Order for more than theological or mathematical reasons: it had practical applications as well. After identifying the mathematical argument, he specifies certain Jesuit missions, within the Torrid zone, where one might experience the regression of the shadows.[47] The greatest detail that Smogulecki provides is consideration of the type of sundial used in the biblical passage, a question which is undoubtedly important because the Sundial of Ahaz is the only miracle testified in Scripture that is also reproducible.[48] In providing such detail in his speculation on the historical form of the sundial, it is apparent that Smogulecki considers contemporary explanations important, whereby the miraculous event could be reproduced not only by mathematics, as previously identified by Clavius, but also with optical illusions, particularly that of the scaphe dial.[49]

Important for his identification of the sundial is his close reading of the biblical text. He argues that it could not have been a horizontal sundial, as Hezekiah was lying ill in his bed and would not have been able to see the sundial. For similar reasons, it likely was not a spherical, hemispherical, or a southern sundial, as with these it would also have been impossible for Hezekiah to see the shadow of the sundial. Instead, he believes it was a vertical column, possibly made of glass, with two gnomons pointed at the appropriate angle, with the hour lines parallel to each other on opposite sides of the sundial.[50]

46 In citing Clavius, he lists chapter 21 of the *Gnomonices*, when it was actually that of the *Fabrica*.

47 The specific locations are, Goa, Calecut, Cananora, Auream Chersonesum, Mexicum Hispaniae nova, Caput Bonae Spei, part of Brasil, Peru, and the major Jova. Smogulecki, *Sol illustratus ac propugnatus*, 98, see also f.n. 17 above.

48 Smogulecki, *Sol illustratus ac propugnatus*, 96.

49 Essentially a scaphe dial is a hemispherical sundial in which the shadow is refracted through the water. Since the sixteenth century this technique was popularly known as a type of Sundial of Ahaz in which the shadow could reverse through the illusion of refraction in water. On the scientific explanation of this, see Dupré. "Galileo, the Telescope, and the Science of Optics," 84–100. On art historical aspects see Chapter 4 in this volume: Andrew Morrall, "Turning back the Sun. Christoph Schisslers's 'Horologium Ahaz' as 'Kunststück'."

50 Smogulecki, *Sol illustratus ac propugnatus*, 102. He spends a significant amount of time weighing the merits of Archimedes and the planetarium globe which he does not rule out altogether, but finds no reason to arbitrate on it either with regard to the biblical text or history itself. On the globe of Archimedes see, Michael T. Wright, "Archimedes, Astronomy, and the Planetarium," in *Archimedes in the 21st Century: Trends in the History of Science* (Cham: Springer International Publishing, 2017), 125–141.

He included an illustration to demonstrate his theory (Figure 3.6).[51] The reconstructed sundial includes half-hours, which Smogulecki justifies based upon the frequent mention of half-hours in the Old Testament. He also surmises that the event occurred around the meridian, an aspect that is in part dependent on his reading of the biblical text – the shadow needed to reverse the same distance that it had progressed, as mentioned in 2 Kings 20. The only way for this to happen would be for the miracle to begin near the meridian. If it began at any other time, the shadow would move in two directions, implying some combination of a progression and regression. Thus in his sundial, the 10 degree reversal begins at the 6 1/2 hour located at the base of the sundial, regressing from point *f* to just beyond point *a*.[52]

While it is difficult to clearly identify whether or not Clavius or Voel thought that the Sundial of Ahaz went against Copernicanism, it is evident within the work of Smogulecki that the event demonstrated the regression of the sun, and as a result one should interpret the passage within the context of the 1616 declaration. This is reiterated in Giovanni Battista Riccioli's *Almagestum novum* of 1651 and its statements about the Sundial of Ahaz and its contribution to the anti-Copernican argument. Riccioli's two volumes, whose frontispiece was previously noted, is an impressive synthesis and catalogue of the Copernican debates, both astronomical and theological, in the middle of the seventeenth century.[53] It is no surprise, then, that Riccioli follows Clavius's tradition: Riccioli explains the miraculous nature of the sundial event, utilising both Clavius's *Fabrica* as well as his *Gnomonices* to explain his points.

Apart from Clavius, Riccioli demonstrates considerable affinity to Smogulecki as well. In particular he includes a long list of the particular biblical citations affirming an immobile earth and a mobile sun, and casts these alongside the condemnations in 1616 by Paul V as well as that of Urban VIII in 1633.[54] Within this, it is notable that the Sundial of Ahaz is listed among the applicable sacred Scriptures that stand against the Copernican opinion.[55] Among the clearest interpretations of the event, which Riccioli includes, the

---

51 Smogulecki, *Sol illustratus ac propugnatus*, 105.

52 Smogulecki, *Sol illustratus ac propugnatus*, 104.

53 Christopher M. Graney, *Setting Aside All Authority: Giovanni Battista Riccioli and the Science Against Copernicus in the Age of Galileo* (Notre Dame, IN: University of Notre Dame Press, 2015).

54 Notably, Riccioli places the account of the event in Ecclesiasticus 48 alongside that of Isaiah 38, both of which comment on the movement of the sun backwards. However, he does not cite 2 Kings 20, which does not mention the sun but mentions the shadow. Riccioli, *Almagestum novum*, 1:51.

55 Riccioli, *Almagestum novum*, 1:290.

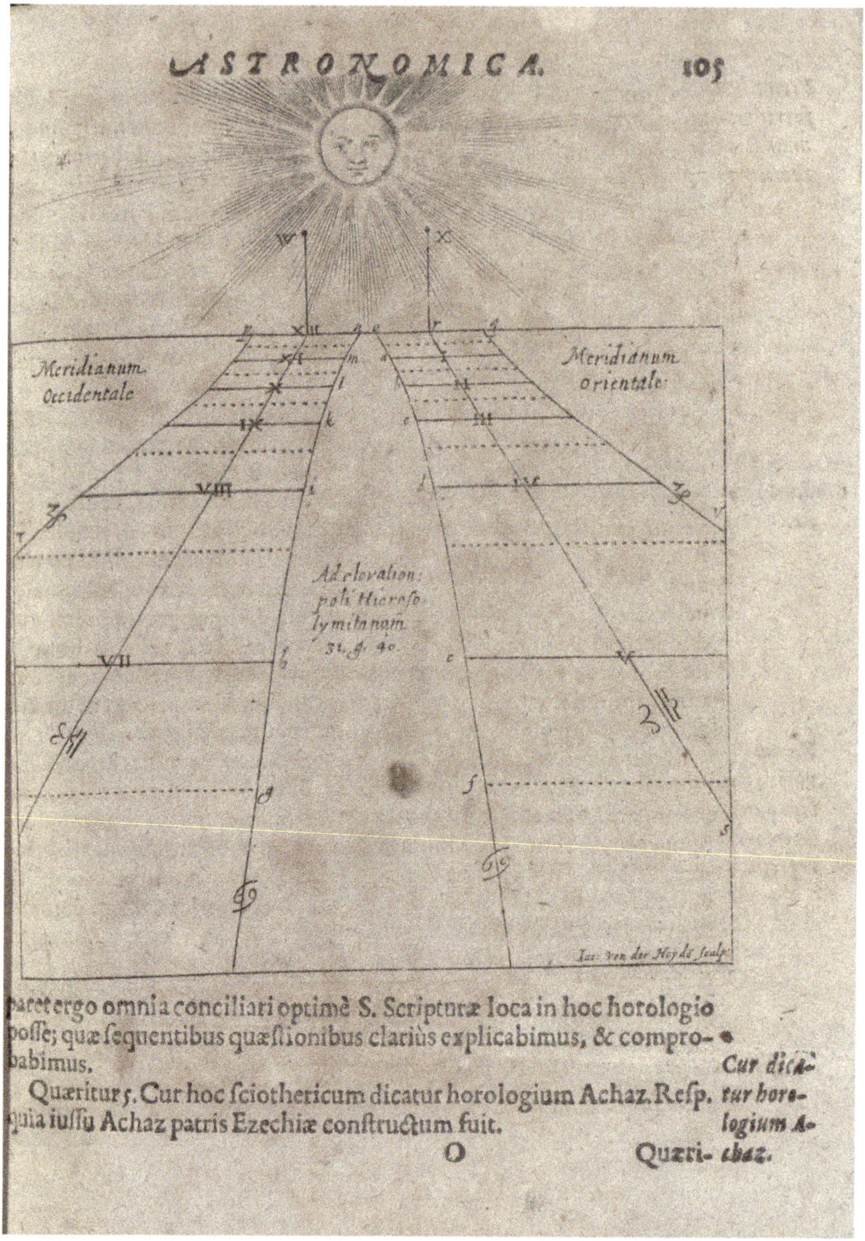
ASTRONOMICA. 105

patet ergo omnia conciliari optimè S. Scripturæ loca in hoc horologio posse; quæ sequentibus quæstionibus clariùs explicabimus, & comprobabimus.

Quæritur 5. Cur hoc sciothericum dicatur horologium Achaz. Resp. quia iussu Achaz patris Ezechiæ constructum fuit. *Cur dicatur horologium Achaz.*

O Quæri-

FIGURE 3.6 *Sundial of Ahaz reconstruction*, in Nicolas Smogulecki, *Sol illustratus ac propugnatus ab illustri ac generoso domino* (Freiburg im Breisgau: Theodorus Meyer, 1626), 105
PHOTO: ETH BIBLIOTHEK, ZURICH/ HTTPS://DOI.ORG/10.3931/E-RARA-3002 / PUBLIC DOMAIN MARK

Sundial of Ahaz stands as a Scripture that specifies the movement of the sun and the immobility of the earth.[56] Riccioli, though, is not as interested in the material condition of the sundial, which Smogulecki or the other Jesuits had already identified. While he includes the possibility of a concave hemisphere sundial, such as a scaphe dial, as one option, he nevertheless follows the theologian Cornelius a Lapide in his depiction of the sundial as plano vertical since such a sundial would be clearly visible to a king who is presumably reclining on account of illness.[57]

In the period after 1616, both the texts of Smogulecki and Riccioli are straightforward in connecting the Sundial of Ahaz to Copernicanism. The moving sun of the biblical text indicates the falseness of Copernicanism. Yet, within this what is important to note is the stated authority upon which such an appeal is made, that of the explicit Papal admonitions in 1616 and 1633 as well as that of the theologians, particularly Cornelius a Lapide. As a result, the reading of their texts relies more on the authority of the theologians and the Papal Decrees in the period after 1616, a context which undoubtedly shaped their textual presentations and gives some indication that the Papal Decrees contributed to the correlation between the Sundial of Ahaz and the anti-Copernican argument. Also evident from the text is that apart from the assertion of the falseness of Copernicus, there was interest in the visible and material reproduction of the event and the type of sundial that was utilised, interests shared by the Jesuits we will consider next.

Mario Bettini and the Hypothetical Sundial

Another important Jesuit to analyse when considering the explanation of the Sundial of Ahaz is Mario Bettini. Bettini studied mathematics at the Collegio Romano during the 1620s and 1630s. His first major publication was the two-volume *Apiaria universae philosophiae mathematicae* (Bologna, 1642). The book itself covers a wide array of topics ranging from mathematics, to optics, to cartography, and represents the flourishing of the visual and mathematical in the Roman Baroque period. It also includes interesting sundials which demonstrate Bettini's mastery of the subject.

What is noteworthy is that Bettini's interpretation of the Sundial of Ahaz does not fall within his analysis of sundials, which is covered in volume one,

56 Riccioli, *Almagestum novum*, 2:220.
57 Riccioli, *Almagestum novum*, 1:484.

but within his section on geography, at the end of the second volume. It is here that Bettini explores the interpretation of the universal sundial within the context of a *Cosmographia*, similar to that of Gemma Frisius.[58] Yet, despite his analysis of the event within a section on geography, Bettini's explanation of the Sundial of Ahaz does not mention the potential mathematical illusion that could occur beneath the Tropic of Cancer, which both Nunes and Clavius addressed. In fact, the only mention of Clavius is at the end, when he points the reader toward Clavius's *Gnomonices*, Book II, proposition 60, where Clavius explains the operation of an antique sundial, which Bettini identifies as most likely the same type as the historical sundial mentioned in the biblical passage.[59]

Bettini provides little explanation as to why he chose to include the Sundial of Ahaz at this point in his book, aside from using the occasion of teaching sundials to instruct on this particularly important biblical event. Among his important points is the antiquity of the Sundial event, in which he follows Clavius in explaining that the biblical text is the earliest historical record of sundials. The inclusion of the sundial was also an opportunity for him to "show that extraordinary miracle of shadow by relenting ten degrees."[60] In addition to this, similar to previous authors, Bettini is interested in countering other contemporary theories on 'hallucinations' and 'magic,' such as the 'scaphe' dial and 'others,' which should not be believed since they operate according to natural means.[61] So while Bettini does not identify Nunes's mathematical problem, he nevertheless evidences a considerable contemporary interest in the reproducibility of the miraculous event.

What is unique in Bettini's explanation is that he not only identifies the probable sundial from the time of Ahaz, likely an unequal equinoctial sundial connected to a pillar, but also introduces a hypothetical 'recent' sundial in which to teach the biblical miracle (Figure 3.7).[62]

Fitting with his broader interest in Gemma Frisius's *Cosmographia*, Bettini's 'recent' sundial is a universal equinoctial sundial, with the geography of landmasses and water as viewed from the northern hemisphere pictured upon it.

---

58 See Peter Apian and Gemma Frisius, *Cosmographia* (Antwerp: Gregorio Bontio, 1550).

59 He refers, as many others had, to the text of Book II, proposition 60, in Clavius's *Gnomonices*. Mario Bettini, "Apiarium XII," in *Apiaria universae philosophiae mathematicae* (Bologna: Io. Baptistae Ferronij, 1642), 2:39.

60 "exhibitum ingens illud miraculum umbrae per denos gradus relabentis." Bettini, "Apiarium XII," 2:37.

61 Bettini, "Apiarium XII," 2:39.

62 "neoterica" Bettini, "Apiarium XII," 2:39.

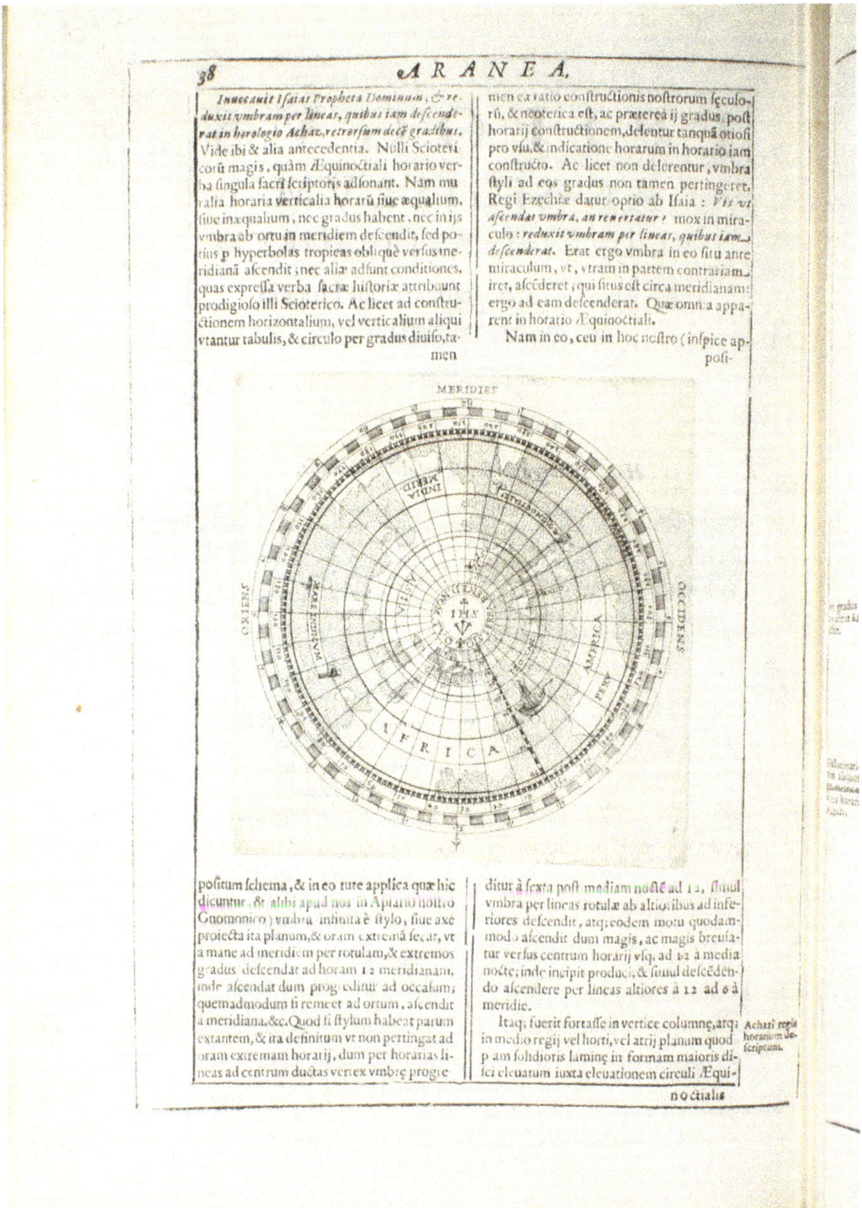

38 ARANEA.

Inuecauit Isaias Propheta Dominum, & reduxit vmbram per lineas, quibus iam descenderat in horologio Achaz, retrorsum decē gradibus. Vide ibi & alia antecedentia. Nulli Scioteri corū magis, quàm Æquinoctiali horario verba singula sacri scriptoris adsonant. Nam mu[ralia] horaria verticalia horarū siue æqualium, siue inæqualium, nec gradus habent, nec in ijs vmbra ab ortu in meridiem descendit, sed potius p hyperbolas tropicas obliquè versus meridianā ascendit; nec aliæ adsunt conditiones, quas expressa verba sacræ historiæ attribuunt prodigioso illi Scioterico. Ac licet ad constructionem horizontalium, vel verticalium aliqui vtantur tabulis, & circulo per gradus diuiso, tamen ea ratio constructionis nostrorum sęculorū, & neoterica est, ac præterea ij gradus, post horarij constructionem, delentur tanquā otiosi pro vsu, & indicatione horarum in horario iam constructo. Ac licet non delerentur, vmbra styli ad eos gradus non tamen pertingeret. Regi Ezechiæ datur optio ab Isaia: *Vis vt ascendat vmbra, an reuertatur?* mox in miraculo: *reduxit vmbram per lineas, quibus iam descenderat.* Erat ergo vmbra in eo situ ante miraculum, vt, vtram in partem contrariam iret, ascēderet; qui situs est circa meridianam: ergo ad eam descenderat. Quæ omnia apparent in horario Æquinoctiali.

Nam in eo, ceu in hoc nostro (inspice appositum schema, & in eo tute applica quæ hic dicuntur, & alibi apud nos in Apiario nostro Gnomonico) vmbra infinita è stylo, siue axe proiecta ita planum, & oram extremā secat, vt a mane ad meridiem per rotulam, & extremos gradus descendat ad horam 12 meridianam, inde ascendat dum progreditur ad occasum; quemadmodum si remeet ad ortum, ascendit a meridiana. &c. Quod si stylum habeat parum extantem, & ita definitum vt non pertingat ad oram extremam horarij, dum per horarias lineas ad centrum ductas vertex vmbrę progreditur à sexta post mediam noctē ad 12, simul vmbra per lineas rotulæ ab altioribus ad inferiores descendit, atq; eodem motu quodammodo ascendit dum magis, ac magis breuiatur versus centrum horarij vsq; ad 12 à media noctę; inde incipit produci, & simul descēdendo ascendere per lineas altiores à 12 ad 6 à meridie.

Itaq; fuerit fortasse in vertice columnę, atq; in medio regij vel horti, vel atrij planum quod p̄ am solidioris laminę in formam maioris disci eleuatum iuxta eleuationem circuli Æquinoctialis

Achazi regis horarium descriptum.

FIGURE 3.7 *Sundial of Ahaz as the 'Recent' dial,* in Mario Bettini, "Apiarium XII," in *Apiaria universae philosophiae mathematicae* (Bologna: Io. Baptistae Ferronij, 1642), 2:38
PHOTO: LINDA HALL LIBRARY OF SCIENCE, ENGINEERING & TECHNOLOGY, KANSAS CITY

In his explanation of how the event occurred, he suggests that it occurred over the course of about 2 1/2 hours, beginning at 10 3/4 and receding backwards to 8 1/2. Although he is not clear as to why those specific hours were chosen, in contrast to Smogulecki's previously discussed explanation, he is clear about the exegetical significance. The benefit for such an interpretation particularly on his sundial in which each hour is divided into quadrants, has to do with the way this time frame correlated to the fifteen years added to Hezekiah's life, described in Isaiah 38:5. To explain this correlation he uses basic mathematics using his sundial. If one added to the 10 degrees the shadow receded, the number of degrees before the meridian when it began to recede, which was 5, then the summation was fifteen – which he notes "produces the number 15 degrees, by the promise of the number of productive years of life to the King."[63] Such a solution, while by no means explicit in the biblical text, nevertheless demonstrates a similar interest to what has been noticed among the previous authors, in which mathematics could be placed within the literal words of Scripture.[64]

In a book which details many optical and visual tricks, Bettini's upholding of the miracle, with such detailed specifics, such as the time and the quarter hours, stands as a sharp contrast to his concerns in the rest of the book with visual illusion, and reinforces the significance of this specificity. His modern equinoctial dial becomes an occasion for teaching the biblical story, which raises the question as to whether there are other uses of sundials for pedagogical or theological purposes within the Jesuit literature at the time. His later comments at the end of *Apiarium XII* about the way in which sundials provide evidence as to how the "salvation of the Lord" was advancing across the globe fits a similar theme as noted above in Clavius, Smogluecki, and Kircher, and indicates that the theory of sundials proved important not only practically, but also theologically. He is also important because his explanation evidences certain contemporary sundials that seem to be labelled as Ahaz's Sundial, such as the scaphe dial. Such a mention in a text published in the 1640s takes on added intrigue alongside the work of our final Jesuit, Athanasius Kircher, in his *Ars magnis lucis et umbrae*, in which the scaphe dial is accorded a prominent place.

---

63 "conficiunt numerum quindecim graduum pro numero annorum vitae productioris Regi promissae." Bettini, "Apiarium XII," 2:39.

64 At the end of Apiarium XII is a table of latitudes, presumably of Jesuit locations. Bettini, "Apiarium XII," 2:43.

Athanasius Kircher and the Multiple Sundials

The final Jesuit mathematician and astronomer of this study, Athanasius Kircher, was an expert in the techniques of natural magic. Early in his career he developed a reputation as a masterful clockmaker, and as a result constructed an intricate sundial in the tower of the Jesuit College in Avignon.[65] By 1633 Kircher had travelled to Rome to become professor of mathematics at the Roman College, succeeding Christoph Scheiner.[66] One of the many intriguing aspects of Kircher's presentation of the Sundial of Ahaz is the way in which he provides many different possible identifications of the type of sundial, and illustrates the many representations of the sundial that existed both within the Jesuit Order and quite likely within the broader culture itself.

For instance, in his first book on sundials, the *Primitiae gnomonicae catoptricae,* which represents the research that Kircher did while at Avignon, Kircher includes a detailed illustration of a Sundial of Ahaz – which he also identifies as a 'Jewish sundial' and an antique sundial. Similar to previous representations of Ahaz's Sundial, his purpose is not to mediate astronomy but instead to emphasise the antiquarian significance of the sundial. Kircher illustrates it as a flat, unequal, sundial with the Hebrew caption located beneath[67] (Figure 3.8). He also mentions that its existence demonstrates how God desired to use time itself in his interaction in the world and proceeds to reconstruct the time of day when the passion of Christ would have occurred based upon the timescale on an antique sundial.[68] Interestingly, he does not explain the biblical text surrounding Ahaz's Sundial in this work, but does so in his most extensive treatment of optics, the *Ars magna lucis et umbrae*. More so than the *Primitiae gnomonicae catoptricae*, the *Ars magna* represents Kircher's skill in the mechanics of sundials, and as noted, there is an iconic image (Figure 3.4) in this book, which shows the Jesuit missions located around the globe, with their geographic location identified with respect to a sundial.[69]

65 Paula Findlen, "The Last Man Who Knew Everything ... or Did He?" in *Athanasius Kircher: The Last Man Who Knew Everything*, ed. Paula Findlen (New York: Routledge, 2004), 11.

66 Findlen, "The Last Man," 15.

67 Athanasius Kircher, *Primitiae gnomonicae catoptricae* (Avignon: Jean Piot, 1635), 170. He also identifies it in the context as a "Jewish sundial," which is a term that is seen in other locations as well, such as in Francis Line's book on sundials. Francis Line, *Explicatio horologii in horto regio Londini* (Liège: Guillaume Henri Streel, 1673), 71.

68 Kircher, *Primitiae gnomonicae catoptricae*, 172–173.

69 This was the only image in the entire book to be changed between the 1646 and 1671 edition, with an update to the sundial portrayal and the orientation of some of the sundials.

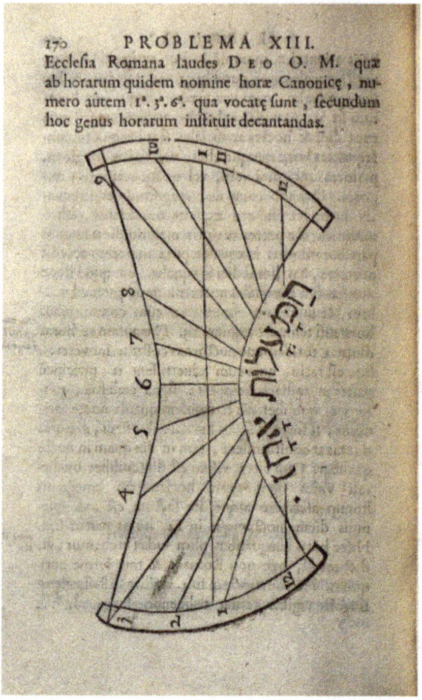
170 PROBLEMA XIII.
Eccleſia Romana laudes DEO O. M. quæ ab horarum quidem nomine horæ Canonicę, numero autem 1ª. 3ª. 6ª. qua vocatę ſunt, ſecundum hoc genus horarum inſtituit decantandas.

FIGURE 3.8
Sundial with Hebrew designation 'ha'malot achaz', in Athanasius Kircher, *Primitiae gnomonicae catoptricae* (Avignon: Jean Piot, 1635), 170
PHOTO: ÖSTERREICHISCHE NATIONALBIBLIOTHEK, VIENNA

Likely owing to his interest and expertise in Oriental languages, Kircher provided an extensive analysis of the Jewish sources behind the Sundial of Ahaz and uses the occasion of Ahaz's Sundial to provide a rather extended explanation as to the theological importance of sundials. Among other things, he explains that the Jewish people must have had a developed understanding of sundials, since there were so many explanations in the Old Testament with regard to specific times.[70] He also speculates that Adam, as the first man, and possessor of all knowledge, must have known how sundials operated.[71] Such a portrayal, while fitting with the historical interest in Ahaz's Sundial that we have previously noted, nevertheless goes considerably beyond the features and mechanics of sundials themselves.[72]

---

70 Kircher, *Ars magna*, 216.

71 Kircher, *Ars magna*, 214.

72 Similarly, it is worth mentioning that Smogulecki stated that the patriarchs of Genesis also would have had a knowledge of astronomy. Smogulecki, *Sol illustratus ac propugnatus*, 100.

Where Kircher deviates from his previous explanation of the sundial in the *Primitiae gnomonicae catoptricae*, it is with regard to identifying how the instrument worked and what it was made from. When identifying what was occurring in Ahaz's Sundial, Kircher begins with a line of interpretation following Jewish rabbis, who had interpreted the sundial as that of a stone pillar sundial, which he accepts as the material and historical basis of the sundial. Such a presentation is similar to previous historical reconstructions, such as in Smogulecki as well as Bettini. Yet, what is interesting in Kircher's interpretation in the *Ars magna* is that after his historical foray, he identifies a second Sundial of Ahaz, which is not intended to be historically accurate, but which is of "recent interest." This sundial, in contrast to his previous historical one, is a scaphe dial, which, he explains operated according to natural means, and thus cannot be a miracle.[73] Such a point is substantiated based upon his previous section, in which he explained the process of refraction in water, which form the basic principles for the scaphe dial (Figures 3.9–10).[74]

Kircher takes a similar approach to the previous Jesuits who applied specifics to the sundial in order to uphold the miracle, except he utilises the specifics of optics to expose the falseness of contemporary optical tricks.[75] Of interest in Kircher's explanation of the Sundial of Ahaz is that he depicts two sundials, both of which could truthfully be identified as Sundials of Ahaz, which nevertheless are different. On the one hand, the historical sundial is unequal and is valuable to identify on account of its antiquarian nature. On the other hand, the contemporary Sundial of Ahaz helps to uphold the miraculous nature of the biblical event, which is also of importance for Kircher. Yet, in neither situation is the question of Copernican astronomy important, indicating that drawing such a connection was not necessary within the context of mathematical or astronomical explanations. While this does not clearly indicate that the Sundial of Ahaz was not an anti-Copernican text for Kircher, it nevertheless throws the clear identification of it as such into question.

More may be said with regard to the identification of the Sundial of Ahaz as a scaphe dial within the Society of Jesus in the middle of the seventeenth

73 One of the oddities when looking into Kircher's interest in the Sundial of Ahaz is that it did not continue in the line of Kircher's reception. It does not show up in the records of the Roman museum by Giorgio de Sepi, nor does it occur within the published works of Athanasius Kircher's student, Gaspar Schott.

74 For the explanation as to how the scaphe dial worked, see Dupré, "Galileo, the Telescope, and the Science of Optics," 84–100.

75 Kircher says that the miracle occurred according to "industriae" (the same term used by Claude Dechales in his *Cursus seu Mundus Mathematicus*, 3:227).

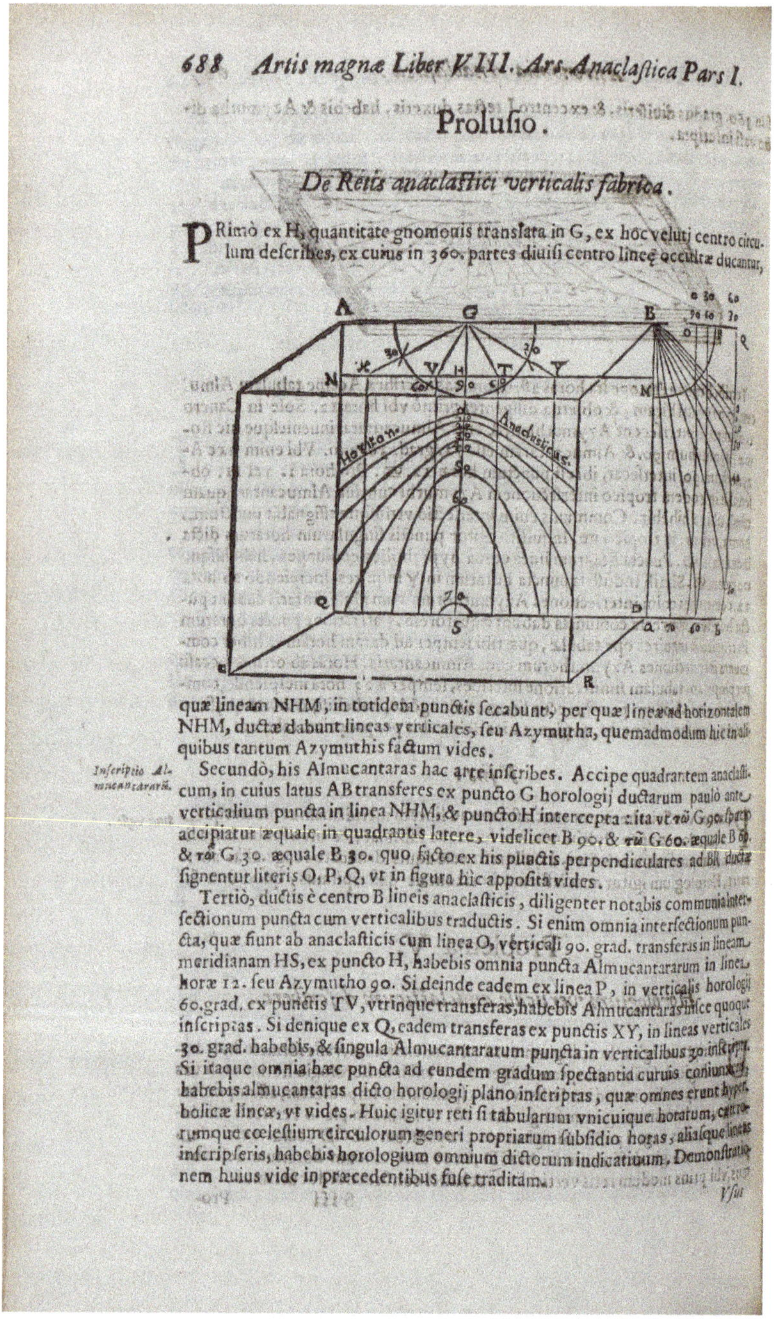

688 *Artis magnæ Liber VIII. Ars Anaclaſtica Pars I.*

Proluſio.

De Retis anaclaſtici verticalis fabrica.

PRimò ex H, quantitate gnomonis tranſlata in G, ex hoc veluti centro circulum deſcribes, ex cuius in 360. partes diuiſi centro lineę occultæ ducantur,

quæ lineam NHM, in totidem punctis ſecabunt, per quæ lineæ ad horizontalem NHM, ductæ dabunt lineas verticales, ſeu Azymutha, quemadmodum hic in aliquibus tantum Azymuthis factum vides.

Inſcriptio Almucantararũ.

Secundò, his Almucantaras hac arte inſcribes. Accipe quadrantem anaclaſticum, in cuius latus AB transferes ex puncto G horologij ductarum paulò ante verticalium puncta in linea NHM, & puncto H intercepta : ita vt τοῦ G 90. [illegible] accipiatur æquale in quadrantis latere, videlicet B 90. & τοῦ G 60. æquale B 60. & τοῦ G 30. æquale B 30. quo facto ex his punctis perpendiculares ad BR ductæ ſignentur literis O, P, Q, vt in figura hic appoſita vides.

Tertiò, ductis è centro B lineis anaclaſticis, diligenter notabis communia interſectionum puncta cum verticalibus traductis. Si enim omnia interſectionum puncta, quæ fiunt ab anaclaſticis cum linea O, verticali 90. grad. transferas in lineam meridianam HS, ex puncto H, habebis omnia puncta Almucantararum in linea horæ 12. ſeu Azymutho 90. Si deinde eadem ex linea P, in verticalis horologij 60. grad. ex punctis TV, vtrinque transferas, habebis Almucantaras hice quoque inſcriptas. Si denique ex Q, eadem transferas ex punctis XY, in lineas verticales 30. grad. habebis, & ſingula Almucantararum puncta in verticalibus 30. inſcripta. Si itaque omnia hæc puncta ad eundem gradum ſpectantia curuis coniunxeris, habebis almucantaras dicto horologij plano inſcriptas, quæ omnes erunt hyperbolicæ lineæ, vt vides. Huic igitur reti ſi tabularum vnicuique horarum, cæterorumque cœleſtium circulorum generi propriarum ſubſidio horas, aliaſque lineas inſcripſeris, habebis horologium omnium dictorum indicatiuum. Demonſtrationem huius vide in præcedentibus fuſe traditam.

Vſus

FIGURE 3.9 *Diagram depicting the refraction of water*, in Athanasius Kircher, *Ars magna lucis et umbrae* (Rome: Hermann Scheus, 1646), 688
PHOTO: HISTORY OF SCIENCE COLLECTIONS, UNIVERSITY OF OKLAHOMA LIBRARIES, NORMAN

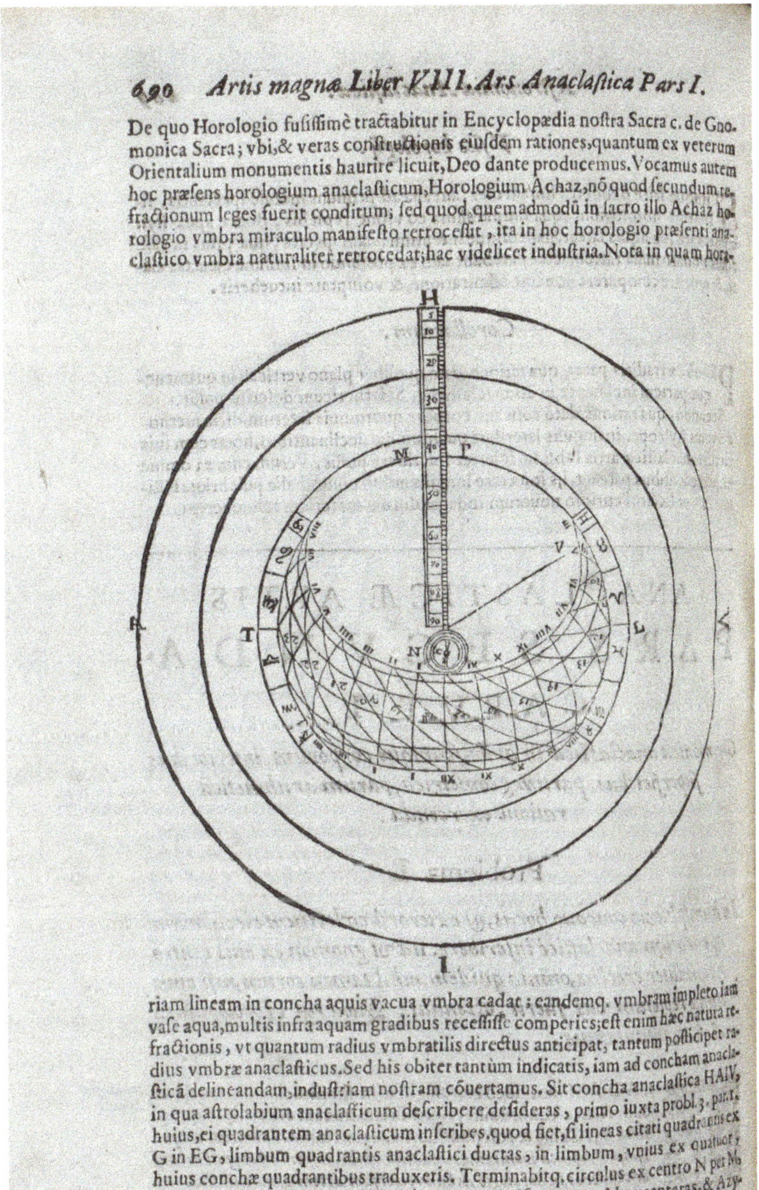

690 *Artis magnæ Liber VIII. Ars Anaclastica Pars I.*

De quo Horologio fusissimè tractabitur in Encyclopædia nostra Sacra c. de Gnomonica Sacra; vbi,& veras constructionis eiusdem rationes,quantum ex veterum Orientalium monumentis haurire licuit,Deo dante producemus.Vocamus autem hoc præsens horologium anaclasticum,Horologium Achaz,nõ quod secundum refractionum leges fuerit conditum; sed quod quemadmodũ in sacro illo Achaz horologio vmbra miraculo manifesto retrocessit , ita in hoc horologio præsenti anaclastico vmbra naturaliter retrocedat;hac videlicet industria.Nota in quam horariam lineam in concha aquis vacua vmbra cadat ; eandemq. vmbram impleto iam vase aqua,multis infra aquam gradibus recessisse comperies;est enim hæc natura refractionis , vt quantum radius vmbratilis directus anticipat, tantum posticipet radius vmbræ anaclasticus.Sed his obiter tantùm indicatis, iam ad concham anaclasticã delineandam,industriam nostram cõuertamus. Sit concha anaclastica HAIV, in qua astrolabium anaclasticum describere desideras , primo iuxta probl.3. par.1. huius,ei quadrantem anaclasticum inscribes,quod fiet,si lineas citati quadrantis ex G in EG, limbum quadrantis anaclastici ductas , in limbum , vnius ex quatuor huius conchæ quadrantibus traduxeris. Terminabitq, circulus ex centro N per M ductus horizontem anaclasticum;intra quem cõsequenter Almucantaras,& Azymutha, & deindè horas inscribes,vt sequitur.

Azy-

FIGURE 3.10 *Sundial of Ahaz as scaphe dial*, in Athanasius Kircher, *Ars magna lucis et umbrae* (Rome: Hermann Scheus, 1646), 690
PHOTO: HISTORY OF SCIENCE COLLECTIONS, UNIVERSITY OF OKLAHOMA LIBRARIES, NORMAN

century. In the works of Smogulecki, Bettini, and now Kircher, scaphe dials have been associated with the Sundial of Ahaz. Alongside this we might place a contemporary of Bettini, Jan Ciermans, who taught mathematics at both Louvain and Antwerp between 1637 and 1640. Ciermans's book on practical mathematics, the *Disciplinae mathematicae*, published in 1640, provides many different explanations and uses for mathematics, of which gnomonics was one. In this section on sundials Ciermans mentions the 'Sundial of Hezekiah' though does not refer to it as miraculous or in relation to Copernicanism. What he does say is that it operated according to refraction in water, thereby implicitly referring to the scaphe dial.[76] While more is needed to substantiate that the Sundial of Ahaz increasingly being identified as a scaphe dial among the Jesuit Order in the middle of the seventeenth century, it is nevertheless a theory that appears in the works of Smogulecki, Riccioli, Bettini, Kircher, and Ciermans.

From the previous explanation, one might think that the Sundial of Ahaz was a common passage included in the early modern Jesuit mathematical and astronomical literature. One important observation from this study is that this was, in fact, not the case. There are numerous examples from Jesuit mathematical and astronomical publications that make no reference to the Sundial of Ahaz, both from the period before the 1616 condemnation as well as afterward.[77] From this it is possible to surmise that the biblical text had only limited appeal among Jesuit mathematicians and astronomers, and an ambivalent application to Copernicanism. In order to reinforce this unclear relationship a few telling points should be made with regard to the Jesuit exegesis of this passage.

An Exegetical Conundrum

Based upon the Jesuit authors analysed, it is clear that they placed great value in interpreting the biblical event according to the literal scriptural text. This is in line with the approach of the Council of Trent where the literal reading of

---

76 Jan Ciermans, *Disciplinae mathematicae traditae anno institutae* (Leuven: Everard de Witte, 1640) in the section titled "Septembris."

77 The following works were surveyed and did not mention the Sundial of Ahaz: Franciscus Aguilonius, *Opticorum libri sex* (Antwerp: Viduam et Filios Io. Moreti, 1613); Christoph Scheiner, *Sol elliptic* (Augsburg: Christopher Mangius, 1615); Georg Schönberger, *Horologiorum novorum radio recto* (Freiburg im Breisgau: Johann Strasser, 1622); Pierre Bobynet, *L'Horographie ingénieuse* (Paris: François Langlois, 1647); Pierre Bobynet, *La cadran des cadrans* (Paris: Mathurin Henault, 1649).

the biblical text was of great importance.[78] This emphasis has been seen in the works previously analysed, for instance, as many of these authors explain why certain types of sundials could not have been used, since an unhealthy, and presumably reclining King could only have seen a sundial attached to a wall or pillar. This interest in a literal interpretation of Scripture also created certain issues when reconciling the three accounts of this miraculous event, which are worth exploring because of their possible bearing upon the interpretation of this passage with respect to Copernicanism.

One of the biggest textual differences between the three accounts of the Sundial of Ahaz involved whether the miracle occurred in the shadow or in the sun. In 2 Kings 20 the biblical text refers to the regression of a shadow, whereas Isaiah 38 and Ecclesiasticus 48 more clearly refer to the regression of the sun. Because of this discrepancy a literal interpretation of the text was troublesome. One either had to prioritise the shadow or the sun, since the former would not necessarily pertain to Copernicanism while the latter would clearly do so. Exegetically, the way forward was to use one text to interpret the other. So, for instance, if the regression occurred in the sun, then one interpreted the regression of the shadow in 2 Kings 20 by utilising Isaiah 38 to show that while 2 Kings 20 explains the phenomenon as a shadow, the miracle nevertheless occurred in the sun. Similarly, if the miracle was in the shadow, then one could explain that the sun was thought to regress because of the essential correlation between the movement of the sun and that of a shadow with respect to the sundial. One of the arbiters in this exegetical debate among the Jesuit commentators is how one interpreted the arrival of the ambassadors of the Babylonian princes, as explained in 2 Chronicles 32:31, "who sent unto him to enquire of the wonder that was done in the land."[79] A local event occurring in the clouds was the only way one could understand why the ambassadors of the Babylonian princes came to the land of Judah to inquire about a miracle, understood as the restored health of the king. Such an interpretation emphases 2 Kings 20 and the way in which the event occurred in the shadow as the prevailing biblical text. According to this interpretation, if it had been a miracle occurring within the sun itself (as Isaiah 38 and Ecclesiasticus 48 indicate) then there would have been no need for the ambassadors to travel to Jerusalem.[80]

---

78 On the allegiance to the post-Trent emphasis on the "literal interpretation" of Scripture see Volker Remmert, "Catholic Biblical Exegesis," 21–25.

79 English Standard Version translation.

80 Here is the way Galileo appeals to the Sundial of Ahaz: "Similarly, in the miracle at the time of Hezekiah, Paul of Burgos thinks that it did not take place in the sun but

The influence of this exegetical line of interpretation played a prominent role in the arbitration of the Sundial of Ahaz as not being an anti-Copernican text in 1615 by the astronomer Galileo Galilei in his *Letter to the Grand Duchess Christina*. As Galileo explained in this important text, similar to the way the sun did not really stand still in the Battle of Gibeon, so also in the Sundial of Ahaz the miracle occurred not in the sun but on the sundial. Of interest for the interpretation of the Sundial of Ahaz among the Jesuits, it has been shown that Galileo borrowed this well-known exegetical explanation from a contemporary of his, the Portuguese Jesuit exegete Cosme de Magalhaens and his 1612 commentary on Joshua, *In sacram Iosue historiam commentarium*.[81] In it Magalhaens explains that the event occurred in the shadow, based not only on the arrival of the Babylonian princes, but also aspects pertaining to its historical documentation. As Magalhaens notes, it would have been odd for the reversal of the sun to have been global on account of the fact that no known records existed with regard to such an odd celestial event.[82] Magalhaens's specific explanations aside, what is important to note is that Galileo's defence of Copernicanism, published just before the Condemnation of 1616, relied extensively upon a Jesuit exegete's critical explanation of the biblical texts.

As already alluded to in the other Jesuits above, Magalhaens's line of reasoning was not the only Jesuit option at the time. Two other important Jesuit commentators, Serarius and Cornelius a Lapide, provide important points of contrast and significantly shaped the opinions of other Jesuits. For instance, in his *In libros Regem et Paralipomenon commentaria posthuma*, Serarius interprets the Sundial of Ahaz in terms of the sun (Isaiah 38), the justification of which is based upon the appearance of the ambassadors of the Babylonian princes.[83] A similar line of interpretation – using Isaiah 38 to interpret the shadows of 2 Kings 20, and then identifying it as global in the context of 2 Chronicles 32 – may be noticed within that of Cornelius a Lapide, one of the greatest Jesuit biblical commentators in the mid-seventeenth century.[84] The

in the clock (sundial)." Galileo Galilei, "Letter to the Grand Duchess Christina," in *The Essential Galileo*, edited and translated by Maurice Finocchiaro (Indianapolis, IN: Hackett Publishing Company, 2008), 135.

81 Goldstein, "Galileo's account of astronomical miracles in the Bible," 3–16.

82 As Henrique Leitão notes, this same line of reasoning continued into the seventeenth century even beyond the Jesuits. See chapter 2 in this volume: Henrique Leitão, "Mathematical Certainty and Biblical Inerrancy. Pedro Nunes and the Retrogradation of Shadows at the Dial of Ahaz."

83 Serarius, *In libros Regem et Paralipomenon*, 352–353.

84 Cornelius a Lapide, *Commentaria in Quatuor prophetas Maiores*, 312. On the importance of Lapide, see Irving A. Kelter, "The Refusal to Accommodate: Jesuit Exegetes and the Copernican System," *The Sixteenth Century Journal* 26, no. 2 (1995): 277–278.

influence of this line of reasoning may be noticed in the works of one of the previously discussed authors, Johannes Smogulecki. As noted, the reason that Smogulecki advocated for the miracle to have occurred in the sun rather than in the shadow had to do with the presence of the ambassadors of the Babylonian princes in Jerusalem, a point in clear contrast to that of Magalhaens, but in accord with both Serarius and Cornelius a Lapide. Smogulecki avoids the issue of the lack of global verification in other regions, which Magalhaens had mentioned, an omission which is quite suspicious since he had verified the retardation of the sun in the reign of Charles V based on other regional reports.[85]

When considering these Jesuit commentators further questions arise, particularly when comparing Serarius's exegesis of the Sundial of Ahaz and that of the Battle of Gibeon. For instance, while Serarius does not comment on the Sundial of Ahaz as an anti-Copernican text within his commentary, he makes a strong argument against the astronomy of Copernicus in the context of his explanation of the Battle of Gibeon in his commentary on Joshua.[86] Such a comparison should raise important questions about the utilisation of this biblical miracle as a vehicle for the anti-Copernican argument. In addition, the exegetical differences of Serarius and Cornelius a Lapide in contrast to Magalhaens suggests the exegesis of this biblical event was not clearly established and was open to at least some debate during the first part of the seventeenth century as Copernicanism was being considered.

When reconsidering some of the Jesuits already analysed, certain aspects of the exegetical debate are noticeable. Some level of exegetical difference is evident from Smogulecki's disputation as well. For instance, when answering the first question, "whether the sun really reversed, or whether a shadow," he identifies two lines of interpretation. The first, which understood the event as having taken place in the shadow, in the tradition of Vatablus, Burgensis, Arias Montanus, and Barniellus.[87] To support this point, he uses the text of 2 Kings 20. In contrast to this, he identifies his own line of interpretation in which the

85 See f.n. 41.

86 Serarius, *In libros Regem et Paralipomenon*, 352. Nicolaus Serarius, *Iosue*, vol. 2 (Paris: Typographica Edmundi Martini, 1610), 238. What is curious is that in Quaestio XVIII, Serarius includes Ecclesiasticus 48, which states that the sun may reverse its course, as an analogue to Joshua stopping the sun in Joshua 10. His point is that the stopping and reversing of the sun are analogous motions, the one reinforcing the other. However, in Quaestio XIV, when Serarius identifies the scriptures against Copernicanism, he does not include any of the references to the Sundial of Ahaz (2 Kings 20; Isaiah 38; Ecclesiasticus 48). Instead he lists the following: Ecclesiasticus 1:4; Psalm 41:1; Psalm 102:5; Psalm 77.

87 "Utrum Sol verè retrocesserit, an tantum umbra." Smogulecki, *Sol illustratus ac propugnatus*, 97.

sun itself reversed according to the traditions of the Sacred Fathers of Cyrill, Procopius, Haymonis, Lyr. Hugonis, Cornelius a Lapide, Serarius, Tornielli, and Barradius.[88] To support this, he uses Isaiah 38 and Ecclesiasticus 48.[89] Thus, despite his agreement with a prevailing line of interpretation, Smogulecki's introduction of the question in his disputation indicates how the exegetical difference had a notable degree of importance in the 1620s.

That the exegetical debate continued into the mid-seventeenth century is testified by a commentary on Joshua by the Jesuit Emmanuel de Naxera from 1651. In this book Naxera interpreted the Sundial of Ahaz only in the context of 2 Kings 20, and explained how the shadow reversed "in the Sundial of Ahaz," and that this occurred prior to anyone else noticing the celestial phenomenon.[90] That it was first noticed in Jerusalem is important for Naxera since the biblical text emphasised the primary importance of the miracle for King Hezekiah himself. When he explains more fully what happened in the miracle, Naxera avoids the moving sun altogether, and yet maintains that "defects in the heavens are noticed by everyone," and by it, particularly those to the East.[91] While there are many curiosities with regard to Naxera's interpretation – particularly what he means by the 'defect' – one wonders whether his exegesis assumes the significance of 2 Chronicles 32 within his interpretation.[92] For, after identifying the 'defect' he then comments on how the ambassadors of the Babylonians came based on what they had seen, an interpretation that demonstrates an awareness of 2 Chronicles 32, which explained the ambassadors arrival based on some phenomenon in the land, rather than due to the illness of Hezekiah, which is what 2 Kings 20 explains as the cause of their arrival. Thus, despite the claims of Smogulecki that strongly support the miracle occurring in the sun as the prevailing exegetical view, Naxera's commentary shows how this was not necessarily the case and that the task of literal interpretation across the various accounts of the Sundial of Ahaz introduced known exegetical conundrums.

While there were clearly theologians, mathematicians, and astronomers who interpreted the Sundial of Ahaz as an anti-Copernican text, it is evident that certain exegetical differences remained among Jesuit theologians.

88 Smogulecki, *Sol illustratus ac propugnatus*, 97.

89 The same line of interpretation exists in Cornelius a Lapide's explanation. Cornelius a Lapide, *Commentaria in Quatuor prophetas Maiores*, 312.

90 Emmanuel de Naxera, *In Iosue Hostilibus redimitum trophaeis Commentarii Litterales, Moralésque* (Lyon: Laurent Anisson, 1651), 188–189.

91 "sublimium defectus ab omnibus observantur." Naxera, *In Iosue Commentarii*, 260.

92 In his text Naxera quotes "P. Gaspar Sanctius" as being in agreement with his explanation, although I have been unable to locate this particular text. Naxera, *In Iosue Commentarii*, 188.

Although more research is needed to substantiate this claim, the known exegetical question makes Galileo's utilisation of Magalhaens's argument less of a poor selection, and perhaps more a shrewd appeal. Behind the scenes of the Copernican question there were exegetical differences among Jesuit authors with respect to the passages utilised in their appeals. As a result more research is needed to understand the history of these exegetical questions among Jesuit commentators in order to better understand whether this shaped the appeal of the Sundial of Ahaz as an anti-Copernican text.[93]

Conclusion

One of the clearest articulations of the tenuous relationship between the Sundial of Ahaz and Copernicanism among early modern Jesuits occurs in a thesis produced at a Jesuit college in Paris in 1693. In the section on sundials, the authors state a few reasons for studying the sundial, one of which is that the sundial affirms the diurnal motion of the stars "which some Copernicans verily deny."[94] The text then explains the process of diurnal motion and how this supports the immobility of the earth and the mobility of the sun, a clearly anti-Copernican argument. The text, however, does not cite the biblical narrative of the Sundial of Ahaz until the appendix.[95] It is there, in a list of engraving titles, that the reader learns that the image at the head of the chapter on sundials is a representation of the Sundial of Ahaz (Figure 3.11). In the section itself the authors make no clear connection between this biblical event and the anti-Copernican argument even though the image is featured prominently. In the context of anti-Copernican explanations, the biblical text is not utilised for its cosmological merits.[96] At a time when many questions were asked of this

93 As an example, when the Jesuit Melchior Cornaeus explains the subtle response to Copernicanism, while he does mention the movement of the stars and sun in the sky, he does not list the Sundial of Ahaz among the Sacred Scriptures that go against Copernicanism. Those he does list are Psalm 103, Ecclesiastes 1, Joshua 10. Melchior Cornaeus, *Curriculum philosophiae peripateticae*, vol. 2 (Würzburg: Elias Michael Zinck, 1657), 534–536.

94 "Quod ne quidem Copernicani negant." Joannes Baptista Thioly and Petrus Taillandier, *Theses opticae, et astronomicae* (Lyon: Pierre Valfray, 1693), 41.

95 Thioly and Taillandier, *Theses opticae, et astronomicae*, 50.

96 See f.n. 93 on Melchior Cornaeus, where similarly sundials were used in an anti-Copernican argument, though when applying biblical texts that negate Copernicanism, none of those pertaining to the Sundial of Ahaz were included.

FIGURE 3.11 *Sundial of Ahaz*, in Joannes Baptista Thioly and Petrus Taillandier, *Theses opticae, et astronomicae* (Lyon: Pierre Valfray, 1693), 41
PHOTO: LINDA HALL LIBRARY OF SCIENCE, ENGINEERING & TECHNOLOGY, KANSAS CITY

biblical miracle, the primary value of it pertained to its historical and miraculous identities.

This does not mean that Jesuits did not appeal to the Sundial of Ahaz as an argument against Copernicanism – as Smogulecki and Riccioli and certain theologians did so – but that such a connection was by no means consistent within Jesuit astronomical, mathematical, or even theological works in the seventeenth century.[97] The more relevant issue was the way in which the biblical event could be understood in the context of competing natural explanations of the text, whether of the mathematics of natural regression introduced by Pedro Nunes or that of the refraction in water with a scaphe sundial. Appeals to anti-Copernican arguments were secondary and, even then, not consistently applied. In a period of religious change, particularly as it related to exegesis and the nature of biblical signs, such an affirmation of the theological importance of symbols and miracles undoubtedly proved important for the Jesuit

97 Another Jesuit who thought the miracle went against Copernicanism was the Jesuit Bartholomew Amicus. Edward Grant, *In Defense of the Earth's Centrality and Immobility: Scholastic Reaction to Copernicanism in the Seventeenth Century* (Philadelphia, Pennsylvania: The American Philosophical Society, 1984), 62.

authors to attest. Yet, despite such a textual affirmation, there were interesting deviations in the way the sundial event was pictured.

I have shown that there was not a specific way in which the miracle of the Sundial of Ahaz was discussed among the members of the Jesuit Order in the seventeenth century, at least among important astronomical and mathematical works. The type of sundial itself was debated: it was suggested that it was a vertical unequal sundial as in Clavius's *Opera Mathematica*, a column with dials on both sides as in Smogulecki, an equinoctial dial as in Bettini, a stone pillar historically as Kircher identifies, and also a scaphe dial as identified by Ciermans and Kircher. From this it is evident that among seventeenth-century Jesuits, there was an interest in engaging with questions pertaining to the physical functionality of the sundial, all of which were also efforts in upholding a literal exegesis of the biblical passage. Even in the case of Kircher, his historical forays into the Jewish rabbinical traditions were woven together with his exegesis of the biblical narratives. To that extent, one might wonder whether their efforts were quite similar to the close interweaving of the mathematical sciences with exegesis among other miraculous events in the sixteenth and seventeenth centuries, such as the deluge and creation itself. Similar to the efforts in those instances, the specific materiality and theory of the sundial were utilised to render the particular biblical narrative intelligible. One might even wonder in the case of sundials, an object of frequent use in Jesuit instruction, whether a tangible articulation, such as in Bettini, was used to bring the sign into the present.[98]

Undoubtedly Galileo's inclusion of this miracle within his defence of Copernicanism intertwined the Sundial of Ahaz with the reception and historiography of Copernicanism, and it is clear at the time of Galileo that others thought this miracle was relevant for such discussions. However, when considering the reception of the event within the Jesuit Order, there is some caution against such a clear and consistent association. Based upon the texts analysed here, which included works by mathematicians, astronomers, and theologians, there are signs that this particular biblical miracle was not as closely correlated with anti-Copernicanism as other biblical passages, particularly that of the Battle of Gibeon in Joshua 10. All of this is a helpful reminder that Copernicanism came about at a time when the Sundial of Ahaz was an important topic, even apart from its scientific and cosmological implications.[99]

98 This would cohere with Mersmann, "Moving Shadows, Moving Sun," 96–123.

99 I wish to thank Volker Remmert, Julia Ellinghaus, and the other conference participants for their feedback during the workshop. Special thanks also to Rienk Vermij, Kerry Magruder, and Samantha Leong for their careful reading and helpful comments.

PART 2

The Horologium Ahaz and Joshua Stopping the Sun on Early Modern Scientific Instruments

CHAPTER 4

Turning Back the Sun

Christoph Schissler's 'Horologium Ahaz' as 'Kunststück'

Andrew Morrall

Abstract

This article examines a sundial by the Augsburg instrument maker, Christoph Schissler (c. 1531–1608) that could re-create the biblical miracle of the Dial of Ahaz by making the sun's shadow go backwards in time. It studies the instrument's status and function through the lens of the "Kunststück," the term Schissler used to describe his creation, and which is defined here as denoting a specific class of artwork, characterised by a high degree of technical, conceptual and thematic ingenuity. Discussing the dial within the wider contexts of Schissler's other precision instruments and the dial's own reception and afterlife, it is argued that the mode of "Kunststück" offered Schissler, the self-styled "geometricus artifex," a means to demonstrate the principles of light refraction; and that, by enacting a biblical miracle, the dial could also provoke questions of theology, astronomy and natural philosophy among mathematically and intellectually sophisticated viewers, whether in the context of a Kunstkammer, an instrument collection, or in the possession of an unorthodox Copernican astronomer.

In a letter of 1572 to August I, Duke of Saxony, the Augsburg instrument maker Christoph Schissler (c. 1531–1608) presented an overview of his creations in the hope of receiving orders. The ensuing list provides a glimpse into the range of specialised goods Schissler was producing for a courtly milieu. It included terrestrial and astronomical globes, astrolabes, planispheres, "unusual and wonderful" clocks, and sundials and compasses that could tell the hours of the day and night. Among these instruments was what he described as a "sonderlich Kunststück," in literal translation, an "especial," or "out of the ordinary" art piece. This was a sundial in the form of a drinking vessel. It was based upon the miracle of the Horologium Ahaz, as recounted in 2 Kings 20:11, and Isaiah 38:10, according to which God caused the shadows on the Dial of Ahaz to retreat a full ten hours backwards in time as a sign to King Hezekiah that he would be brought back from the brink of death, and indeed given a further fifteen years

 | DOI:10.1163/9789004677654_006

of life.[1] The next "item" listed, presumably a similar instrument, references the episode in Joshua, 10:12–15, where the sun stood still in its course through the heavens for a full day, as a sign to Joshua of the Lord's favour.[2]

A version of Schissler's *Horologium Ahaz*, signed and dated 1578, survives today in a partial and restored state in the collection of the American Philosophical Society of Philadelphia.[3] It consists of two parts, an upper concave bowl and a circular base. A presumed stem that originally supported the bowl was missing in 1895, when the dial was restored by Julius Sachse. He replaced it with three brass rods, which have since been disassembled (Figure 4.1).[4]

---

1 I would like to thank Volker Remmert and Julia Ellinghaus for the invitation to speak at the workshop at the Bergische Universität Wuppertal in August, 2019 from which this volume grew, and for their valuable editorial suggestions and incisive comments in reading my text.

Hauptstaatsarchiv Dresden, Loc. 8679, fols. 44 and 45. In the relevant part of the letter, Schissler asks whether the duke would be interested in any of the following: "globus Coelestis et Terrestis, Spheram materialem, Astrolabien, Planispherien oder zu seltsamen wunderbarlichen horologis, die sich inclinierien und declinierien, auch frembden Compassen, daß die magnet zeugen, die stunden weist; item mancherley horas, in- und außwendig bei Sonn- und mondschein, die gewiße Zeit und stunde bey tag und nacht zu erkundigen; item ein sonderlich kunststück ein trinkgeschür zu machen, auf den Text des 20. Cap. etc. im 4. buech der künigen von dem horologium Achaz: da der schatten zuruckgelauffen etc. Item auf den Text deß 10. Capitels Josua, da die Sonnen ist stil gestanden, Item ein Instrument zu machen, durch wölches die 4 ort deß himmels als ortum, meridiem, occasum et septentrionem an allen bekannt und unbekannten enden on allen Sonn- mon- und sternschein on alles licht bei stückfinsterer nacht, wöllich instrument villeicht E(uer) G(naden) dienstlich sein möcht, ein fürgenomen ort in hölzern oder wäldern zu finden." Quoted in Maximilian Bobinger, *Christoph Schissler der Ältere und Jüngere* (Augsburg: Verlag Die Brigg, 1954), 34–36.

2 The phrasing is ambiguous but appears to describe two separate instruments, each designed around a separate biblical miracle, and each announced, like other instruments in the list, by the term "Item." Inaccurately translated in Peter Plaßmeyer, "Christoph Schissler. The Elector's Dealer," in *European Collections of Scientific Instruments 1550–1750*, ed. Giorgio Strano et al. (Leiden: Brill, 2009), 20.

3 Inv. no. 58. 66. It has been restored at least twice: by Julius F. Sachse in 1895 and most recently, and to its present state, by Owen Gingerich and Philip M. Sandler. See Julius F. Sachse, "Horologium Achaz," in *Proceedings of the American Philosophical Society* 34, no. 147 (January 1895): 21–30; Louis Winkler, "Pennsylvania German Astronomy and Astrology X: Christopher Witt's Device," *Pennsylvania Folklife* 24, no. 2 (Winter 1974–1975): 36–39; Owen Gingerich and Philip Sadler, "Christopher Schissler's Wonderful 'Bowl of Ahaz' of 1578," *Bulletin of the American Astronomical Society* 21 (1989): 1218; Philip M. Sadler, "An Ancient Time Machine: The Dial of Ahaz," *American Journal of Physics* 63, no. 3 (1995): 211–216.

4 Sachse also removed a short cone-shaped gnomon that had been placed on the bottom of the dial a few centimetres from the centre, and fashioned a new gnomon by stretching a wire from the tip of the staff of the figure on the bowl's rim to a point within the scales of the more northern scale of the bowl. He set the gnomon at a 40° angle to match the local altitude of

FIGURE 4.1
Julius Sachse's 1895 reconstruction of the possible original arrangement of the refractive sundial by Christoph Schissler, known as the "Dial of Ahaz," signed and dated 1578, brass and alloy, American Philosophical Society, Philadelphia, inv. no. 58.66
PHOTO: SACHSE, "HOROLOGIUM ACHAZ," PLATE 1

The upper part consists of a shallow basin made of gilded brass, engraved with a web of lines that allowed one to tell the time at different latitudes and during different seasons of the year. A string, now missing, would have extended from the staff held by the turbaned figure mounted on the rim to the centre of the bowl to form a gnomon. The dial is divided into two separate sets of scales engraved on opposite sides of the bowl that allow it to operate within two ranges of latitude, a more southerly from 41° to 46°, and a more northerly from 47° to 52°. The rim, engraved with the signs of the zodiac, is moveable, such that the figure can be rotated to adjust the gnomon to the requisite local latitude (Figure 4.2). The southerly dial is marked with two scales: one for "common" or equal hours, another for so-called "Italian" hours, made up of twenty-four equal hours, reckoned from 0 at sunset and ending with 24 at the same point the following day. The northerly dial is also graduated in two different scales: one for "common" hours, another for "planetary" or unequal hours, adjusted to accommodate daylight hours as they grow longer in summer and

Philadelphia, but outside the bowl's actual range for the variety of latitudes for which the instrument was originally designed. Sachse, "Horologium Achaz," 21–23.

shorter in winter. Markings for the twelve planetary houses are also included on this scale; there is also a planetary table that indicates which planet rules over each hour of each day of the week. Such information was important for astrological readings, used, for instance, by medical practitioners and others whose activities were guided by the stars.[5]

A second, smaller, slightly concave, circular dial forms the reverse side of base, whose upper side is decorated in relief with aquatic imagery (Figure 4.3).[6] The surface of this base dial is divided into five engraved sections; two make up the dials, graduated for different latitudes of between 44–46°degrees and 47–49°respectively, each with a Latin inscription identifying its range (latitudes that collectively would extend roughly from Genoa to Augsburg).[7] A thin strip running around the outer perimeter, its function unclear, is marked out into four equal bands calibrated into degrees running from 0–90. Two further inner fields contain engraved scenes of Isaiah's miracle and Hezekiah's cure. In one, the prophet Isaiah stands before King Hezekiah, who languishes in his sickbed, and gestures towards the retrograde movement of the sun's shadow on the Dial of Ahaz behind; the matching scene shows King Hezekiah in his recovery, reaching for the paste of figs prescribed to him as a divine cure to his afflicted parts (2 Kings 20:7). A helix in the centre contains a Latin inscription, which gives the key to the instrument's operation. Translated, it reads:

> This hemicyclum-like shell denotes the miracle of the 38th chapter of Isaiah. For if you fill it to the brim with water, the shadow of the sun is borne backwards by ten or twenty degrees. Besides this, it indicates the sign and degree of the sun, that is, the common hour of any day whatever, together with what they call the planetary hours.[8]

---

5 Sachse, "Horologium Achaz," 23. Bobinger, *Christoph Schissler*, 113–119, esp. 115–118; Winkler, "Pennsylvania German Astronomy," 37–38.

6 This relief decoration is identical with that found on the base of an earlier armillary sphere by Schissler, dated 1569, today in Munich, Bayerisches Nationalmuseum, inv. no. Phys 27.

7 "Horologii Achaz hydrographica declinatio ad elevat: poli · 44 · 45 · 46 · gradu." and idem "Horolgii [sic] Achaz hydrographica declinationes ad elevat: poli · 47 · 48 · 49 · grad:" For latitude ranges, see Penelope Gouk, *The Ivory Sundials of Nuremberg 1500–1700* (Cambridge: Whipple Museum of the History of Science, 1988), 15 and 42. See fig. 39.

8 "Notat concha isthac hemiciclea capitis 38 Esaia miracvlvm: nam hanc si aqva labrvm vsque impleveris: vmbra solis ... 10 ... imo: zo ... gradibvs retrorsvm fertvr: signvm ac gradvm solis: quin etiam horam diei vvlgarem qvamcvnque vna cvm planetarvm qvas vocant horis denuncians."

FIGURE 4.2 Christoph Schissler, *Refractive sundial*, 1578: Bowl, 6 in. (incl. figurine) x 12 in. (diam.)
PHOTO: AMERICAN PHILOSOPHICAL SOCIETY, PHILADELPHIA, INV. NO. 58.66

The purpose of this second dial remains mysterious and has not been satisfactorily explained in the existing literature. As Julia Ellinghaus and Volker Remmert point out in their essay for this volume (Chapter 5), its eccentric position obviates its function as a working dial; they raise the plausible possibility that it was originally detachable and capable of being used independently. Yet

FIGURE 4.3 Christoph Schissler, *Refractive sundial: The reverse side of the base*, 1578
PHOTO: AMERICAN PHILOSOPHICAL SOCIETY, PHILADELPHIA, INV. NO. 58.66

positioned beneath the base, it would have served to conceal the narrative imagery and explanatory inscription and required that the cup be inverted and scrutinised in order for its second, 'miraculous' narrative function to be revealed. Such a conceit or trick is fully consistent with the instrument's peculiar status as a 'Kunststück,' as will be discussed below. Such a position would also fit with what can be inferred from the existing visual elements: that the inscriptional reference to a "hemicyclum" matches the shape of the 'upper' bowl, but not the 'lower' dial; and that the size and shape of this latter dial precisely fits the decorative base, which was demonstrably made in the Schissler workshop; even the relationship between the proportions of the 'upper' bowl and 'lower dial' leave one with the strong presumption that the two elements were indeed part of the same instrument.

In 1989, Owen Gingerich and Philip Sandler were able to experiment with the bowl, and demonstrated that the addition of water to the upper bowl did indeed make the shadow move backward by between ten and twenty degrees, corresponding to forty to sixty minutes on the calibrated surface, at different times of the afternoon.[9] On the one hand, therefore, the instrument could operate as a regular sundial, capable of delivering a considerable amount of astronomical data, and on the other hand, as a re-creation of the mysterious biblical Dial of Ahaz, with the miraculous ability to make time retreat.

What was Schissler's interest in creating this dial? What prompted an instrument maker to engage with this particular biblical subject? In seeking answers to such questions, this article attempts to establish the instrument's status, character and function by examining the criteria implicit in the technical, formal, and symbolic choices involved in its making. Particular attention is given to the term "sonderlich kunststück" that Schissler used to describe his instrument. For in specifically describing his invention thus, by making it perform a narrative function, and by fashioning it as a drinking vessel ("Trinckgeschür"), Schissler clearly distinguished it from the other instruments he offered the duke, taking it out of a strictly practical context and, as will be argued, implicitly grouping it with objects that were similarly described as 'Kunststück' in collection inventories and contract agreements.

The term is difficult to translate, especially so, as precise and consistent use of terminology was rare in the period. In general usage, 'Kunststück' often connoted a form of trickery, or deceit, even magic or sorcery, as when Luther described the Devil's lies and deceptions as "sein meister- und kunststücklin," or "works of arch-artistry."[10] Yet in the context of craft production, the word seems to have been used as a descriptor of works of especial ingenuity and high artistic accomplishment, sometimes synonymous with the more general

9 Gingerich and Sadler, "Wonderful 'Bowl of Ahaz,'" 1218; Ivars Peterson, "Turning Back Time: An Antique Sundial Simulates a Biblical Miracle," *Science News* 137, no. 6 (February 1990): 91; and especially Philip M. Sadler, "An Ancient Time Machine: The Dial of Ahaz," *American Journal of Physics* 63, no. 3 (1995): 211–216; available online at https://aapt.scitation.org/doi/10.1119/1.17956.

10 "der teufel ... gehet mit eitel lügen und betrügen umb ... das ist sein meister- und kunststücklin," Luther *Tischreden*. 320 (60, 132 *Irm.*), quoted in Jacob and Wilhelm Grimm, "KUNSTSTÜCKLEIN," in *Deutsches Wörterbuch von Jacob und Wilhelm Grimm* (Leipzig, 1873), 11:2731. Available online at http://www.woerterbuchnetz.de/DWB?lemma=kunststuecklein.

'Kunstwerck,' but retaining the sense of unusual contrivance, of something out of the ordinary.[11] It could also carry a sense of demonstration or show, somewhat akin to the term 'Kabinettstück' that combined these connotations with those of the collectible and of display. Certainly, in the context of the Kunstkammer, the term appears to have denoted a quite specific genre of artwork, as further examination will suggest. A second aim of the article therefore is to draw attention to a previously unstudied contemporary descriptive category for a particular kind of artistic creation.

'Kunst' and 'Kunststück'

To understand the particular semantics of the word 'Kunststück,' one must briefly unpack contemporary usages of its root term 'Kunst' and its various cognates. 'Kunst' itself was a capacious term that was used as a general descriptor of any qualified activity, discipline, craft or trade. It was applied to practices ranging from medicine to husbandry, as well as to all the disparate fields within the artisanal world, from painting, sculpture, and goldsmithing, to cloth dyeing, embroidery, metal stamping, bell casting and so on. Furthermore, it denoted mastery of the particular skill-set of that profession, including not only technical prowess, but also a grasp of theoretical principles. Johann Neudörffer (1497–1563), the Nuremberg humanist, mathematician and writing master, articulated this essentially artisanal attitude to the arts, characteristic of Nuremberg craftsmen in the 1540s, in his *Notes on Artists and Craftsmen from the Year 1547*.[12] Typical is his account of the famous goldsmith, Wenzel

11 Less specific usages of the term do exist. In the autumn of 1565, when Paul Buchner, an agent of the Duke of Saxony, was sent to Nuremberg to seek out "etzlichen Kunststücken" in Nuremberg as suitable New Year presents, he suggested some "künstlich Instrumentlein" ("artfully made small instruments"), a set of three measuring rods (rulers) and a compass, which the goldsmith Wenzel Jamnitzer had originally made for the emperor and was willing to duplicate. See Cornelius Gurlitt, "Aus den sächsischen Archiven. Jamnitzer und der kursächsische Hof," in *Kunstgewerbeblatt. Monatsschrift für Geschichte und Litteratur der Kleinkunst*, ed. Arthur Papst (Leipzig, 1885), 51–53. Available online at https://digi.ub.uni-heidelberg.de/diglit/kunstgewerbeblatt1885/0058. The fact that they were "small instruments" might suggest the kind of artful contrivance – beyond mere utility – that made them suitable as courtly gifts.

12 Johann Neudörffer, *Des Johann Neudörfer Schreib- und Rechenmeisters zu Nürnberg Nachrichten von Künstlern und Werkleuten daselbst aus dem Jahre 1547 nebst der Fortsetzung des Andreas Gulden nach den Handschriften und mit Anmerkungen*, ed. G. W. K. Lochner (Vienna: Baumüller, 1875).

Jamnitzer and his brother Albrecht, whom he characterises as equals in "artistic invention" ("Erfindung der Kunst").[13] 'Kunst' here is to be understood on one hand in terms of practice, for it is essentially technical accomplishments that Neudörffer then enumerates: the skilful carving of coats of arms and seals in silver, stone and iron; glass enamelling in the most beautiful colours, etching in silver, and in particular, casting small animals from the life. But this characterisation also contained an important theoretical dimension: for Neudörffer recognised also their proficiency in perspective and measurement/geometry ("haben der Perspectiv und Messwerk einen grossen Verstand").[14] Elsewhere in his biographies, Neudörffer frequently marks out for praise a grounding in similar mathematical principles. Peter Flötner and Georg Pencz are but two of many artists commended for their grasp of "Perspectiv und Masswerk,"[15] the former also for his skill in composing narrative histories to adorn goldsmiths' work; the latter for his understanding ("Verstand und Geist") of the principles of narrative composition.[16] "Perspectiv" is what characterises the work of Georg Glockendon the Elder, woodblock cutter, printer and illuminator, whose son published a book on the subject.[17] The sculptor, Johann Teschler "created [...] whole images according to such painstaking and correct proportion, that it was wonderful to see." ("machte [...] ganze Bildnisse von solcher lieblichen und gerechten Proportion, dass es wunderbarlich zu sehen war").[18] The goldsmith Jacob Hoffman was highly sought after by kings, princes, and noblemen for his protean technical skills. Yet it was his great understanding of "Symmetria" that gave his large output its particular distinction.[19] Perspective, measure, proportion, symmetry: these were the principles that Neudörffer valued most highly in the work of his contemporaries. This held true among the humanist theorists and advanced artisanal circles of Nuremberg and Augsburg in the 1540s, for whom the theoretical component, frequently characterised by the term 'Verstand' ('understanding'), became an increasingly important criterion for any successful 'Kunstwerk.'

13 Neudörffer, *Nachrichten von Künstlern,* 126.
14 Neudörffer, *Nachrichten von Künstlern,* 126.
15 Neudörffer, *Nachrichten von Künstlern,* 115.
16 Neudörffer, *Nachrichten von Künstlern,* 137.
17 Neudörffer, *Nachrichten von Künstlern,* 140; likely referring to Jean Pèlerin and Albrecht Glockendon, *Von der Kunnst Perspectiva* (Nuremberg, 1540).
18 Neudörffer, *Nachrichten von Künstlern,* 116.
19 Neudörffer, *Nachrichten von Künstlern,* 127.

For Neudörffer's contemporary and acquaintance, Walther Ryff, medical doctor, editor and compiler of humanist texts and the first translator of Vitruvius into German, 'Verstand' contained an important ethical component. For him, it meant quite specifically the exercise of the mind, for only through its use was it possible for humans to recover their rational understanding, lost at the Fall, and which differentiates them from the animals. In practice, for both Ryff and Neudörffer, this meant primarily the use of mathematical principles. Indeed, for Ryff, the pursuit of true understanding was best satisfied by a special attention to the mathematical disciplines, which he enjoined the layman to study in his free time through books or learned discourse.[20] Ryff in fact addressed his book to "Kunstliebhaber," literally, "lovers of art."[21] These were understood, not as admirers of the fine arts in today's narrow sense, but rather as lay non-practitioners who were cultural all-rounders, whose interests encompassed all the disciplines enfolded under the term 'Kunst.'[22]

It is clear that with this expanded understanding of 'Kunst' that was current among humanist commentators, elite craftsmen, and the "Liebhaber" or "Philologi" who were their patrons, a work such as Schissler's sundial would have satisfied the necessary criteria of technical finesse, formal inventiveness, mathematical and conceptual understanding and aesthetic beauty for it to qualify as a 'Kunstwerk' in the highest sense.

Yet among all these cognate uses, the term 'Kunststück' appears to have been employed in a quite narrowly specific way. In the documentary instances where it is found, it almost invariably denotes an object type or instrument, whose primary purpose went beyond the merely functional to encapsulate a specific idea or concept. For example, Thomas Rucker, a master sword-maker from Dresden, who was a resident of Augsburg in the 1570s and who worked

---

20 Marcus Pollio Vitruvius, *Vitruvius Teutsch. Nemlichen des aller namhafftigisten und hocherfarnesten / Römischen Architecti / und Kunstreichen Werck oder Bawmeisters / Marci Vitruuij Pollionis / Zehen Bücher von der Architectur und künstlichem Bawen*, ed. and trans. Walther Hermann Ryff (Nuremberg: Johann Petreius, 1548), fol. 30v, 58–59v and Walther Hermann Ryff, *Der furnembsten, notwendigsten der gantzen Architectur angehörigen Mathematischen und Mechanischen Künst eygentlicher Bericht* (Nuremberg: Johan Petreius, 1547), VIII Preface.

21 Ryff, *Der furnembsten, notwendigsten der gantzen Architectur*, title page, VIII, fol. 113.

22 Ryff explained the term quite exactly in translating the words 'Philologus' and 'Philotechnus,' which he adopted from Cesare Cesariano's earlier Italian translation and commentary of Vitruvius, *Di Lucio Vitruvio Pollione de Architectura libri dece* [...] (Como: Gottardo da Ponte, 1521). The 'Philologus' is the "Liebhaber der Künste und Disziplinen," the lay person, widely read in "the arts and disciplines"; the 'Philotechnus' is a practitioner, one qualified in a craft, and a maker of "Kunstreicher Arbeit," works of great artistry. Ryff, *Vitruvius Teutsch*, fol. 193.

competitively with Schissler for commissions within the orbits of the courts, referred to a monumental iron throne that he made specifically as a "Kunst stühck" in a letter of 1574, addressed to the Emperor Rudolf II. Destined for the imperial collection, the throne contained in every member – legs, arms and back rails – small, chiselled scenes that described the entire course of Roman history, from Aeneas and the heroes of Troy to the sixteenth-century present, framed within an allegory of the Four World Empires, based on the Book of Daniel. Drawing on multiple historical and biblical sources it created an allegory of Habsburg imperial dominion over time and its place within divine providence. The effect was such that the emperor, once enthroned upon it, would have assumed into his own person the entire course of Roman imperial history.[23]

Similarly, a further documentary source refers to a work by Schissler as "ein klein kunststücklein zu dem heiligen Christ," "a little art work of the blessed Christ," presented to the Saxon court in 1572, perhaps as a new year's gift.[24] The nature of this piece is unclear, but an implied connection between its narrative or representational character and its designation as 'Kunststück' is consistent with the other examples.

Indeed, the common denominator among these disparate pieces is that they are all instruments or objects of use into which some kind of concept or narrative idea has been incorporated, not merely as a decorative or ornamental gloss, but as fully integrated, such that their forms, materials, imagery, and, in certain cases, instrumentation all work collectively to embody a single, primary meaning or concept. The 'Kunststück,' used in this specific sense, therefore seems to involve an ontological shift from a functional to an expressive or representational object, its compositional strategies more akin in intention to easel painting or independent sculpture.

A vivid illustration of this shift is provided by Johannes Kepler, in his well-known – and ultimately unsuccessful – determination to present his patron, Duke Friedrich I of Württemberg, with a three-dimensional model of his cosmological theory, set out in his *Mysterium Cosmographicum* of 1596.[25] This

---

23 Staats- und Stadtbibliothek Augsburg, "Thomas Rucker an den Kaiser," Cod. Halder 2, Cod. MS 231 (1574), quoted in Zentralinstitut für Kunstgeschichte und Städtische Kunstsammlungen Augsburg, *Welt im Umbruch. Augsburg zwischen Renaissance und Barock* (Augsburg: Augsburger Druck- und Verlagshaus, 1980), 2:497.

24 Dresden, Hauptstaatsarchiv, "Ein klein kunststücklein zu dem heiligen Christ und einem Newen Jahresgeschenk," Loc. 8679, fols. 44–45 (1572). Bobinger, *Christoph Schissler*, 132.

25 Johannes Kepler, *Prodromus dissertationum cosmographicarum, continens mysterium cosmographicum* [...] (Tübingen: Georg Gruppenbach, 1596).

posited a correspondence between the planetary intervals and those of the Platonic solids when nested within a sphere in a particular order.[26] A first idea was to produce such a model in the form of a drink-dispensing cup, or "Credentzbecher," a form of courtly joke for the table, by which different liquids would pour from different planetary spheres. It probably approximated the well-known illustration, the "Tabula III," published in the *Mysterium Cosmographicum* (Figure 4.4).[27] When difficulties with the goldsmith commissioned to make the piece caused this to be abandoned, Kepler produced a completely new design. This would have introduced a clockwork mechanism intended to demonstrate not merely Kepler's intervals between the planets, but also their *motions*, such that they would turn in proportionate ratios with each other in a manner consistent with the principles of Copernicus's universe.[28] Crucial to this new design was the fact that the duke intended to invest a considerable sum in the piece and to display it in his Kunstkammer. This changed Kepler's conception of the piece, for as he wrote, such a setting would now give him the opportunity to demonstrate Copernican theory not only to cognoscenti but also to a broad spectrum of non-specialist visitors. Movement, he said, was crucial to this ambition, for if the bodies were simply static, the ordinary viewer would see only a model of the five Platonic bodies and entirely miss the significance of Copernican principles on display. For this, he wrote, and for the project to be worth the considerable financial outlay, the work, "*had* to be a 'Kunststück' ("das Werckh soll ein kunststuck sein") so that it would be seen as such, not just by the 'Pythagoreans' but by the many."[29]

26 "Intereâ verò dum Tubingâ absum, nescio, quae me puerilis an fatalis cupiditas Principibus placuisse viris, Stuccardiam adegerit, ubi duo agere volebam [...] deinde instrumentum sive Sphaeram, vel imaginem mundj Principi offerre, tanquam rem novam dignam visu a Principe, primo mortalium inspiciendam, sicut ego primus sim inventor, haec in supplicatione egj." Johannes Kepler, *Gesammelte Werke*, ed. Max Caspar and Walther von Dyck (Munich: Bayerische Akademie der Wissenschaften, 1945), 13:71. Quoted in Adam Mosley, "Objects of Knowledge: Mathematics and Models in Sixteenth-Century Cosmology and Astronomy," in *Transmitting Knowledge. Words, Images and Instruments in Early Modern Europe*, ed. Sachiko Kusukawa and Ian Maclean (Oxford: Oxford University Press, 2006), 202, f.n. 24.

27 For an account of the complicated production history of the Tabula III engraving, see Noam Andrews, "Tabula III: Kepler's Mysterious Polyhedral Model," *Journal for the History of Astronomy* 48, no. 3 (August 2017): 281–311.

28 For the second design, see Henry Charles King and John R. Milburn, *Geared to the Stars: The Evolution of Planetariums, Orreries, and Astronomical Clocks* (Toronto: University of Toronto Press, 1978), 92–93.

29 (Emphasis mine.) Kepler, *Gesammelte Werke*, 13:220. "3. Das deren so dem Copernico beyfallen, nicht gar vil, und gmeinglich nur hin und her zerstrewete mathematicj seind.

The addition of clockwork and moving parts would create an altogether different order of object. While the static object would have provided merely an explanation, the appeal of the 'Kunststück' to a non-specialist audience lay in its peculiar rhetorical power: to inspire admiration by means of its technical ingenuity and to persuade via the spectacle of movement – of bodies smoothly turning in perfectly proportionate concinnity.

As these sources indicate, each of these various works was made in the context of elite patronage and destined for display in a collection. As such, they formed a subset of the class of *artificialia* or man-made things, one of the broad categories by which contemporary Kunstkammers were organised. This distinguished them from the *naturalia*, the things of nature, including the prodigious (*mirabilia*), and indeed from *scientifica*, a category that denoted practical instruments.[30] On a broader point of historiography, the role of these *artificialia* has not so far been very clearly established. Modern scholarship has tended to regard them mainly in terms of their technical skill alone and explained their presence in collections largely in terms of an Art/Nature paradigm – as demonstrations of human *techne* made in rivalry to the "wonderful" creations of nature.[31] All too often, moreover, they have been dismissed, quite uncritically – and often pejoratively – as objects of luxury, as "purely decorative" or as "princely playthings."[32] Their expensive materials, elaborate ornamentation and preciosity were clearly important aspects of their appreciation and reflect the social and artistic contexts in which they were designed,

---

4. Das man villeicht noch vil lange Jahr zuebringen werde, biß dise widersinnige meinung in die Gelehrten zubringen. […] Weiln dan nun nicht zu hoffen, das irer sehr vil in Irer Fürl: Gnaden Kunstkammer khommen werden, die dem Copernico beyfallen, so mueß folgen, das die Spectatores und besichtiger das innere Werckh für nichts anders, als für die 5 Geometrische Corpora ansehen, und vermeinen werden, es sey unvonnötten gewest, das man den costen drauff wende. […] Nu wol an gedachte ich, ob wol sollicher stefft nichts zur kunst taugt, sondern fabricae corporum elegantj zuewider, jedoch, weil es ja sein muesse, khönte man eben durch diß mittel einen motum darein bringen. Ist derowegen und schlieslich diß mein meinung, wel das Werckh soll ein kunststuck sein, damit es dafür angesehen werde, nicht allein von Pythagoraeis, sondern auch von meniglichen, so begehrte ich nitt nur proportionem, sondern auch Motum caelorum darein zubringen." Quoted in Mosley, "Objects of Knowledge," 204, f.n. 31.

30 On the organisation of the Kunstkammer of the Saxon Court, for instance, see Helen Watanabe-O'Kelly, *Court Culture in Dresden. From Renaissance to Baroque* (Basingstoke: Palgrave, 2002), 71–83.

31 Especially influential in situating *artificialia* within a philosophical context of "Wonder" has been Lorraine Daston and Katherine Park, *Wonders and the Order of Nature, 1100–1750* (New York: Zone Books, 1998).

32 Jasmin Mersmann, "Moving Shadows, Moving Sun. Early Modern Sundials Restaging Miracles," *Nuncius* 30, no. 1 (January 2015): 121.

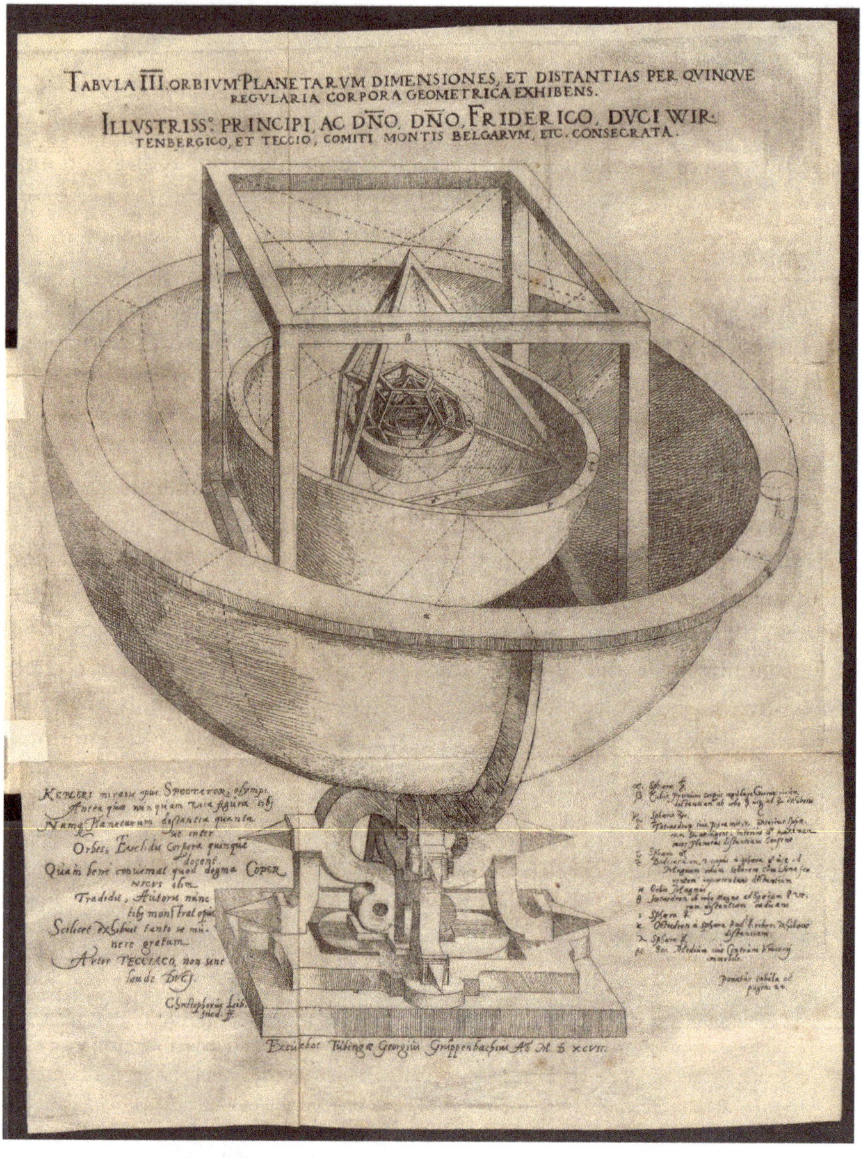

FIGURE 4.4 *First edition of tabula III* from Johannes Kepler, *Prodromus dissertationum cosmographicarum, continens mysterium cosmographicum* [...] (Tübingen: Georg Gruppenbach, 1596)
PHOTO: ETH BIBLIOTHEK, ZURICH, RAR 1367: HTTPS://WWW.E-RARA.CH/DOI/10.3931/E-RARA-445 PUBLIC DOMAIN MARK

constructed and commissioned. Yet to regard them purely in such terms or for their technical skill alone is to obscure the fact that many of these craft works were expressive of carefully meditated themes: their materials, forms and iconography often embodied or addressed significant areas of human knowledge and experience, enjoining their owners to regard them as objects of contemplation, as material interventions into the world of the intellect and ideas, intended to enlighten, instruct and inspire.

Schissler's 'Kunststück': Design and Iconography

Schissler's description of his Dial of Ahaz as a 'Kunststück' therefore invites one to look at its character and functioning in this light: he has effected a similar transformation of a technical instrument into an art work: more than simply a form of historiated ornament, the miracle is physically embodied by the dial itself: that is to say, the gnomon, dial plate and corpus are all organised so as to participate in recreating King Hezekiah's miracle. Like Kepler's projected working model of the heavens, it reenacts its subject performatively, in this case introducing its own kind of motion by harnessing the natural movement of the sun's shadow. Furthermore, its status as a "mathematical" instrument gives the representation of the miracle an added rhetorical authority quite unique to its instrumental nature – differentiating it from representations of the miracle in paintings, prints or other media.

Examination of the dial's individual parts makes clear how each contributes to this overall concept. In the first instance, the viewer is led by the figurative imagery: it is in the playful nature of the 'Kunststück' that the two scenes, hidden from view on the base dial, must be "discovered" in order that the subject around which the piece is constructed be revealed (Figure 4.3). This differs from other, near-contemporary dials that incorporate an image of the miracle of Ahaz. The scene occurs for instance in the upper corners of a printed diagram of a sundial, dated 1561, by Georg Hartmann (1489–1564), Nuremberg pastor, mathematician, and maker of both metal and printed paper instruments (Figure 5.1).[33] In the upper left, Isaiah gestures to a dial,

33 See Suzanne Karr Schmidt, "Georg Hartmann, *Bowl of Ahaz* Sundial," in *Prints and the Pursuit of Knowledge in Early Modern Europe*, ed. Susan Dackerman, exh. cat., Harvard Art Museums, Cambridge; Mary and Leigh Block Museum of Art, Northwestern University, Evanston (New Haven: Yale University Press, 2011), 296–297, cat. no. 69.

while on the right King Hezekiah lies in his bed, his hands raised in prayer.[34] The design is for a concave, semi-cylindrical column sundial, which operated by using the two outer edges of the semi-cylinder as gnomons, such that the sun's shadow, cast by the right-hand edge, would move from left to right for six hours until noon, at which point the left-hand edge would take over as the gnomon and the shadow would travel backwards in the opposite direction, the directional change thereby recalling, if not quite reenacting, the biblical miracle.[35] By contrast, a magnificent brass sundial by the Nuremberg goldsmith, Wenzel Jamnitzer (1508–1585), dated 1578, the same year as Schissler's dial, pairs a vivid representation of the miracle, showing a similarly gesturing Isaiah and a bed-bound King Hezekiah, with an equally dramatically drawn scene of Joshua standing beneath the arrested sun on Mount Gibeon, a sign of God's favor (Joshua 10:14) (Figure 5.8).[36] These two Old Testament scenes of divine astronomical intervention are set within a dense pictorial program across front and back, that includes allegories of the elements, the seasons, two ancient astronomers, and Atlas and Hercules (Figure 5.7). As Volker Remmert has shown, these latter mythological figures, traditional bearers of the world's sphere, had been understood since antiquity as the first teacher and student of Astronomy (Atlas as an astronomer king and Hercules, who took over his burden of the earth, as pupil), and by an extension of the tradition, as the founders of the discipline.[37] A similar meaning can be attributed to the miracle of

34 Hartmann's paper model forms part of the "Collectio figurarum." Bayerische Staatsbibliothek München, "Collectio figurarum," Georg Hartmann, MS Rar. 434, fol. 139. It is reproduced in Georg Hartmann, *Hartmann's Practika. A Manual for Making Sundials and Astrolabes with the Compass and Rule*, ed. and trans. John Lamprey (Bellvue, Co: John Lamprey, 2002), 297.

35 For a detailed explanation and reconstruction of its operation see Chapter 5 in this volume by Julia Ellinghaus and Volker Remmert, "Manipulating the Sun in Depictions on Early Modern Scientific Instruments. An Iconographic Study."

36 Gilded brass, 51.7 x 12 cm; Bibliothèque de l'Observatoire, Paris, inv. no. 7, currently on loan to the Bibliothèque Humaniste in Sélestat. For an analysis of the imagery see Chapter 5 in this volume by Julia Ellinghaus and Volker Remmert. Also see Denis Savoie, "Le cadran solaire de hauteur de Wenzel Jamnitzer de l'Observatoire de Paris," *Cadran Info* 36 (October 2017): 136–151, available online at https://hal.sorbonne-universite.fr/hal-02048624. Another version of the dial was formerly in Dresden (lost during World War II), for which see Max Engelmann, "Wenzel Jamnitzers Dresdner Meßscheibe," in *Der Kunstwanderer: Zeitschrift für alte und neue Kunst, für Kunstmarkt und Sammelwesen*, ed. Adolph Donath (Berlin, September 1919 – August 1920), 311–314, available online at https://digi.ub.uni-heidelberg.de/diglit/kunstwanderer1919_1920/0315.

37 See Volker Remmert, *Picturing the Scientific Revolution: Title Engravings in Early Modern Scientific Publications* (Philadelphia: St. Joseph's University Press, 2011), 127–165; Volker

Hezekiah's cure. For Ahaz, as the maker of the only sundial to be mentioned in the Bible, was likewise considered an ur-exemplar of his profession, sanctioning and ennobling it by his ancient scriptural precedent.[38] One function of the scenes on the Jamnitzer dial was thus to dignify the instrument by tying it to the long tradition of astronomy that extends across time, myth and miracle to the very origins of the discipline.[39] Yet, as Julia Ellinghaus and Volker Remmert argue convincingly in Chapter 5 of this volume, the pairing of two great Old Testament rulers, each the recipient of Divine favour and miraculous intervention, makes them also exemplars of righteous godly rulership, intended to redound flatteringly upon the virtues of the dial's intended – probable ducal, even imperial – owner.

The figurative imagery on Schissler's instrument is directed to a somewhat different end. For the two scenes of the miracle work in dialogue with the centrally positioned Latin inscription. This cites the biblical chapter but without actually indicating its content, stating merely: "This hemicyclum-like shell demonstrates the miracle of the 38th chapter of Isaiah."[40] The images thus amplify the textual reference by unfolding the actual story; indeed, in visual terms, the images appear to flow centripetally out of the text (Figure 4.3). This text-image relation is derived from the popular picture Bibles of the day, such as Jean de Tourne's influential *Quadrins Historiques de la Bible*, illustrated by Bernard Solomon, or, within Germany, the *Biblische Figuren* such as those illustrated by Virgil Solis: both are structured by means of an abbreviated textual reference to a Bible passage, explicated by an image (Figure 4.5).[41]

Remmert, "Analyse des Imageries des Instruments Scientifiques de L'Époque modern (XVIe–XVIIe siècle)," Images des Mathématiques, CNRS, last modified May 29, 2019, https://images.math.cnrs.fr/Analyse-des-imageries-des-instruments-scientifiques-de-l-epoque-moderne-XVIe.html. Atlas and Hercules appear similarly positioned on an astrolabe from 1568 by Johannes Practorius, today in the Staatliche Kunstsammlungen Dresden, Mathematisch-Physikalischer Salon.

38 See Chapter 3 by Brent Purkaple in this volume, "Appealing to Ahaz. The Jesuits and the Sundial of Ahaz in the Era of Copernican Astronomy."

39 The dignifying use of origin stories was widespread within every aspect of contemporary civic, scientific, artistic and literary culture. For its use on other instruments, see Remmert, "Analyse des Imageries." For a broader use among publishers and artisans, see Andrew Morrall, "Virgil's Flute: the Art and Science of 'Antique Letters' and the Origins of Knowledge," in *The Primacy of the Image in Northern European Art, 1400–1700. Essays in Honor of Larry Silver*, ed. Debra Taylor Cashin, Henry Luttikhuizen, and Ashley D. West (Leiden: Brill, 2017), 193–206.

40 See f.n. 8 for the quotation in full.

41 Claude Paradin, *Quadrins Historiques de la Bible* (Lyon: Jean de Tournes, 1560), with woodcut illustrations by Bernard Solomon; and Virgil Solis, *Biblische Figuren des*

Schissler's design adopts this emblem-like relation but goes further: for the text's content also acts as a connecting thread that explicitly links the images of the miracle to the operations of the upper dial ("This hemicyclum-like shell demonstrates the miracle of the 38th chapter of Isaiah. For if you fill the basin altogether with water, the shadow of the sun is borne backward by ten or twenty, degrees").[42] The inscription is thus key to structuring the viewer's experience of the piece: as the viewer turns the instrument to grasp the words in their spiraling helix, the narrative meaning of the upper dial is literally unravelled. In a dynamic activity of eye and mind, the work's meaning is revealed in a series of visual and textual connections, as each element builds on the last and recognition of one prompts interpretation of another. The dial's other figurative element, the cast turbaned figure that directs the gnomon on the upper bowl, although lacking a specific identity, conveys a generalised sense of exotic antiquity, and in the context of the imagery on the dial below, contributes to an aura of ancient astronomical practice.[43]

By these means, the upper dial's working parts (i.e. gnomon, dial plate and corpus) are integrated into the biblical story such that these ordinarily neutral, mechanical parts assume a figurative role in the narrative. As Jasmin Mersmann has argued of *skaphe*-, or bowl-dials, their hemispherical shape could be understood figuratively as a model of the dome of the heavens, the gnomonic grid acting as a projective image of the network of parallels and meridians, while the shadow moving over it functioned as an index of the sun's movement.[44] Mersmann was writing about the only other known surviving northern European sixteenth-century Dial of Ahaz, made by Georg Hartmann (1489–1564) in Nuremberg, and a possible prototype for Schissler's more fully

---

Alten und Neuen Testaments (Frankfurt: Sigmund Feyerabend, 1560). Other examples include: Sebald Beham, *Biblische Historien* (Frankfurt: Christian Egenollf, 1533); Hans Holbein the Younger, *Historiarum veteris instrumenti icones* (Lyon: Melchior and Caspar Trechsel, 1538). See chapter 1 by Walter Melion in this volume, "The Prayer of Joshua and the Dial of Ahaz. Paratactic Scriptural Illustration in the Lempereur, Vorsterman, and Van Liesveldt Bibles." Also, Max Engammare, "Les Figures de la Bible: le destin oublié d'un genre littéraire en image (XVIe–XVIIe siècles)," *Mélanges de l'École française de Rome* 106 (1994): 549–591.

42 See f.n. 8.

43 Schissler used the same figure to somewhat similar semantic effect on a horizontal sundial dated 1562, today in Dresden: Dresden, Staatliche Kunstsammlungen, Mathematisch-Physikalischer Salon, inv. no. D I 37. See Peter Plaßmeyer and Christoph Emmendörffer, eds., *Weltenglanz: Der Mathematisch-Physikalische Salon Dresden zu Gast im Maximilianmuseum Augsburg* (Berlin: Deutscher Kunstverlag, 2009), 86–87.

44 Mersmann, "Moving Shadows, Moving Sun," 98.

FIGURE 4.5 Bernard Solomon, *Woodcut illustration of 2 Kings 20*, in Claude Paradin, *Quadrins Historiques de la Bible* (Lyon: Jean de Tournes, 1560)
PHOTO: GALLICA.BNF.FR / BIBLIOTHEQUE NATIONAL, PARIS

representational dial (Figure 4.6).[45] As she wrote, "While armillary spheres, astrolabes or clockwork-driven celestial globes simply *represent* those constellations on a small scale, sundials exploit their own indexical relationship with the Sun and exhibit that relationship on their surface."[46]

Mersmann thus implicitly comes down on one side of an older scholarly debate as to the nature of early modern 'scientific instruments.' While James A. Bennett and others have argued that instruments such as astrolabes, armillary spheres and globes functioned simply and straightforwardly as calculating

45 There are three surviving specimens: (1) Cambridge, Harvard, Collection of Historical Scientific Instruments, inv. no. 7397; cf. Stephanie Dick, in *Art, Printmaking, and Science* (Cambridge MA: typescript, 2010), 35–36, and Karr Schmidt, "George Hartmann," 296–297. (2) Toledo, Museo de Santa Cruz, inv. no. 5045; and (3) Madrid, Museo Nacional de Cienca y Tecnologia, inv. no. 1999/023/0001; cf. Lino Colombo, "Refractive Sundials in Italy," accessed February 16, 2021, https://www.academia.edu/39913916/Gli_orologi_solari_a_rifrazione_in_Italia.

46 Mersmann, "Moving Shadows, Moving Sun," 99.

FIGURE 4.6 Georg Hartmann, *Scaphe Dial of Ahaz*, for the latitude of 43 degrees 10', signed and dated 1548, brass
PHOTO: HARVARD, COLLECTION OF HISTORICAL SCIENTIFIC INSTRUMENTS, CAMBRIDGE, INV. NO. 7397

devices,[47] others have drawn attention to the use of such mathematical instruments as models of the cosmos, even, as Adam Mosley has shown as regards Kepler's planned instrument, for embodying and communicating novel cosmological theories.[48]

This argument is particularly borne out by instruments deliberately crafted as 'Kunststücke,' in which the purpose was as much narrative as computational. Even more than Hartmann's dials, this is true of Schissler's Dial of Ahaz,

47 See James A. Bennett, "Geometry in Context in the Sixteenth Century: The View from the Museum," *Early Science and Medicine* 7, no. 3 (2002): 214–230 and James A. Bennett, "Knowing and Doing in the Sixteenth Century: What were instruments for?" *British Journal for the History of Science* 36, no. 2 (June 2003): 129–150.

48 Mosley argued that because Kepler's proposed moving model of the Copernican universe also incorporated the ability to determine planetary movement relative to the earth, the astronomer saw no intrinsic incompatibility between a model and a calculating device, between determining planetary positions instrumentally and representing the true form of the cosmos. Mosley, "Objects of Knowledge," 204–205.

where, through text and figuration, the craftsman has exploited the idea of the dial as a celestial stage and the setting for his biblical drama. In this way, the sun's moving shadow becomes the crucial player, whereby its motion, incorporated into the design of the dial, reenacts celestial phenomena on a terrestrial scale. The sundial, quite unlike other kinds of artwork had the ability to harness the immense, ineffable, motions of the heavens, to bring them down to earth, and to manipulate them within its tiny compass. One might say that Art ['Kunst!'] has channelled the forces of Nature to its own purposes: in this case, paradoxically, to reproduce an aberration from Nature's norms. This was spectacle indeed!

Schissler's Technical and Mathematical Background

What then does this instrument tell us about Schissler as a practitioner and about the nature of his thinking in fashioning his dial? Was the trick of light refraction merely the clever solution to a technical problem of an accomplished technician? Or did Schissler have a deeper interest in making commentary that touched on natural philosophical or theological questions regarding the sun's movement, and by extension, the nature of miracles? In the almost total absence of documentation, one might look to the material evidence of Schissler's instruments themselves for what can be inferred about his mathematical knowledge and involvement with astronomy.

The range of Schissler's instruments runs between the two poles of practical utility and the representational art object. While some had the status of works of art or collectors' objects, and others were utilitarian surveying or measuring instruments, a great many had the character of consumer objects, that is to say, instruments such as portable dials, where elaborate decoration or shape served as fashionable attributes superimposed onto, but secondary to the actual practical function. In his early career Schissler was particularly adept at creating compact, multi-dial, precision instruments that could perform multiple purposes (Figure 4.7).[49] These found a fashionable market, and while novelties, they did not create anything genuinely new.

Such works nonetheless demonstrate Schissler's mastery of the sundialist's art of projecting the sun's movement through three dimensions onto a flat dial by means of a perspectival grid. In the case of the *Horologium Ahaz*,

49 Christoph Schissler, *Astronomical Compendium*, signed and dated 1557, gilt brass, 77 x 77 x 15 mm, British Museum, London, inv. no. MLA 1913, 7–10.1.

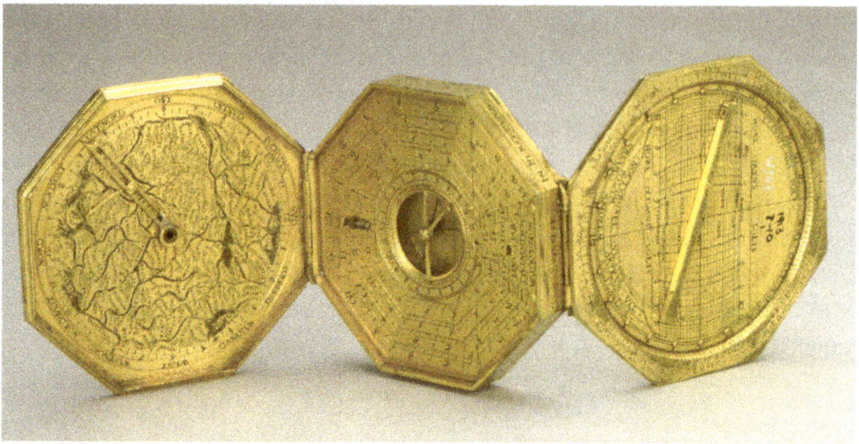

FIGURE 4.7 Christoph Schissler, *Astronomical compendium*, signed and dated 1557, gilt brass; 77 x 77 x 15 mm
PHOTO: BRITISH MUSEUM, LONDON, INV. NO. MLA 1913, 7-10.1

Schissler's ability to calibrate precise temporal markings upon an *ellipsoidal* surface, while taking into account different latitudes and seasonal variations, and to order them within a satisfying design of elegantly intersecting lines, was a remarkable feat of applied mathematics, based on a set of complicated astronomical calculations and an advanced grasp of perspectival drawing.[50]

Very little, however, is known of the degree of Schissler's own mathematical training or astronomical knowledge or of his collaborations with mathematicians and astronomers. A measure of his own grasp of the mathematics necessary in the production of his instruments can be taken from his knowledge of the current printed mathematical and technical literature (some in

50 How he achieved this is unclear. Philip Sadler has suggested that the calibration of this device "probably relied on use of large astrolabes, which model the path of the sun, since the mathematics required to model the sun's motion and project it onto a dish had not yet been fully developed." Sadler used a combination of spherical trigonometry and a technique of "ray tracing" to reconstruct the original positioning of the string gnomon, such that it would correctly indicate the backward movements of the sun's shadow when the bowl is filled with water. He established that a bead attached to a gnomon string, that led from the staff held by the turbaned figure to the centre of the dial, suspended 3.5 cm above the bottom and 3 cm from the centre of the bowl, would produce a shadow with the correct degree of recession, between 12.00 and 3.00 pm at any of the degrees of latitude marked on the dial. Philip Sadler, "An Ancient Time Machine. The Dial of Ahaz," *American Journal of Physics* 63, no.3 (March 1995): 211–216, esp. 213–214.

Latin) that, as Bobinger's analysis revealed, underpins the construction of many of his instruments. These sources included Johann Dryander's *Sphaerae materialis gründliche Ausslegung* (German edition, Marburg, 1539), Oronce Finé's *Protomathesis* (Paris, 1532), Walther Ryff's (Gualtherus Rivius) *Der furnembsten … Mathematischen und Mechanischen kuenst eygentlicher bericht* … (Nuremberg, 1547), Philipp Apian's *De utilitate Trientis* (Tübingen, 1586), and numerous works by Peter Apian (1495–1552), such as his *Instrumentbuch* (Ingolstadt, 1533), the *Cosmographicus liber* (Landshut, 1524), and the great *Astronomicum Caesareum* (Ingolstadt, 1540).[51] Schissler referred obliquely to his knowledge of such works in a manuscript instruction-booklet for a surveying instrument he provided for the Duke of Saxony in 1569, stating that he "had so far not found much of use in the writings of earlier authors on this art, or that had been already applied."[52]

Schissler and Mathematical Circles in Augsburg

From such sources, Schissler derived basic designs and templates for dials and globes and their accompanying tables that he could adapt to his own use. Yet the larger significance of this wide immersion in printed sources is that it situates him squarely within the dynamic culture of scientific and technical humanism that flourished across northern Europe through a network of theoretical and technical interchange conducted among a wide constituency of interested parties. In southern Germany, it extended back to at least a century earlier, when the mathematician and astronomer, Johannes Regiomontanus (1436–1476) settled permanently in nearby Nuremberg, established his own printing press, dedicated to publishing ancient and modern mathematical and astronomical texts, and set up a workshop to design and produce more accurate mathematical instruments and better astronomical tables.

51 Bobinger, *Christoph Schissler*, 128–129.

52 "bißher bey anndern Authorn so von dieser Khunst geschrieben nit vil befunden und im gebrauchgewest." Christoph Schissler, *Geometria, das ist die Distantia der Stätt und Flecken und was dergleichen zu messen ist* (Augsburg, 1569), inv. 1587, no. 107 (destroyed in WWII), dedication. Quoted in Herbert Wunderlich, *Das Dresdener 'quadratum geometricum' aus dem Jahre 1569 von Christoph Schissler der Ä., Augsburg* (Berlin: VEB Deutscher Verlag der Wissenschaften, 1960), 47. The instrument was his *Quadratum Geometricum*, today in the Physikalisch-Mathematischer Salon, Dresden, inv. no. CI1. See below.

Regiomontanus famously chose Nuremberg because of its local expertise in instrument making.[53] His legacy in both areas persisted, indeed flourished, throughout the following century. By 1567, the French mathematician, logician, and educational reformer, Petrus Ramus (1515–1572), could praise the mathematical and technical achievements of Germany in his *Prooemium mathematicum* and specifically "the senate and citizens of Nuremberg," for encouraging the mathematical arts and doing their best "to always have [new] Regiomontani."[54]

A similar mathematical and scientific culture had grown up in Augsburg. Ramus encountered it at first hand during a stay in the city between 1569 and 1570. He recounts meeting the Danish astronomer, Tycho Brahe (1546–1601) ('Tycheo Bracheo, adolescente nobili Dano') at a dinner of a local 'astronomy circle,' which comprised a group of learned citizens centred around the brothers Paul (1527–1581) and Johann Baptist Hainzel (1524–1581), wealthy and enlightened patricians and members of the city council. Those present included Hieronymus Wolf (1516–1580), erstwhile student of Melanchthon and Camerarius in Wittenberg, historian, and collector and translator of Greek texts, who was the private secretary and librarian to Johann Jakob Fugger between 1551–1557, and from 1557, the Rector of the St. Anna Gymnasium; and Johannes Maior (c. 1540–1615), who from 1560 onwards was teacher and later also Rector of the St. Anna Gymnasium and maintained an extensive epistolary exchange with Tycho Brahe. Other members of the circle included the Augsburg goldsmith Lorenz Denn (1522–1599), who carried out alchemical experiments with his son-in-law, the doctor, Balthasar Minderer of Esslingen, and with whom Tycho stayed during his Augsburg stay between 1569 and 1570.[55]

53 In a letter to the mathematician M. Christianus in Erfurt, in 1471, he wrote: "Nuperrime Norinbergam mihi delegi domum perpetuam, propter commoditatem instrumentorum maxime astronomicorum, quibus tota sideralis innititur disciplina, tum propter universalem conversationem facilius habendum cum studiosis viris ubicunque vitam degentibus, quod locus ille perinde quasi centrum Europae propter exoursum mercatorum habeatur." Quoted in Friedrich Wilhelm Ghillany, *Geschichte des Seefahrers Ritter Martin Behaim, nach den ältesten vorhandenen Urkunden bearbeitet* (Nuremberg: Bauer und Raspe, 1853), 23, f.n. 1.

54 "Senatus enim populusque Noribergensis operam dedit, ut perpetuos Regiomontanos haberet." Petrus Ramus, *Prooemium mathematicum* (Paris: Wechel, 1567), 278. Quoted in Isabelle Pantin, "New Philosophy and Old Prejudices: Aspects of the Reception of Copernicanism in a Divided Europe," *Studies in the History and Philosophy of Science* 30, no. 2 (1999): 242, f.n. 27.

55 Inge Keil and Helmut Zäh, "Tycho Brahes Aufenthalt in Augsburg (1569–1570 und 1575)," in *Beiträge zur Astronomiegeschichte*, ed. Wolfgang R. Dick and Jürgen Hamel (Frankfurt: Verlag Harri Deutsch, 2004), 7:45–98.

At the same dinner, Ramus was shown the "Great Quadrant" that had been commissioned by Paul Hainzel to Brahe's specifications and erected in the grounds of the patrician's country house. At the time it was the largest and most accurate astronomical instrument in Germany.[56] Brahe had come to Augsburg precisely to seek out craftsmen with the technical expertise to make astronomical instruments of a sufficient size and accuracy for his exacting needs. While there is no evidence to connect Schissler with this particular instrument, a letter from Johannes Maior to Brahe in 1576, a year after the latter's second Augsburg stay, mentions him explicitly in familiar, proprietary terms as "Schissler, our sundial maker," implying his close involvement with this group of cognoscenti, and perhaps Brahe's own familiarity with the instrument maker.[57] Particularly striking is Maior's use of the Greek, γνωμονοποιον [*gnomon* (sundial) and *-poios* (maker)] to describe him. An apparent neologism that used the elevated "*-poios*," normally reserved for learned or literary forms of making, the term both acknowledged Schissler's status as a kind of "artifex doctus," and by the choice of Greek, associated him with an ancient Greek tradition of sundialists and astronomers. Its usage forms a neat linguistic parallel to the images of ancient astronomers with which the craftsmen ornamented their instruments, as discussed above.[58] And it confirms that the

56 Petrus Ramus, *Defensio pro Aristotele adversus Jac. Schecium* (Lausanne: J. Probus, 1571), 109–118. For Tycho's observations at Augsburg (116–117): "Interea ab Joā Baptista Hainlelio, senatore primario agnitus et liberalissime tractatus sum adhibitis in convivium eruditis variae doctrinae convivis, sed imprimis Tycheo Bracheo adolescente nobili Dano, cum quo post prandium in suburbanam Hainlelii villam deductus matimaticā observandis syderibus machinā libenter aspexi." The Quadrant's great size, with a radius of 543 cm, allowed measurements of such exactitude that Tycho subsequently incorporated them into his later astronomical tables. He published a description and illustration of the instrument in his *Astronomiae Instauratae Mechanica* of 1598.

57 Adam Mosley, in a consideration of Brahe's use of the Atlas-as-globe-bearer motif, suggests that he might have been influenced by Schissler's Atlas-bearing Armillary sphere of 1569, today in the Bayerisches National Museum in Munich (inv. no. Phys. 27), seen during a visit to his workshop in that year. Adam Mosley, *Bearing the Heavens. Tycho Brahe and the Astronomical Community of the Late Sixteenth Century* (New York: Cambridge University Press, 2011), 18.

58 *Poios* was normally used to designate a poet or writer working in one or another genre, for instance, *logopoios* (prose writer) or *mythopoios* (myth maker, or fabulist in the negative sense). My thanks to Caspar Meyer for his assistance with these issues of translation. A striking, near-contemporary usage of the term is found in Henri de Monantheuil's 1599 Latin translation of the *Mechanics*, attributed to Aristotle. Monantheuil characterised God as a mechanic and the cosmos as a machine and described the divine maker as "sapientissimus, optimus, potentissimus mechanikos et mechanopoios." Henri de Monantheuil, *Aristotelis Mechanica Graeca, emendate, Latina facta, & Commentaris illustrate* (Paris: Jérémie Perier, 1599), Dedication to King Henry IV of France. For other

elevated status Schissler conferred upon himself when he signed his own works (including the Dial of Ahaz), "geometricus et astronomicus artifex" was not mere hubris, but widely recognised in the learned community. Maior, recounting a conversation with Schissler in his workshop, recommended him to Brahe as a craftsman who possessed the necessary expertise and who had expressed himself willing to supply him with a large, lightweight, and perfectly spherical celestial globe made of copper that was to be easily transportable. This was on the basis of a similar copper globe that he had fashioned the previous year and which had been sent to Spain under the auspices of the merchant, Marcus Fugger.[59]

A copper globe, signed by Schissler and dated 1575, today in the Palaçio Nacional de Sintra in Portugal (inv.no. PNS 3457), conforms closely to this description, and although no documentary evidence can be found to link it to Maior's reference, it demonstrates Schissler's competence in the exacting task of transcribing detailed and fresh astronomical data onto a sphere with the utmost accuracy (Figure 4.8). Samuel Gessner's detailed study of the globe has shown how the representations of the constellations on its surface are based on a much earlier printed paper globe of 1532 by the Cologne mathematician, Caspar Vopel de Medebach (1511–1561), but with significant adjustments.[60] One is the inclusion of entirely new constellations; the other is that the star positions have been shifted from those on Vopel's globe to take into account the precession of the equinoxes for the then current epoch of 1575, using calculations that had been established by Copernicus. This concern for the greatest possible astronomical accuracy accords well with Brahe's requirements for a working instrument, as well, more generally, with the interests of Hainzel, Wolf, Maior, and the scholars and mathematicians working within

---

meanings and applications of the word *gnomon*, see Thomas Little Heath, *A History of Greek Mathematics,* 2nd ed. (New York: Dover Publications, 1981), 78–79.

59 "Porro te ignorare nolo, mi Domine TYCHO, SCHISLERUM γνωμονοποιον [sundial maker] nostrum confecisse globum ex cupro fabrefactum secumdum omnes dimensiones perfectae, veluti tuum Automaton, rotunditatis cuius diameter quinquepedalis. Globus autem tali arte est fabrefactus, ut in 8 partes dissolve et rursus in integrum restitui possit, necessitate exigente, quem in Hispaniam unâ cum Astrolabio quantitates eiusdem per MARCUM FUGGERUM est misurus." Tycho Brahe, *Tychonis Brahe, Dani, Opera Omnia*, ed. John Louis Emil Dreyer (Copenhagen, 1924), 7:36; original letter in Österr. Nationalbibliothek, Handschriften u. alte Drucke, Cod. 10686/66, fol. 74r–75r. Quoted in Samuel Gessner, "'Geometricus et astronomicus faber': Chr. Schissler aus Augsburg," in *Weiter sehen/Seeing further*, ed. Jürgen Hamel and Michael Korey (Frankfurt am Main: Verlag Harry Deutsch, 2012), 137–138.

60 Gessner, "Geometricus et astronomicus faber," 149–151.

the orbit of the St Anna Gymnasium.[61] Gessner's meticulous reconstruction of the theoretical basis, the sources, and the process by which the new star alignments were calculated and applied to the globe, reveals the nature of the collaborative pooling of intellectual and practical resources necessary to the instrument's construction: on the one hand, the trained mathematician's grasp of the complicated astronomical theory required to determine the new, updated star alignments themselves; and on the other hand, the technical finesse and more modest mastery of conventional sexagesimal arithmetic necessary to work from available astronomical tables, which would have to have been within Schissler's competence.[62] The globe underscores the still murkily understood interdependency for knowledge and expertise between the skilled instrument maker and the various different parties invested in the study and practice of astronomy during this period – among them the mathematicians and professors of astronomy at the *Gymnasium*, educated patricians, clerics, and medical doctors.[63]

The degree of Schissler's expertise is no less apparent in his land surveying instruments. The most elaborate of these is the *Quadratum Geometricum*, a geometric square, made in 1569 for the Duke of Saxony (Figure 4.9) that was accompanied by an instruction book, cited above, in manuscript form, illustrated with elaborate watercolour drawings of surveying techniques.[64] In a striking statement in the first of two prefaces, Schissler wrote that his explicit aim in constructing it was to advance the art of measurement with new and

61 Karl-August Keil and Inge Keil, "Astronomie am Gymnasium bei St. Anna," in *1531–1981: Festschrift 450 Jahre Gymnasium bei Sankt Anna in Augsburg*, ed. Johannes Rettenberger and Helmut Schmid (Augsburg, 1981), 12–23.

62 Gessner, "Geometricus et astronomicus faber," 141–148, 153–155.

63 The eighteenth-century archivist and historian of Augsburg, Paul von Stetten, presented a vivid record of these intertwining interests as well as the presence of mechanical and astronomical instruments in local Kunstkammer collections, including those of the Hainhofer, Hopfer, Staininger, and Bimmler families: "Birk, Wolf, die Höschel und Henisch, Rectores und Professores an unserm Gymnasium, Bayr ein Rathskonsulent, und vornämlich die rechtschaffenen und gelehrten Rathsglieder Johann Baptist und Paulus Hainzel, waren Beförderer dieser Wissenschaft, die selbsten wichtige Beobachtungen anstellten, und also auch die Werkzeuge dazu nötig hatten. [...] Ueber dieses gab es auch damals Kunstsammlungen, in welchen man allerley mechanische Maschinen und mathematische Werkzeuge für grosse Zierden hielte, wie auch dergleichen in der Hainhoferschen, Hopferischen, Stainingerischen und Bimmlischen Kunstkammer zu finden waren. Dergleichen Gelehrte und Liebhaber beschäfftigten die Künstler, und die ersten wussten ihnen auch sehr viel Gutes an die Hand zu geben." Paul von Stetten, *Kunstgewerb und Handwerks Geschichte der Reichs-Stadt Augsburg* (Augsburg, 1779), 168–169.

64 Christoph Schissler, *Geometria, das ist die Distantia der Stätt und Flecken und was dergleichen zu messen ist*, Augsburg, 1569, inv. 1587, no. 107 (destroyed in World War II).

FIGURE 4.8 Christoph Schissler, *Celestial globe*, signed and dated 1575, copper, Palaçio Nacional de Sintra, inv. no. PNS 3457
PHOTO: JOAQUIM ALVES GASPAR, WIKIMEDIA COMMONS, CC BY-SA 4.0

useful methods and especially "by means of this fine and ingenious instrument" which, he continued, "can measure all heights, depths, distances, lengths and breadths, all from one place, without running to and fro, and all without the need to make special calculations."[65] As Michael Korey has demonstrated, Schissler applied infinitely more finely calibrated scales to what was a traditional instrument, which allowed readings of an unprecedented precision; he added slanting transversal lines near the edges of the square that allowed greater accuracy in reading the scales; and he engraved computational tables on both sides that enabled the user simply to look up the answer to the particular surveying problem at hand without the need to actually carry out any

65 "durch dises schöne und kunstliche Instrument welliches Quadratum Geometricum gennent wurdt. Zue messen ist alle Höhe, Thüffe, Weyte, Lenge und braitte, Diss alles an einem Ortt, on alles hin- und hergehn, auch one alle soundere Rechnung." Quoted in Bobinger, *Christoph Schisssler*, 58–59.

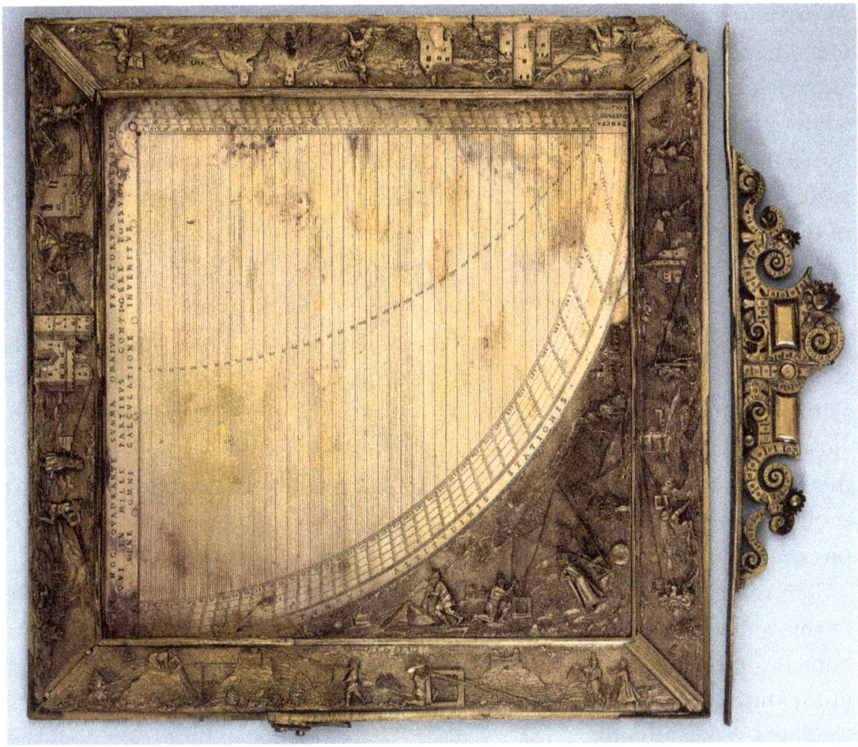

FIGURE 4.9 Christoph Schissler, *Quadratum geometricum*, signed and dated 1569, gilt brass, 37,5 x 37,5 cm
PHOTO: STAATLICHE KUNSTSAMMLUNGEN DRESDEN, MATHEMATISCH-PHYSIKALISCHER SALON, DRESDEN, INV. NO. C I 1

elaborate mathematical calculations.[66] All this necessitated an expert grasp of the mathematics of triangulation and its practical applications. It bears out his proud title of "master of geometrical and astronomical instruments" ("Geometrischen und Astronomischen Werckhmaister"), with which he signed his booklet.

This deepening involvement in the production of highly innovative, cutting-edge instruments for astronomical mapping dates from the late 1560s onwards and appears to mark a new phase in Schissler's output, reflected in the type

66 For a precise explanation of the working of the instrument, see Michael Korey, *The Geometry of Power. Mathematical Instruments and Princely Mechanical Devices from around 1600* (Munich: Deutscher Kunstverlag, 2009), 17–19. See also Wunderlich, "Das Dresdener 'quadratum geometricum.'"

of instruments produced by his workshop. These included a series of odometers that he made for August I, Duke of Saxony and for the Emperor Rudolf II, which he presented in person in Prague, travelling via Vienna in 1583.[67] At a later point, his expertise in using such instruments is borne out by the commission given to him by the civic authorities of Augsburg to undertake a large-scale survey of the city between 1598 and 1602, followed by a similar survey of the bailiwicks around the city, which occupied him until his death in 1608.

Schissler's professional trajectory from the late 1560s onwards, therefore, demonstrates an increasing immersion in the theoretical aspects of his craft. While his own avowed approach to "advance the science of measurement" was that of the skilled technician solving practical problems with new methods ("etlichen neuen und nützlichen Praktiken"), it is important to register how his close collaboration with professional astronomers and mathematicians in the production of advanced technological tools became an important part of his professional persona, and an extension of his practice into what today would be called joint scientific enterprises.

Returning to the Horologium Ahaz in this light, it becomes more possible to think that Schissler was using the mode of the 'Kunststück' to engage, as both Regiomontanus and Georg Hartmann had before him, in solving a set of theoretical and technical problems, namely, the nature of the sun's – or its shadow's – movements across a dial, the optics of water refraction, and the whole possibility of recreating the effect of retrograde solar movement. Both Regiomontanus and Hartmann were deeply interested in the science of optics and there is evidence that the former also constructed refractive dials already in the fifteenth century, although none have survived.[68] Hartmann inherited a manuscript copy of Ptolemy's *Optics* that had belonged to Regiomontanus, which discussed the optics of refraction and provided tabulations of angles of refraction. Sven Dupré has suggested that their common interest in Ptolemy's work was precisely because the ancient Greek astronomer's tabulations presented a means to understand the operations of a refractive dial and the means to calculate the angles of refraction of light passing through water.

67 Bobinger, *Christoph Schissler*, 69–70.

68 See Sven Dupré, "The Dioptrics of Refractive Dials in the Sixteenth Century," *Nuncius* 18, no. 1 (2003): 39–67, esp. 42–45. Dupré cites an anecdote of the Italian mathematician Mutio Oddi how Christopher Clavius had told him in Rome that Regiomontanus had made one for a German Prince: "& un giorno parlandone io col Padre Christoforo Clauio in Roma, mi disse, che Giovanni da Monteregio n'havea fatto uno ancor lui, per vn Principe d'Alemagna." Mutio Oddi, *De gli Horologi Solari Trattato* (Venice: Per il Ginammi, 1638), 100.

The Ptolemaic tradition had also been passed down through medieval natural philosophers such as John Pecham, (d. 1292), an edition of whose *Perspectiva Communis*, Hartmann produced in 1542,[69] and Witelo (1230–1314), who had formulated his own tables of refraction for air-water, glass-air, glass-water based on Ptolemy's calculations. His writings on illusions of perception, flat and parabolic mirrors, the problem of the refraction of light on the border of centres of diverse optical density and meteorological optics, including the rainbow, remained influential into the sixteenth century.[70] It was only after Thomas Harriot (1560–1631) in around 1600 and later, Willebrord Snellius (1580–1626) in c. 1620, independently discovered the sine Law of Refraction that a systematic method of calculation could be employed.[71]

While Regiomontanus and Hartmann were both university-trained and deeply vested in ancient learning, Schissler, with his practical training, would have lacked direct access to such ancient texts, while sharing their mathematical interests. His contribution lay in his ability to grasp this kind of quantitative knowledge, mediated to him by learned associates like Johannes Maior and others within the astronomical circles of the St Anna Gymnasium, and to translate it into working instruments.

On a fundamental level, therefore, Schissler's motivation in constructing a Dial of Ahaz was theoretical: like Hartmann, he sought a way to demonstrate the abstract mathematics of light refraction in tangible and visible form. Given the conditions of patronage and of the markets and the modes in which he was working, the best, perhaps only, category of work at his disposal was the

69 John Pecham, *Perspectiva communis. Summa cura et diligentia emendate ...*, ed. Georg Hartmann (Nuremberg: I. Petreius, 1542). See John Pecham, *John Pecham and the Science of Optics. Perspectiva Communis,* ed. and trans. David C. Lindberg (Madison: The University of Wisconsin Press, 1970), 56–58.

70 He was studied by both Nicolaus Copernicus and Johannes Kepler, who wrote a treatise on optics entitled *Ad Vitelonem paralipomena*.

71 Dupré, "The Dioptrics of Refractive Dials," 45–51, for ancient and medieval knowledge of refraction and methods of construction of refraction dials. On Witelo, see Clemens Baeumker, *Witelo. Ein Philosoph und Naturforscher des XIII. Jahrhunderts* (Münster: Aschendorff Buchhandlung, 1908). On Harriot, see Johannes Lohne, "Thomas Harriot (1560–1631), the Tycho Brahe of Optics," *Centaurus* 6 (1959): 113–121; Johannes Lohne, "Zur Geschichte des Brechungsgesetzes," *Sudhoffs Archiv für Geschichte der Medizin und der Naturwissenschaften* 47, no. 2 (June 1963): 152–172. On Snellius, see Klaus Hentschel, "Das Brechungsgesetz in der Fassung von Snellius: Rekonstruktion seines Entdeckungspfades und eine Übersetzung seines lateinischen Manuskriptes sowie ergänzender Dokumente," *Archive for History of Exact Sciences* 55, no. 4 (2001): 297–344; Cornelius de Waard, "Le manuscrit perdu de Snellius sur la réfraction," *Janus* 39 (1935): 51–73; more generally, Liesbeth Cornelia de Wreede, "Willebrord Snellius (1580–1626), a Humanist Reshaping the Mathematical Sciences" (PhD diss., University of Utrecht, 2007).

'Kunststück.' Using this mode, he could overlay his fundamental technical interest with layers of narrative, conceptual and aesthetic sophistication that would meet the expectations of a wealthy and intellectually astute patron like the Duke of Saxony. In the speculative context of a Kunstkammer, it would be entirely natural for these combined elements to provoke questions of theology, natural philosophy and astronomy and prompt discussion and debate (just as Kepler's projected instrument was intended to do).

As regards the choice of theme, it is also important to recognise the craftsman's own social investment in his invention as an important conditioning factor of its conception. From the inclusion of a similar instrument, listed as among his own inventions, in the letter to the Duke of Saxony, with which this article opened, and from the frequency and familiarity with which other instrument makers used the theme in the decoration of similar works (discussed above), there is no reason to doubt that the choice of the Miracle of Ahaz was Schissler's own and that he had no need of the participation of a patron or other "advisor." Indeed, there is an undeniable self-referentiality in choosing a subject that redounds cleverly upon his own field of expertise as a dial maker. Schissler, in effect, was presenting himself as a modern-day Ahaz, the ur-exemplar of his profession, if not surpassing him in replicating the very miracle that played out on the biblical dial. Unusual enough in the oeuvre of a maker of predominantly practical instruments, one might see Schissler's 'Kunststück' as a statement of artistic ingenuity and uniqueness, the grandiloquent gesture of a craftsman, aware of his position as one of *the* premier instrument makers of his day, sought after by emperors and dukes. It matches other expressions of his professional and social status evident in his works, notably his own coat-of-arms engraved on the Copper Globe of 1575, his use of Latin inscriptions, and above all, his proud signature, 'geometricus et astronomicus artifex/faber/opifex' that, as discussed above, aligned him with a learned, Latinate and mathematical culture.[72]

Reception and Afterlife of the Horologium Ahaz

As for ideas or broader theories that might underlie Schissler's use of the biblical theme? While the miracle of the Dial of Ahaz was to become a flashpoint

72 "Werkmaister": ms. *Geometria* 1569; "faber": Celestial globe 1575; "artifex": Horologium Ahaz, 1578; quadrant, 1579; pair of compasses 1579, and 1580; "optifex": terrestrial globe 1597, "Geographicorum et astronomicorum instrumentorum fabricator": large armillary sphere 1605. Gessner, "Geometricus et astronomicus faber," 148–150.

in the arguments for and against a Copernican worldview, one cannot point to any references to helio- or geocentrism in any of Schissler's works, nor any indication that this particular instrument was intended as a commentary on this debate (notwithstanding his awareness of Copernican precession in modelling the 1575 globe and his likely association with Tycho Brahe).[73] Since the publication of Copernicus's *De Revolutionibus* in 1543, astronomers of all religious persuasions had been content to use Copernicus's model for its accuracy in plotting celestial movements, while treating its revolutionary cosmological theory as no more than a useful thought experiment.[74] Neither does Schissler's Dial overtly comment upon current disputes among theologians of differing confessional standpoints as to the nature of miracles and of God's intervention in the world.[75] And yet, once out in the world, an instrument that demonstrated the workings of solar retrogression as described in the Bible must have elicited a considerable interest that went beyond its entertainment value or technical ingenuity and tantalisingly gestured towards questions that would become explicit and polarising only a few years later.

Contemporary mathematicians were already pondering such questions. Already in the mid-sixteenth century, the Portuguese mathematician Pedro Nunes proved by predictive mathematics that solar regression was in fact a natural phenomenon.[76] And in the early seventeenth century, Galileo, in his letter to the Grand Duchess Christina of Lorraine (1615), pointed out that Paul

73 Gessner, "Geometricus et astronomicus faber," 137–138 and 145.

74 Andreas Osiander's infamous, anonymous preface, the *Ad lectorem* (To the reader) added in the first edition of the *De Revolutionibus*, set the scene for such a reading: "It is the duty of an astronomer to record the motion of the heavens with diligent and skillful observations, and then he has to propose their causes, or rather hypotheses, since he cannot hope to attain the true reasons. [...] these hypotheses need not be true nor even probable; it is sufficient if the calculations agree with the observations." See Robert S. Westman, "The Melanchthon Circle, Rheticus, and the Wittenberg Interpretation of the Copernican Theory," *Isis* 66 (1975): 165–193; and Robert S. Westman, "Three Responses to the Copernican Theory: Johannes Praetorius, Tycho Brahe and Michael Maestlin," in *The Copernican Achievement*, ed. Robert S. Westman (Berkeley: University of California Press, 1975), 285–345; Isabelle Pantin, "New Philosophy and Old Prejudices: Aspects of the Reception of Copernicanism in a Divided Europe," *Studies in the History and Philosophy of Science* 30, no. 2 (1999): 237–262.

75 For an account of the different approaches of Protestants and Catholics towards miracles, see Peter Dear, "Miracles, Experiments, and the Ordinary Course of Nature," *Isis* 81, no. 4 (December 1990): 663–683.

76 See Peter Lynch, "Pedro Nunes and the Retrogression of the Sun," *Bulletin of the Irish Mathematical Society* 81 (Summer 2018): 23–30; and especially Chapter 2 by Henrique Leitão in this volume, "Mathematical Certainty and Biblical Inerrancy. Pedro Nunes and the Retrogradation of Shadows at the Dial of Ahaz."

of Burgos (c. 1351–1435) had long ago suggested that Hezekiah's miracle did not take place in the sun, but locally, affecting only those rays that produced the narrow shadow on the dial.[77] At much the same time, as Volker Remmert has shown, Catholic theologians were invoking the sun's backward movement in the miracle of Hezekiah to demonstrate the biblical truth of a traditional Ptolemaic, geocentric cosmology.[78]

There is one, albeit later, source that offers a valuable perspective on how a refractive dial such as Schissler's could prompt speculations of this kind. It is found in the travel journal of the celebrated scientist and traveller Zacharias Conrad von Uffenbach (1683–1734), that recorded his "intellectual tour" with his brother, Johann Friedrich Armand von Uffenbach (1687–1769), of collections and libraries across North-West Europe between 1709–1711.[79] During their stay in the university town of Helmstedt, the brothers visited the collection of Johann Andreas Schmidt (1652–1726), abbot of the nearby Marienthal monastery, which served as a school and teacher training college. Schmidt taught natural science and had built a teaching collection of instruments, which he described in a printed, illustrated catalogue.[80] Uffenbach's journal records the instruments they saw, including a number of sundials, among which was, a "Polyedro," which "could make the shadows retreat by a certain amount of time."[81] This instrument can be precisely identified because the phrase

77 See Bernard R. Goldstein "Galileo's Account of Astronomical Miracles in the Bible: a Confusion of Sources," *Nuncius* 5, no. 1 (January 1990): 6–12. For an English translation of the letter to Duchess Christina, see Maurice Finocchiaro, ed. and trans., *The Essential Galileo* (Indianapolis: Hackett Publishing Company, Inc., 2008), 109ff.

78 Volker Remmert, "Picturing Jesuit Anti-Copernican Consensus: Astronomy and Biblical Exegesis in the Engraved Title-page of Clavius's *Opera mathematica* (1612)," in *The Jesuits II: Cultures, Sciences, and the Arts, 1540–1773*, ed. John W. O'Malley, Gauvin Alexander Bailey, et al. (Toronto: University of Toronto Press, 2006), 291–313.

79 Zacharias Conrad von Uffenbach, *Merkwürdige Reisen durch Niedersachsen, Holland und Engelland,* 3 vols. (Ulm, 1753–54). See also Peter de Clercq, "The travel journals of Balthasar de Monconys (1608–1665) and Zacharias Conrad von Uffenbach (1683–1734)," *Bulletin of the Scientific Instrument Society* 128 (2016): 2–13.

80 Johann Andreas Schmidt, *Collegii Experimentalis Physico-Mathematici Demonstrationes Singulis Semstribus in Academia Iulia Curiosis B.C.D. Exhibendae,* 4th ed. (Hamm, 1721).

81 Uffenbach, *Merkwürdige Reisen*, 1:251–252: "Was die in der Erklärung gemeldete, und in denen Figuren 107. bis 115. abgebildete vielerley Sonnen-Uhren betrifft, so sind selbige nicht unbekannt. Jedoch muss von der letzten, nemlich dem Polyedro Fig. 115. was dabey gesetzet, nemlich quod in eo certo tempore umbra retrogrediatur, erkläret werden, daß solches auf diese Weise geschehe, daß nemlich solches sich nach den Aequinoctiis verändere, und in dem einen der Schatte auf die eine Seite in dem andern aber auf die andere falle. Jedoch muss hievon Georg Varenius in seiner *Geographia generali*, aus welchem Herr Abt Schmid diese Sonnen-Uhr genommen, nachgeschlagen werden."

Uffenbach used to describe the instrument, "quod in eo certo tempore umbra retrogrediatur" is taken exactly from Schmidt's catalogue description (the "a" of "retrogrediatur" notwithstanding), and he explicitly references the illustration of the instrument, "Fig. 115." in Schmidt's volume (Figure 4.10).[82] Most usefully, Uffenbach also recorded the abbot's opinion of the piece, that:

> this particular invention of a sun-dial might well give atheists and those people who did not believe in miracles the idea that the sundial upon which the shadow moved backward on itself as a sign to King Hezekiah was fashioned in somewhat the same, or in a closely conforming manner, which the prophet Isaiah observed, and precisely on that account, proposed as a sign to the king.[83]

The polyhedral dial illustrated is quite unlike Schissler's and operates according to a different gnomonic system.[84] Yet the abbot's comments are revealing in a number of ways. They show how naturally a refractive dial of any kind instantly called to mind the biblical miracle, even without an explicit reference, and could prompt musings about the theological implications it raised. In this case it caused an orthodox Catholic churchman and natural philosopher to ponder the heresy of attempting to replicate a divine miracle. The abbot conceived the dial not as an act of human mimesis of a natural wonder, made in the spirit of pious emulation, as some recent commentators have suggested, but as a work of atheist trickery.[85]

---

82 Schmidt, *Collegii Experimentalis* [...], fol. D., and Tabula VIII, Figs 106–115. The relevant passage, under the section *de Caelo*, reads: "Quae vel *universalia* sunt, *annulus universalis*, duobus circulis meridian & aequinottalia constans, non uno elaborates modo: *crux horologia* vniversalis: *horologia aquinoctalia* universalia alia: cylindri horodictici &c. vel *particularia*, tum *horizontalia*, tum *inclinata*, tumalia. His subjicitur *polyedrum*, diversa habens horologia: & *aliud*, in quo certo tempore umbra retrogreditur."

83 Uffenbach, *Merkwürdige Reisen*, 1:252: "Es möchte diese besondere Erfindung von einer Sonnen-Uhr den Atheisten, und denjenigen, so keinen Glauben an Wunderwerken haben, wohl die Gedanken geben, als seye die Sonnen-Uhr, daran der Schatte dem Könige Hiskia zu einem Zeichen hinter sich zurück gegangen, auch auf eine etwan dergleichen, oder ihr sonst nahe kommende Manier verfertigt gewesen, daran der Prophet Esaias solches vermerkt, und deßwegen eben dieses Zeichen dem König vorgeschlagen haben. 2. Kön. 20, 8. u. f."

84 Sachse, "Horologium Achaz," 24–25, who references Uffenbach's memoirs, but without apparent knowledge of Schmidt's catalogue, mistakes the Helmstedt instrument for Schissler's actual Dial of Ahaz. Uffenbach states that Schmidt had taken his refractive sundial from an example found in Georg Varenius's *Geographia generali* (Amsterdam, 1650).

85 See Anthony J. Turner, "A Biblical Miracle in a Renaissance Sundial," *Bulletin of the Scientific Instrument Society* 61 (1999): 11–14: "What man could do by natural means was,

FIGURE 4.10 *Plate VIII*, from Johann Andreas Schmidt, *Collegii Experimentalis Physico-Mathematici Demonstrationes Singvlis Semstribus in Academia Ivlia Curiosis B.C. D. Exhibendae a Io. Andr. Schmidt D.P.P* (5th edition, Helmstedt: Schnorrius, 1727)
PHOTO: HERZOG AUGUST BIBLIOTHEK, WOLFENBÜTTEL, H: 510 HELMST. DR. (24), CC BY-SA

of necessity, possible to God and the act of mimesis itself became a form of worship." Also Samuel L. Bethell, *The Cultural Revolution of the Seventeenth Century* (London: D. Dobson, 1951), 44–58.

It is interesting, too, that the instrument held no apparent astronomical implications of either the helio- or geocentric system for its owner (or his lay visitors); at issue, rather, was the sacred nature of miracles and the hubris in trying to emulate or disprove them by human means. This is all the more striking because the abbot was evidently comfortable with using heliocentric systems in his teaching: the instrument that immediately precedes the refractive dial in Uffenbach's account of Schmidt's instruments, and figure 106 in Schmidt's catalogue (Figure 4.10), is described as a "Plani-planeticum armillare" that laid out the Copernican system, and which could, by a simple rearrangement of parts, also be made to demonstrate the Tychonic universe.[86] Clearly neither the abbot nor the Uffenbachs regarded the polyhedral dial as a means to prove or disprove the sun's movement. In the index to Uffenbach's journal, there is a reference to "Sundial, – Hiskia, Where the Shadow turns back," listed simply as "Curiosus."[87]

By contrast, some thirty years before the Uffenbach's visit to Helmstedt, Schissler's actual Horologium Ahaz can be traced to the possession of an astronomer and dissenting Lutheran minister, Johann Jakob Zimmermann (1642–1693).[88] As an astronomer, Zimmermann embraced the heliocentric universe of Copernicus, while as a theologian, he developed a set of radical millennial and mystical beliefs, which led to his eventual removal from the Lutheran ministry in 1685 and forced him and his family into a peripatetic exile from his native Württemberg. Settling finally in Hamburg, he published two works, the pseudonymous *Exercitatio theoricorum Copernico-coelestium* (1689)

86 Uffenbach, *Merkwürdige Reisen*, 1:249–251: "Nicht weniger merkwürdig und nützlich ist das hernach folgende Plani-planeticum armillare Fig. 106., wie es im jetzt ermeldten Kupferstücke vorgestellet wird, so zeiget es den Copernicanischen Welt-Bau. Will man nun geschwinde und so zu sagen ohne Mühe einem den Tychonischen Welt-Bau zeigen, so rücket man es nur, wie die Figur N. 25 ausweiset. Fig XXV." A lengthy explanation of the functioning of the instrument follows.

87 Quoted in Sachse, "Horologium Achaz," 24.

88 The historical record is retrospective and based on a strong local tradition associated with the American Philosophical Society in Philadelphia. Zimmermann was the *magister* or leader of a group of Rosicrucian mystics in Hamburg, whose passage across the Atlantic to Pennsylvania in search of religious freedom in 1693 he organised. At his untimely death at the outset of their voyage across the Atlantic, Schissler's dial, along with his other instruments, passed to his successor, Johannes Kelpius (1667–1708), who took over leadership of the group and was one of the founding members of a hermetic community who settled in the Wissahickon woods near Philadelphia in the late seventeenth century. The Dial came thence to the American Philosophical Society through Christopher Witt, an associate of that community of hermits and friend of Benjamin Franklin, around 1765. Sachse, "Horologium Achaz," 28–29.

and *In Scriptura Sacra Copernizans* (1690, reprinted, 1706, 1736) that reconciled these two aspects of his thought in a singular cosmological vision.[89]

At the core of his thinking was a form of speculative metaphysics that strove to integrate a heliocentric universe that drew on Copernicus, Kepler and, by this time, Descartes, with a vitalist conception of nature, inspired by the ideas of Giordano Bruno and Jakob Boehme. From this, as Mike A. Zuber has shown, Zimmermann developed a form of cosmotheism, which set out to explain the workings of "the invisible being of God" within the physical universe.[90] Drawing on Boehme's mystical conception of an antagonism between light as an active principle and darkness as its negative, passive opposite, Zimmerman equated light with the Divine, and the Copernican sun, spinning on its own axis at the centre of the planetary system, with the triune Godhead and the motive power of the entire universe. Emissions of light particles from this spinning central force field set in motion the secondary planets, which, composed of sluggish dark matter, were impelled to turn with it, seeking its warmth. In this way, Zimmermann explained both the axial rotation of the planets and their elliptical orbits as the result of divine power.[91]

In his second volume, *Scriptura Sacra Copernizans,* Zimmermann demonstrated how the writings of Scripture bore witness to the truth of this form of Copernicanism.[92] In part two, he dealt systematically with each of the biblical references to miracles that involve the movement of the sun and which had

89 Johann Jakob Zimmermann, *Scriptura S. Copernizans seu potius Astronomia Copernico-Scripturaria Bipartita* [...] (Hamburg: Christian Reymers, 1690); and Johann Jakob Zimmermann, *Exercitatio theoricorum Copernico-coelestium mathematico-physicotheologica* (Philadelphia, 1689).

90 Zimmermann, *Exercitatio theoricorum*, 4. The phrase, cited in Greek, is derived from Romans 1:20. See also 101, where Zimmermann translated the phrase as "Gottes unsichtbahres Wesen." Quoted in Mike A. Zuber, "Copernican Cosmotheism: Johann Jacob Zimmermann and the Mystical Light," *Aries* 14, no. 2 (2014): 219, fn. 19.

91 See Zuber, "Copernican Cosmotheism," 215–245.

92 The book's full title summarises its intention: *Scriptura S. Copernizans seu potius Astronomia Copernico-Scripturaria Bipartita Das ist: Ein gantz neu- und sehr curioser Astronomischer Beweißthum, Des Copernicanischen Welt-Gebäudes aus Heil. Schrifft: Worinnen I. Beydes der Sonnen sam[m]t andern Fixsternen räumlicher Stillstand, als auch der Erdkugel samt anderer Planeten natürlichen Umlauff buchstäblich angewiesen; II. Die vermeintlich widersprechende Gegen-Sprüche aus dem Hebräischen und Griechischen Text deutlich erörtert werden; Um, so wohl die Hoheit, Fürtreffligkeit und Autorität des H. Prophetischen Worts, wider die heuttägige Verkleinerung der Spinosisten, Naturalisten, Atheisten .... zu verwahren; Nebens untermengter Erläuterung einiger Carthesianischen und Chymischen Geheimnissen, auch etlicher ... Glaubens-Articuln.*

been used by orthodox Catholics to endorse the Ptolemaic world view.[93] In his commentary on the miracle of Hezekiah, he begins by refuting the conventional Ptolemaic (and Catholic) position that the sun itself moved backwards 10 degrees across the sky with the argument, well known to sceptics from Galileo onwards, that both the texts of 2 Kings 20:9–11 and Isaiah 38:8 clearly state that it was the shadow on the dial, not the sun that moved backwards. He also refutes the "followers of Spinoza," along with "naturalists" and "atheists," mentioned on the title page, who asserted that the retrogression of the shadow was a purely local phenomenon, caused by a highly unusual effect of light refraction on this single dial.[94] For how was it, he asks, citing 2 Chronicles 32:31, that embassies from distant Babylon had come to Jerusalem to ask what had caused the solar phenomenon that they too had witnessed from their own distant vantage points?[95]

Zimmermann's answer lay in his mystical Copernican system: though stationary in space at the centre of the planetary system, it was the sun's *axial* movement "from West to East" that God reversed. As the sun was "the well-spring of all movement" this had the effect – through its solar emissions ("ausgehende Effluvia solaria") of reversing the orbits of all the lesser dark planets, including the earth, such that the entire planetary system was reversed by ten degrees.[96] Only a Copernican universe could explain

---

93 Mike A. Zuber, "Johann Jacob Zimmermann and God's Two Books: Copernican Cosmology in Lutheran Germany around 1700," in *Knowing Nature in Early Modern Europe*, ed. David Beck (London: Routledge, 2015), 83– 99. Zuber focuses on Zimmermann's interpretation of Joshua's miracle on Mount Gibeon as his primary example and does not address the miracle of Hezekiah.

94 Zimmermann, *Scriptura S. Copernizans*, 43: "Das aber des Sonnen-Zeigers Ahas eben in specie gedacht wird / hat es nicht der Meinung / als ob dieses Wunder des Phoenomeni nur allein an demselbigen Orte geschehen seye; wie die Spinosisten es für eine blosse äusseriche Apparenz auslegen / meldende: es ware nur durch eine starcke und sondersame Refraction am Sonnen-Zeiger Ahas effectuirt worden."

95 "Howbeit in *the business of* the ambassadors of the princes of Babylon, who sent unto him to enquire of the wonder that was *done* in the land, God left him, to try him, that he might know all *that was* in his heart."

96 Zimmermann, *Scriptura S. Copernizans*, 42–43: "Wolte aber Gegentheil par force erzwingen / dass die Sonne selbsten zurück gegangen seye / so hat sich doch solches bey der Hypothesi Copernicana werckstellig machen lassen. Dann / dieweil die Sonne im Centro dieses Vorticis, da sie sich von Westen gen Osten gyriret/ hiermit ein Quellbronn aller Bewegung ist / und vermittelst solcher ihrer Central-Gyration auch die Planeten und unsere Erd-Kugel nach gleicher Plaga zum Umlauff stimuliret / und aber alhier die Sonne übernatürlicher weise im Centro ihres Vorticis ware von Osten gen Westen/ und also in plagam oppositam so lang zurück gezogen worden / so würde auch die Erde / durch die aussgehende effluvia solaria in orbem magnum, veranlasset / dass sie sich auch nach

the biblical text in literal terms and leave the harmony of the cosmos undisturbed.[97]

It is not known how Zimmermann came to acquire Schissler's dial or how he might have used it in the light of his cosmology. But certainly, the dial offered him a material analogue of his particular proof of the cosmotheist truth of the biblical account, as outlined above. After all, to reverse the sun's shadow by water refraction was in effect to unshackle the shadow from its origin such that it no longer implied the sun's corresponding regressive motion through space. This reasoning was of course also that of the "Spinozans, naturalists and atheists," who also rejected the Ptolemaic basis of the miracle, as Abbot Schmidt's comments about his own refractive dial made clear. All the same, Schmidt's and Zimmermann's opposing views provide interesting examples of reception and ownership of a refractive, miracle-inducing dial, even if they date from a century or so after Schissler. In the intervening period, astronomical knowledge had made significant advances and the ideological lines around the Ptolemaic/Copernican debate had widened considerably: Zimmermann's radical cosmotheism was but one idiosyncratic form of Copernicanism among many; nor was the debate along simple Catholic/Protestant lines; Zimmermann, after all, was seeking to convince a conservative *Lutheran* leadership who still espoused a Ptolemaic worldview.[98] Nonetheless, the views of an orthodox abbot and a radical nonconformist together indicate how, once out in the world, a showpiece of craft ingenuity could be implicated very differently in the struggles of individuals and institutions to come to terms with new systems of knowledge and to assimilate them into older patterns of theology and natural philosophy.

---

solcher Plaga hat wenden müssen / vermöge dessen [...] Wordurch dann der Schatten / so lange dieser Gyrus Retrogradus der Sonnen selbsten gewähret / hätte müssen zurück gehen / nemlichen zehen Stuffen weit. Welches abermals den buchstäblichen Verstand klärlich behielt / und zumalen die Natur selbsten an Hand giebet."

97 The point is clearly and polemically made in Zimmermann's explanation of Joshua's miracle: "Woraus klar /dass er einen solchen Stand begehret / durch welchen die Harmonye des gesammten Planetologii nicht zustöret / sondern mit der Sonnen auch der Mond und andere Planeten dieses Vorticis, und also dieser gantze Vortex oder Würbel so lang ruhen möchte; welches Josua nicht von der Ptolomaischen / wol aber von der Copernicanischen Sonne erlangen mögen / und kann man dahero nach der vulgaren Astronomia diese Biblische Histori gantz nicht/ wol aber nach Copernici Sätzen buchstäblich erklären, da alles verwunderlich zusammen stimmet." Zimmermann, *Scriptura S. Copernizans*, 38.

98 Zuber, "Johann Jacob Zimmermann," 96.

Conclusion

For Schissler in the late sixteenth-century, however, the use of his sundial to espouse or deny a rational scepticism of miracles or to embrace or reject Copernicanism was not at issue: it is certainly impossible to attribute to him any such views. His interest in reconstructing a miracle by means of a refractive dial is best seen in the light of his professional practice and interests, in the skill and the sheer inventive variety he brought to his sundials, surveying tools and astronomical instruments: that is to say, in the recognition that his creative imagination was essentially mathematical and technical. In that sense, his professional passions were fully consonant with the European-wide spirit of the new cartography and astronomy that sought to map and measure the world and the heavens in the abstract language of mathematics, and which encouraged the growing belief that the tools of the mathematician rather than the texts of ancient philosophers or indeed the Bible, were the correct ones to use to understand the natural world.[99] The Horologium Ahaz demonstrates this principle several years before Galileo's discoveries began to make the natural world mathematically legible: for Schissler has used a technical instrument to explain a miracle of nature, normally the province of theology, in the abstract terms of refractive geometry.

In that sense, perhaps the largest claim that can be made for Schissler's Horologium Ahaz and for the category of the 'Kunststück' as a creative mode, operating within the collecting and commercial conventions of the day, is that it formed one small, tentative, part of a larger movement that swept over the cultural and intellectual landscape of northern Europe in the sixteenth century and laid the foundations for the "Scientific Revolution" in the next: what historians of science have called the "mathematization of nature."[100] As the curious afterlife of Schissler's instrument bears witness, it was a process that developed only slowly and uncertainly. Nonetheless, it becomes possible to follow a thread that connects a dial maker's ingenious experimentation with the sundial's normal rules of operation and of the Kunstkammer as a place of speculative discussion, to the development, by the mid-seventeenth century, of an entire

99 See for instance, Lesley B. Cormack, Steven A. Walton, and John A. Schuster, eds., *Mathematical Practitioners and the Transformation of Natural Knowledge in early Modern Europe* (Cham, Switzerland: Springer International Publishing, 2017).

100 See H. Floris Cohen, *The Scientific Revolution: A Historiographical Inquiry* (Chicago: The University of Chicago Press, 1994), 59–97; and H. Floris Cohen, *How Modern Science Came into the World: Four Civilizations, One 17th-Century Breakthrough* (Amsterdam: Amsterdam University Press, 2010), xvii–xviii, 201.

experimental philosophy. Schissler's Horologium Ahaz and the 'Kunststück' in general, stood at the threshold of an age that Henry Power (1623–1668), English physician and experimenter, writing in 1664, could describe as "the days that must lay a new Foundation of a more magnificent Philosophy [...] that will Empirically and Sensibly canvass the Phenomena of Nature, deducing the Causes of things from such Originals in Nature, as we observe are producible by Art and the infallible demonstration of Mechanicks."[101]

---

101 Henry Power, *Experimental Philosophy in Three Books* (London, 1664), 192.

CHAPTER 5

Manipulating the Sun in Depictions on Early Modern Scientific Instruments

An Iconographic Study

Julia Ellinghaus and Volker Remmert

Abstract

Scientific instruments of the early modern period can be decorated with various kinds of figurative scenes. In this article we examine instruments that bear depictions of the biblical miracles of Hezekiah's life prolonged (2 Kings 20:1–11; Isaiah 38:1–8) and Joshua stopping the sun (Joshua 10:10–18), either solo or combined. These depictions adorn devices of different types and purposes, though all of them may be used for time measuring: a paper sundial by Georg Hartmann (1561), a torquetum by Johannes Praetorius (1568, Nuremberg), a refractive sundial by Christoph Schissler (1578, Philadelphia), a measuring instrument by Wenzel Jamnitzer (1578, Paris) and an astronomical clock by Hans Gruber (1583, Stuttgart). Analysing the pictorial programmes of these instruments brings to light the different meanings underlying the scenes, that are closely tied to either the way the items operate or to the purchaser or addressee, showing that none of them is explicitly linked to an anti-Copernican message.

Quite a few early modern scientific instruments are adorned with imagery reflecting a more or less complex pictorial programme. Sometimes the imagery points to the various uses of the instruments, but usually it is not relevant to actual work with the instrument.[1] Often the imagery addresses popular themes related to the structure and the understanding of the cosmos in the early modern period as in depictions of the seven planets, the four seasons, or the cycle of the twelve months of the year. However, some instruments carry less standard subjects; among these, the representations of the two biblical miracles of the retrograding sun (Horologium Ahaz, 2 Kings 20:1–11; Isaiah

1 Research for this paper was kindly supported by the *German Research Foundation* in the framework of the project *Ikonographie auf wissenschaftlichen Instrumenten der Frühen Neuzeit* (*Iconography on early modern scientific instruments*). We are grateful to Andrew Morrall for discussing various aspects with us and for cross-reading this text.

 | DOI:10.1163/9789004677654_007

38:1–8) and the sun standing still over Gibeon (Joshua 10:10–18) – both briefly paraphrased in the introduction of this volume – are of particular interest. In this chapter we focus on how these two miracles played out on early modern scientific instruments and analyse them in the relevant iconographic contexts and programmes. Images of these two miracles could in some cases carry clear anti-Copernican undertones by the late sixteenth and early seventeenth centuries. A case in point are the two vignettes on the frontispiece of Christopher Clavius's *Opera mathematica* of 1612.[2] However, in what follows we address the question of whether such anti-Copernican undertones would automatically be associated with the two images or whether there are other messages they were intended to convey – possibly even commenting on the instrument itself and its usages.

Depictions of both miracles can be seen on Wenzel Jamnitzer's measuring instrument of 1578 (Paris) and on Hans Gruber's astronomical clock of 1583 (Stuttgart). In 1572 the goldsmith and instrument maker Christoph Schissler mentioned a pair of instruments dedicated to the two miracles in a letter to August I, elector of Saxony. While we only know of one isolated depiction of Joshua stopping the sun on scientific instruments at the moment,[3] the Horologium Ahaz appears without its counterpart on a paper instrument by Georg Hartmann (1561), a torquetum by Johannes Praetorius of 1568, a refractive sundial by Christoph Schissler (1578), and on a polyhedral paper sundial by Georg Brentel[4] (1619), which is a slightly varied remake of Hartmann's paper instrument. In what follows we will first discuss the significance of the

2 Volker Remmert, *Picturing the Scientific Revolution: Title Engravings in Early Modern Scientific Publications* (Philadelphia: St. Joseph's University Press, 2011), 25–45. For a different kind of view see Chapter 3 in this volume by Brent Purkaple, "Appealing to Ahaz. The Jesuits and the Sundial of Ahaz in the Era of Copernican Astronomy."

3 It is a rather simple depiction on a polyhedral dial dated around 1650, attributed to Johann Melchior Volckmair and made for an unnamed Austrian archduke, today kept in the Kunsthistorische Museum Wien, inv. no. 728, cf. Kunsthistorisches Museum Wien, *Katalog der Sammlung für Plastik und Kunstgewerbe, II. Teil: Renaissance* (Vienna: Kunsthistorisches Museum Wien, 1966), 124, cat. no. 403; online collection, acceessed July 20, 2022: https://www.khm.at/objektdb/detail/86927/. The instrument came to our knowledge only after finishing this article.

4 Georg Brentel, *Kurtzer Bericht Und Gebrauch deß beygelegten in Kupffer gezeichneten Sonnenuhr-Stöckleins* (Lauingen: Jacob Senfft, 1619). The print is mentioned in Suzanne Karr Schmidt, "Georg Hartmann, Bowl of Ahaz Sundial," in *Prints and the Pursuit of Knowledge in Early Modern Europe*, ed. Susan Dackerman, exh. cat., Harvard Art Museums, Cambridge; Mary and Leigh Block Museum of Art, Northwestern University, Evanston (New Haven: Yale University Press, 2011), 296–297, cat. no. 69. We will not deal with this remake extensively in this article, as the main features are identical to Hartmann's paper instrument.

Horologium Ahaz when depicted by itself and then proceed to the meaning of its combination with Joshua stopping the sun. To conclude we will discuss the question of whether and how the Copernican debate plays out in the material we have analysed.

Before we proceed, a remark on our use of the term *scientific instrument* is in order. It began to be used as a generic term in the nineteenth century. Our use expands on Gerard Turner's conceptualisation of (metal) scientific instruments as *ideas made of brass* to include instruments irrespective of material – for example ephemeral devices of wood or paper – as *materialised and shaped ideas*. In the early modern period scientific instruments were rather called mathematical, optical or philosophical instruments.[5]

Solo: The Horologium Ahaz

Hartmann's Paper Instrument of 1561

As far as we know, the earliest scientific instrument with a depiction of the Horologium Ahaz is a do-it-yourself or interactive print for a paper sundial designed by the instrument maker Georg Hartmann (1489–1564) (Figure 5.1).[6] After studying theology and liberal arts in Cologne Hartmann spent several years in Italy before settling in Nuremberg in 1518, where he held the position of vicar of St. Sebaldus from 1518 to 1544. He was well acquainted with artists and astronomers, humanists and printers in Nuremberg, which was an intellectual and cultural hub at the time.[7] Beyond Nuremberg he counted Philipp Melanchthon and Georg Joachim Rheticus amongst his correspondents. Rheticus repeatedly travelled to Nuremberg, where he saw Copernicus's *De revolutionibus* to print in 1542, and he met Hartmann there in 1538/39.[8] While this does not directly play out in our story, it is worth mentioning that Rheticus was

5 Cf. Liba Taub, "On Scientific Instruments," *Studies in History and Philosophy of Science* 40 (2009): 337; cf. Jim Bennett, "Early Modern Mathematical Instruments," *ISIS* 102 (2011): 697–705; Liba Taub, "Introduction: Reengaging with Instruments," *ISIS* 102 (2011): 689–696.

6 On Hartmann see Hans Gunther Klemm, *Georg Hartmann aus Eggolsheim (1489–1564): Leben und Wirken eines fränkischen Mathematikers* (Forchheim: Ehrenbürg-Gymnasium, 1990); Suzanne Karr Schmidt, *Interactive and Sculptural Printmaking in the Renaissance* (Leiden: Brill, 2017), 205–240; Lucille B. Ritvo, "Hartmann, Georg," Complete Dictionary of Scientific Biography, accessed March 17, 2021, https://www.encyclopedia.com/science/dictionaries-thesauruses-pictures-and-press-releases/hartmann-georg.

7 On Nuremberg as a cultural hub see the classic, Jeffrey Chipps Smith, *Nuremberg. A Renaissance City, 1500–1618* (Austin: University of Texas Press, 1983).

8 Cf. Kurt Pilz, *600 Jahre Astronomie in Nürnberg* (Nuremberg: Verlag Hans Carl, 1977), 194–195.

extremely knowledgeable of the exegetical reverberations of the Copernican debate and wrote a booklet on the topic.[9] Hartmann, besides being theologically informed, was a master in the artistic and mathematical design of sundials ("magister horologiorum"[10]) and produced scientific instruments made of metal and other durable materials (e.g. ivory) as well as paper. His paper instruments had a double purpose, serving as templates for instrument makers as well as being suitable to be pasted on cardboard or wood, thus serving as inexpensive, but workable instruments.[11] Collections of drawings and prints of Hartmann's instruments are extant in various libraries.[12] In our context the collection of ca. 110 printed instruments or instrument designs in the Bavarian State Library in Munich is of particular interest, as it contains two prints of Hartmann's sundial of 1561.[13]

The story of the Horologium Ahaz is represented in the two upper corners of the rectangular print above the dial and explicitly referred to in the bottom line. On the left, the prophet Isaiah points to a vertical sundial mounted on a tower. On the right, the bedridden King Hezekiah is seen clasping his hands. Given the fact that Hartmann affirmed that he designed, etched, and printed his instruments himself, it seems probable that the images on the sundial are of is his own design as well. However, we have not yet been able to identify his pictorial sources.

While illustrations of the scene in illuminated Bibles or universal chronicles (Weltchroniken) were quite frequent, it is only from 1529/30, with the publication of the *Durlach Bible*, that they began to appear in printed Bibles.[14] Here,

9 Cf. Reyer Hooykaas, ed. and trans., *G. J. Rheticus' Treatise on Holy Scripture and the Motion of the Earth with translation, annotations, commentary and additional chapters on Ramus-Rheticus and the development of the problem before 1650* (Amsterdam: North-Holland Publishing Company, 1984).

10 Klemm, *Georg Hartmann*, 17.

11 On Hartmann's paper instruments see Suzanne Karr Schmidt, "Georg Hartmann and the Development of Printed Instruments in Nuremberg," in *Prints and the Pursuit of Knowledge in Early Modern Europe*, ed. Susan Dackermann, exh. cat., Harvard Art Museums, Cambridge; Mary and Leigh Block Museum of Art, Northwestern University, Evanston (New Haven: Yale University Press, 2011), 268–279; Karr Schmidt, *Interactive*.

12 Cf. the volumes listed and characterised in Karr Schmidt, *Interactive*, 215–221.

13 Bayerische Staatsbibliothek München, "Collectio figurarum," Georg Hartmann, MS Rar. 434, fol. 139. (http://daten.digitale-sammlungen.de/bsb00028171/image_265). All in all the album contains ca. 200 prints. Cf. Pilz, *600 Jahre Astronomie*, 174; Karr Schmidt, *Interactive*, 215f.; Nicola Severino, "Georg Hartmann, Il primo grande gnomonista della Rinascenza," last modified December 2008, https://www.academia.edu/39912162/George_Hartmann_Il_primo_grande_gnomonista_della_Rinascenza.

14 The Luther Bible partly printed by Valentin Kobian in Karlsruhe-Durlach, partly by Wolfgang Köpfl in Strasbourg. Martin Luther and Leo Jud, trans., *Die gantz Bibel Alt vnnd*

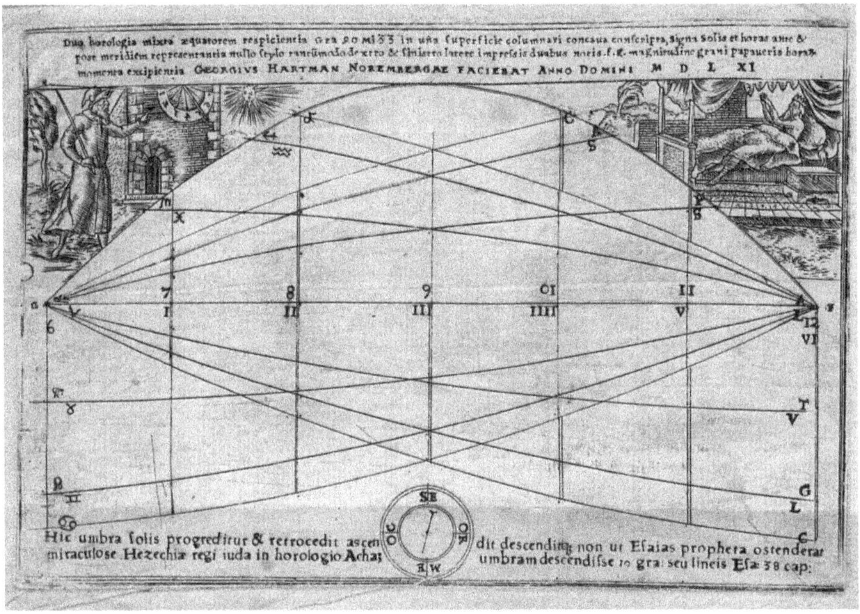

FIGURE 5.1 Georg Hartmann, *Paper sundial*, 1561
PHOTO: BAYERISCHE STAATSBIBLIOTHEK, MUNICH, MS RAR 434, FOL. 139R

as in most of the earlier illuminations, Isaiah is shown standing close to King Hezekiah in his sickbed. The Horologium Ahaz itself is not presented, however, Isaiah is shown facing the sun, which indicates the miracle at the core of the story.[15]

A woodcut from the *Looßbuch zu ehren der Römischen, Ungarischen und Böhmischen Künigin* (1546), a booklet dealing with divination and geomantic practice, follows this representation.[16] In the rare cases where we find depictions of the Horologium Ahaz it is usually rendered as a vertical sundial. To our knowledge, it was first represented as a vertical wall clock in 1470 in a woodcut by an unknown artist in the *Defensorium inviolatae virginitatis Mariae* by Franz

Neüw Testament: Register – weiset alle Historien vnd fürnehme sprüch uber beyde Alt vnd Neüw Testament. Jtem auch mitt Zweyhundert Figuren mehr dann vor hien nie jm Truck [...] *Verteütscht durch Doctor M. L.* [*und Leo Jud*] (Strasbourg: Wolff Köpfl, 1530), 2: fol. XCVIIIr.

15 A limited list of examples can be found in Hans Bardtke, "Ezechias," in *Lexikon der christlichen Ikonograhie*, ed. Engelbert Kirschbaum SJ (Freiburg: Herder, 1968), 1: Col. 714–715.

16 Paul Pambst, *Looßbuch zu ehren der Römischen, Ungarischen und Böhmischen Künigin* (Strasbourg: Balthasar Beck, 1546), 37.

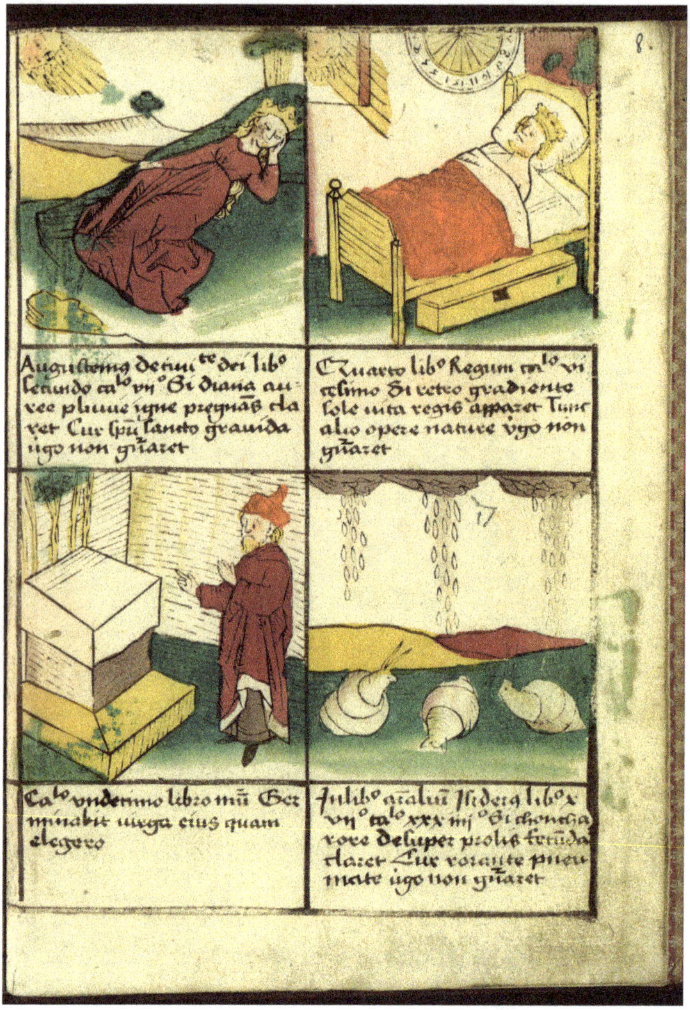

FIGURE 5.2 *King Hezekiah and the Horologium Ahaz* (upper row on the right), in Franz von Retz, *Defensorium inviolatae virginitatis Mariae* (Nördlingen: Friedrich Walther, 1470), fol. 8r
PHOTO: LMU-UNIVERSITÄTSBIBLIOTHEK, MUNICH, CIM. 49

von Retz (Figure 5.2).[17] The bedridden King Hezekiah and the wall clock are in the same room, the sun shining in through a window.

17 Franz von Retz, *Defensorium inviolatae virginitatis Mariae* (Nördlingen: Friedrich Walther, 1470), fol. 8r.

Hartmann's design is different as he splits the scene into two parts, situating the prophet not at Hezekiah's bed, but in front of the tower with the vertical sundial. Above the two scenes and the dial itself Hartmann gave a short description of the instrument, dated 1561, a concave cylindrical sundial.[18] We need not attend to the details of his construction here, but it is obviuos, that he must have assigned a particular importance to the imagery, which he explained at the bottom to the left and right of a compass rose, stating that "here the shadow of the sun moves forward and goes back, climbs up and descends, not like the prophet Isaiah showed in a miraculous way to Hezekiah, king of the Jews, that the shadow on the dial of Ahaz descended 10 degrees or lines."[19] The link between the printed instrument and the dial of Ahaz lies in the movement of the shadow within a concave cylindrical sundial, the axis of which is aligned with the polar axis. Before noon it moves from the left edge to the right edge, first going down in the hemispheric cylinder and then going up again. At solar noon, the shadow jumps back by six hour lines and repeats its path. Thus the analogy to the dial of Ahaz, while being plausible is still rather superficial as it is not the intention of the paper instrument to re-enact or replicate the biblical miracle, which Hartmann had achieved earlier in his refraction dials made of brass.[20]

Praetorius's Torquetum of 1568

A more elaborate cycle about Isaiah and King Hezekiah is engraved on a torquetum designed by Johannes Praetorius in 1568, currently in the Germanisches Nationalmuseum in Nuremberg (Figure 5.3).[21] A torquetum is a complex instrument for measuring and converting time and angles, mechanically illustrating

18 We are grateful to Denis Savoie who examined the paper sundial after our inquiry. Cf. Denis Savoie, "Cadran de Georg Hartmann," *Cadran Info* 43 (May 2021): 181–184.

19 "Hic umbra solis progreditur & retrocedit ascendit descenditq[ue] non ut Esaias propheta ostenderat miraculose Hezechiae regi iuda in horologio Achaz umbram descendisse 10 gra: seu lineis Esa: 38 cap:." Hartmann, "Collectio figurarum," fol. 139.

20 On the hemispherical brass dials by Hartmann and Schissler that worked as refraction dials and were called Horologium Ahaz, see Jasmin Mersmann, "Moving Shadows, Moving Sun. Early Modern Sundials Restaging Miracles," *Nuncius* 30, no. 1 (January 2015): 111–122; Sven Dupré, "The Dioptrics of Refractive Dials in the Sixteenth Century," *Nuncius* 18, no. 1 (January 2003): 39–67, esp. 42–45; and especially Chapter 4 in this volume by Andrew Morrall, "Turning Back the Sun. Christoph Schissler's 'Horologium Ahaz' as 'Kunststück.'"

21 Cast Brass, engraved, embossed and gilded, 28,5 cm high, base plate: 17,0 x 17,0 cm, inv. no. WI 33, on loan from the City of Nuremberg; Johannes Willers, Gerhard Bott, and Wolfgang Pülhorn, *Focus Behaim-Globus*, exh. cat., Germanisches Nationalmuseum Nürnberg, vol. 2 (Nuremberg: Verlag des Germanischen Nationalmuseums, 1992), 644–645, cat. 1.128. See

the relative position of the astronomical great circles: horizon, equatorial, and ecliptic.[22] It could be used for astronomical observations as well as for surveying tasks, e.g. to measure the height of a building. Naturally, it could also be used for astrological purposes as Regiomontanus had emphasised.[23] In principle the instrument had a wide range of applications in astrology, astronomy, and geodesy, but due to a certain top-heaviness, static problems impaired the exactness of measurements in the field.

The torquetum at hand is gilded and richly decorated. It was designed by the astronomer and instrument maker Johannes Praetorius (Johannes Richter, 1537–1616). Educated at the University of Wittenberg, Praetorius lived and worked in Nuremberg between 1562 and 1569.[24] While Praetorius signed and dated the instrument on the back, nothing is known about the monogrammist ww who richly engraved the instrument, literally covering every vacant spot not needed for technical data with ornaments and images. The precious instrument had been commissioned by the wealthy physician Melchior Ayrer (1520–1579), a descendant of an old Nuremberg merchant family who had been knighted in 1561. Ayrer's coat of arms, a deer with a feathered arrow protruding from its breast, is prominently displayed on the back just above the signature of Praetorius. Ayrer had studied in Erfurt (Baccalaureus) and Wittenberg (Master) where he had been a student of Melanchthon. He went on to Bologna earning a PhD in medicine.[25] Returning from Italy he practiced medicine in Nuremberg from 1549, beginning in the local hospital, but constantly moving up to more prestigious positions. In 1570 the countess palatine Dorothea of Denmark, widow of prince-elector Frederic II, appointed him her personal physician. Moving up the social ladder in the 1560s he ordered a number of

---

the entry on the torquetum from Christian Forster that is to be published in the digital collection on www.stadtbibliothek.nuernberg.de.

22 On the torquetum see Ernst Zinner, *Deutsche und niederländische astronomische Instrumente des 11.–18. Jahrhunderts* (Munich: C.H. Beck, 1979), 177–183; Giles M. Hudson, "Torquetum," in *Instruments of Science. An Historical Encyclopaedia,* ed. Robert Bud and Deborah Jean Warner (New York: Garland, 1989), 623–626; cf. Richard L. Kremer, "How Did the Turketum (or Torquetum) Get Its Name?" in *Scientific Instruments between East and West,* ed. Neil Brown, Silke Ackermann, and Feza Günergun (Leiden: Brill, 2019), 80–107.

23 Johannes Schöner, *Scripta clarissimi mathematici M. Ioanis Regiomontani* (Nuremberg: J. Montanus & U. Neuber, 1544), fol. 1r–13v; Regiomontanus wrote *De Torquetum* in 1469.

24 On Praetorius see Menso Folkerts, "Johannes Praetorius (1537–1616) – ein bedeutender Mathematiker und Astronom des 16. Jahrhunderts," in *History of Mathematics, States of the Art,* ed. Joseph W. Dauben and Christoph Scriba (San Diego: Academic Press, 1996), 149–169.

25 On Ayrer see Doris Wolfangel, *Dr. Melchior Ayrer (1520–1579)* (Würzburg: Baehr, 1957).

FIGURE 5.3 Johannes Praetorius, *Torquetum*, 1568
PHOTO: GERMANISCHES NATIONALMUSEUM, NUREMBERG, INV. NO. WI 33 / MONIKA RUNGE

scientific instruments and globes from Praetorius: a polyhedral dial (1562), a chalice dial (1563), a sundial compendium (1566), a pair of celestial and terrestrial globes (1566), an astrolabe (1568), the torquetum (1568), and a large sun quadrant (1571).[26] While not much is known about Ayrer's personal or intellectual preferences the large number of expensive scientific instruments points

26 Cf. Willers, Bott, and Pülhorn, *Fokus Behaim*, 2:637–654. Apart from the torquetum the adorned instruments and the globes were engraved by the goldsmith Hans Epischofer. On some new attributions of scientific instruments to the Ayrer collection see Christine Sauer, "Die Memorabilien in der Stadtbibliothek," in *Wunderkammer im Wissensraum. Die*

not only to a clear strategy of social distinction, but also to a strong interest in astrological and astronomical issues.[27]

Indeed, astronomical instruments had a natural place in the toolkit of every early modern physician and Ayrer would certainly have been exposed to a certain amount of instruction in astrology/astronomy in Bologna, including how to operate astronomical instruments.[28] Astrological knowledge was part and parcel of early modern medical practice.[29] Instruments such as those Ayrer acquired from Praetorius had a clear double function as usable tools as well as statements of status and "symbols of higher education" vis-à-vis the patients.[30] Thus it is not surprising that Ayrer commissioned his instruments when he was climbing the social ladder and setting up his own medical practice in the early 1560s. The instruments were displayed as showpieces or would be shown on request as Sebald Welser, a patient of Ayrer, noted in 1577: "Saw Dr Ayrer's globes."[31]

Given the complexity of the torquetum as an instrument we will first describe the composition of the instrument and then analyse the imagery:

(1) The base of the instrument/base of the horizon (tabula horizontis) is mounted on four lions.

(2) The equatorial plane (tabula equinoctialis) is hinged to the base of the horizon at its North side. A free-rotating equatorial disk (basilica) is mounted on the equatorial plane.

(3) The circular ecliptic plane (tabula orbis signorum) is elevated by two supporting triangles.

(4) A rotating alidade is fixed at the centre of the ecliptic plane. From the same centre the circle of latitude (christa) extends vertically.

Before discussing the pictorial programme of the torquetum we will briefly survey the elaborate imagery on the torquetum beginning at the bottom.

Memorabilien der Stadtbibliothek Nürnberg im Kontext städtischer Sammlungskulturen, ed. Christine Sauer (Wiesbaden: Harrassowitz, 2021), 17–95.

27 Cf. Paul Freher, *Theatrum virorum eruditione clarorum*, vol. 2 (Nuremberg: Hofmannus, 1688), 1277.

28 Cf. Darin Hayton, "Instruments and Demonstrations in the Astrological Curriculum: Evidence from the University of Vienna 1500–1530," *Studies in History and Philosophy of Biological and Biomedical Sciences* 41, no. 2 (June 2010): 125–134.

29 On the interaction of astronomy and medicine in practice see e.g. Michael Stolberg, *Gelehrte Medizin und ärtzlicher Alltag in der Renaissance* (Berlin: De Gruyter, 2020), esp. 154–159.

30 Cf. Willers, Bott, and Pülhorn, *Fokus Behaim*, 2:637.

31 Ursula Koenigs-Erffa, "Das Tagebuch des Sebald Welser aus dem Jahre 1577," *Mitteilungen des Vereins für Geschichte der Stadt Nürnberg* 46 (1955): 290; cf. the excerpt in Wolfangel, *Dr. Melchior Ayrer*, 22–23.

(1) The presentation on the base of the horizon is partitioned into two parts by a tower in the middle that is being measured by a surveyor with a cross-staff. The astronomer on its right is shown observing a celestial object with a horoscopion, a universal time measuring instrument. His left arm is resting on a geometric quadrant positioned on an icosahedron. These images have been copied from the title page of Peter Apian's *Instrument Buch* (1533).[32]

The part on the left varies from Apian's template in that his two astronomers have been replaced with an astrologer measuring distances on a celestial globe with a compass.[33] He has been copied from the title-page of the edition of Mashallah ibn Athari's *De Scientia Motus Orbis* edited by Johann Stabius in 1504. The text by the Persian astronomer and astrologer Mashallah ibn Athari (ca. 740–815), author of a tract on the astrolabe, had first been translated into Latin by Gerard of Cremona (c. 1114–1187). The title-page, probably from the entourage of Dürer's workshop, presented an imaginary portrait of the author, widely known as an astrologer in the fifteenth and sixteenth centuries.[34] There was no need to be able to recognize Mashallah ibn Athari as a person on the torquetum. The essential point was that the astrologer on the left stood in stark contrast to the astronomer on the right. His throne, composed of the seven spheres of the planets and the fixed stars accompanied by the zodiac, indicated that he represented the study of the cosmic influences on individuals, while the astronomer on the right stood for the mundane art of measuring time and angles.

(2) Surveying scenes, partly traceable to Apian's *Instrument Buch*, can be seen on the reverse side of the equatorial plane, and the inner and outer

32 Peter Apian, *Instrument Buch* (Ingolstadt: Peter Apian, 1533); cf. Fleur Richter, *Die Ästhetik Geometrischer Körper in der Renaissance* (Stuttgart: Verlag Gerd Hatje, 1995), 11.

33 The three men on the torquetum have been identified as an astrologer, an astronomer, and a surveyor in Paul Adolf Kirchvogel and Bernhard Sticker, *documenta astronomica. Eine Ausstellung historischer Instrumente und Dokumente zur Entwicklung der astronomischen Messkunst im Museum für Völkerkunde und Vorgeschichte Hamburg* (Wiesbaden: Steiner, 1964), 40.

34 Mashallah ibn Athari, *De Scientia motus orbis* (Nuremberg: Johann Weissenburger, 1504); on its attribution see Matthias Mende, Rainer Schoch and Anna Scherbaum, *Albrecht Dürer. Das druckgraphische Werk*, vol. 3 (München: Prestel, 2004), 524, cat. A 39. According to the interpretation in Susan Dackerman, ed., *Prints and the Pursuit of Knowledge in Early Modern Europe*, exh. cat., Harvard Art Museums, Cambridge; Mary and Leigh Block Museum of Art, Northwestern University, Evanston (New Haven: Yale University Press, 2011), 366, cat. 91, the portrait shows Ptolemy.

sides of the triangles supporting the ecliptic plane.[35] Scholars with instruments (sun-dials, a celestial globe and an astrolabe), dressed in Oriental as well as European garments, adorn the outer segments of the freely rotating equatorial disk.[36] The inner field shows an anthropomorphic sun, in correspondence with an anthropomorphic moon on the bottom side of the circular ecliptic plane. The latter's outer segments show four stereometric bodies (pyramid, cube, cone, and irregular Hexahedron).

(3) The cycle of scenes of Isaiah and King Hezekiah is on the upper side of the circular ecliptic plane (Figure 5.4). The rim of the plate carries two rings of scales, one for the twelve months and one for the twelve signs of the zodiac. The inner circle is divided into four parts separated by columns. The cycle runs anti-clockwise and begins with a depiction of the bedridden Hezekiah praying. To his left a wailing woman gazes after the prophet Isaiah who is leaving the room. The next quadrant shows Isaiah receiving instructions from God, who is situated on a cloud. In the third image Isaiah conveys the divine message to Hezekiah, pointing to a sundial mounted on a tower. Finally, Hezekiah can be seen sitting and stretching out his leg, which is being tended to by Isaiah and a servant offering him a fig poultice. This order follows the story as told in Isaiah 38:1–8 and verse 21 ("For Isaiah had said, Let them take a lump of figs, and lay it for a plaister upon the boil, and he shall recover").[37]

The engraver of the torquetum used printed material as templates for most of the images on the torquetum as described in (1), (2), and (4). Considering this, it seems plausible that he would also have used a printed template for the four scenes of the cycle about Isaiah and King Hezekiah. As of now we have not been able to find such printed material. The second scene, Isaiah receiving instructions from God, could

35 Three of the scenes had already been printed in Peter Apian's *Quadrans Apiani Astronomicus* [...] (Ingolstadt: Peter Apian, 1532) and reprinted in his *Horoscopion Apiani generale* [...] (Ingolstadt: Peter Apian, 1533). The figure with the cross-staff can be found in his *Introductio Geographica Petri Apiani* [...] (Ingolstadt: Peter Apian, 1533). The scene on the back side of the equatorial plane is taken from Walther Ryff, *Der furnembsten, notwendigsten der gantzen Architectur angehoerigen Mathematischen und Mechanischen Kuenst eygentlicher Bericht* (Nuremberg: Johan Petreius, 1547), third book, chapter XV.

36 The scholar with an astrolabe may have been inspired by a print of Hartmann's. Bayerische Staatsbibliothek München, "Collectio figurarum," Georg Hartmann, MS Rar. 434, fol. 91, http://daten.digitale-sammlungen.de/bsb00028171/image_179.

37 In the story as told in 2 Kings 20:1–11, the fig poultrice is applied before the miracle on the Horologium Ahaz takes place.

FIGURE 5.4 Johannes Praetorius, *Torquetum*, 1568, ecliptic plane: Monogrammist W.W., *The story of Isaiah and King Hezekiah*
PHOTO: GERMANISCHES NATIONALMUSEUM, NUREMERG, INV. NO. WI 33 / ROLAND SCHEWE

potentially have been inspired by Hans Holbein's visual rendering of Isaiah 6:1–3 (the vision of Isaiah) in the *Icones* of 1538.[38] There Isaiah can be seen approaching from the right, while God on his heavenly throne is shown on the upper left side (Figure 5.5). An angel reaching out from the clouds touches the prophet's lips with a piece of glowing coal as a sign of his vocation (Isaiah 6:6–9). Holbein's composition was often copied in later illustrated Bibles, at times even as a portrait of the

38 Hans Holbein the Younger, *Historiarum Veteris Testamenti icones* (Lyon: Melchior and Caspar Trechsel, 1538). On this image and its dissemination see Manfred Kästner, *Die Icones Hans Holbeins des Jüngeren. Ein Beitrag zum graphischen Werk des Künstlers und zur Bibelillustration Ende des 15. und in der ersten Hälfte des 16. Jahrhunderts*, vol. 1 (Heidelberg: Esprint, 1985), 303–304, W.73.

FIGURE 5.5 Hans Holbein the younger, *The vision of Isaiah* (Is 6:1–3), in *Historiarum Veteris Testamenti icones* (Lyon: Melchior and Caspar Trechsel, 1538)
PHOTO: BAYERISCHE STAATSBIBLIOTHEK, MUNICH, ESLG/4 GERM.G. 90, S. 306

prophet at the beginning of the book of Isaiah.[39] The engraving on the torquetum is very similar to this image, the main difference being that there are no angels.

(4) The two sides of the circle of latitude are lavishly illustrated as well. A sighting arm with a pivoting vertical half circle, used to measure elevation angles, stands out from one side. This side is partitioned into four

39 Cf. the woodcut at the beginning of the book of Isaiah in the *Biblia Iconibus artificiosissimis, quo Lectoris memoriae consulatur, tanquam emblematis quibusdam exornata* (Antwerp: Johan Steels, 1541–1542); the woodblock was reused in *Den Bibel: Inhoudende het oude ende nieuwe Testament* (Antwerp: Hans de Laet, 1556), as well as in the later editions from 1560 and 1565. On this see Bart Rosier, *The Bible in Print: Netherlandish Bible Illustration in the Sixteenth Century*, vol. 1 (Leiden: Foleor, 1997), 39–40, 247, N° 74, 297, N° 82.

quadrants, each showing one of the four seasons based on a series of etchings by Virgil Solis.[40]

Parts of the engravings on the back side also refer to prints from Virgil Solis's large Nuremberg workshop. Most of the upper semi-circle is filled by Ayrer's coat of arms. On its right is a woman resting her left arm on a celestial globe while pointing to the stars with the right hand. At her feet is a compass on a cube. She is either a personification of astronomy as one of the artes liberales or of Urania, muse of astronomy. On the left side two scholars can be seen at a table with an armillary sphere, one writing busily, the other pointing to the sign of the planet Mercury. The scene is part of a cycle by Virgil Solis depicting the children of the planets.[41] Merchants and thieves were subjugated to Mercury, as were scholars and physicians, Ayrer's own profession. The lower semi-circle presents two women in a landscape. The one on the left plays a string instrument, while the one on the right looks at an open music book and holds a banderole. Both were copied from a series of the *Nine Muses* by Virgil Solis and thus can be identified as Polihymnia (on the left) and Calliope (on the right).[42]

The Pictorial Programme of the Torquetum

The programme of the torquetum is clearly tripartite in that it covers the three fields of application of the instrument, namely geodesy/surveying, astronomy, and astrology, rising from bottom to top. On the base of the instrument we find the three subjects represented by one figure each as we have seen above (1): the astronomer on the right, the surveyor in the middle, and the astrologer on the left (Mashallah ibn Athari on his planet throne). The parts of the programme pertaining to geodesy – the surveyor at work and the stereometric bodies pointing to the rules of geometry – are placed in the lower part of the instrument (1, 2). Rising from the measuring of the earth towards the next level of measuring time and the heavens, the programme proceeds to astronomy on the free spinning equatorial disk and the circular ecliptic plane (2, 3). The sun

40 Virgil Solis, *Four seasons*, etchings, ca. 6 x 4,8 cm, in Friedrich Hollstein, *Hollstein's German engravings, etchings and woodcuts*, vol. 64 (Rotterdam: Sound & Vision Publishers, 2004), 73–74, no. 369–372.

41 Virgil Solis, *Mercury*, etching, 5 x 8,4 cm, in Hollstein, *Hollstein's German engravings*, 64:95–96, no. 408.

42 Virgil Solis, *Nine Muses*, etching and engraving, 4,6 x 17,5 cm, in Friedrich Hollstein, *Hollstein's German engravings, etchings and woodcuts*, vol. 63 (Rotterdam: Sound & Vision Publishers, 2004), 220, no. 283.

and the moon stand for the heavenly bodies, which are fundamental for measuring time, as do the scholars with time measuring devices such as sundials. The programme then ascends to more heavenly spheres. Firstly, to the cycle about Isaiah and King Hezekiah, where another sundial, the Horologium Ahaz, and the associated miracle of the retrograding sun take centre stage. Secondly, at the top level of the instrument, on the circle of latitude (4), astrology takes the place of pride. Here the influence of the stars on terrestrial affairs is symbolised in various ways: the four seasons point to the cosmic influences on humans, corresponding to the four elements, air, fire, earth, and water, crucial for humoral pathology, and its four humours, blood, phlegm, yellow and black bile, as well as to the four temperaments, melancholic, phlegmatic, sanguine, and choleric. The influence of the planets also plays out in the depiction of Mercury and his children on the reverse side.

Only the two muses, Calliope, muse of eloquence and epic poetry, but also of philosophy and scholarship, and Polyhymnia, muse of "many hymns," do not easily fit into the programme. They are placed along with Ayrer's coat of arms and may refer to his own interest in music.

What has become clear up to this point is that the programme makes no reference to the discussion about the world systems. Rather it is concerned with possible fields of application of the instrument and is finely tuned to Ayrer, who commissioned it. It is to Ayrer and his profession as a physician that the prominent cycle about Isaiah and King Hezekiah refers. The recovery of Hezekiah, and Isaiah's role as mediator between God and Hezekiah as well as his medical knowledge take centre stage, with the astronomical miracle itself only playing a minor role. The torquetum's programme is highly appropriate for a wealthy and successful physician. Old testament prophets were often incorporated into legitimisation strategies of the medical profession, as there are various reports about their healing powers (e.g. 1 Kings 17:17–22, 2, Kings 4:32–36, 5:10–14), even though the application of the fig poultice by Isaiah is one of the rare passages of the Bible where a specific remedy is mentioned.[43] Isaiah's medical expertise had also been highlighted in early modern exegetical literature.[44] The miraculous recovery of Hezekiah was often referred to along

43 On the prophets as powerful and spiritual healers see Johannes Hempel, *Heilung als Symbol und Wirklichkeit im biblischen Schrifttum* (Göttingen: Vandenhoeck & Ruprecht, 1965), 294.

44 See e.g. Georgius Wicelius [Georg Witzel], "Am Dornstag nach dem Aschermitwoch. Prophetische Lection aus Esaie. Cap. XXXVIII," in *Qvadragesimale Catholicvm. Gegrundte Auslegung aller Episteln, oder Lectionen und der Evangelien so die Kirche Gottes von alters durch die Tage der heiligen Fasten zu Lesen und Predigen Christlich und löblich verordnet hat* (Cologne: Peter Quentel, 1546).

with the praise of the physician in the book of Sirach (chap. 38) when arguing for the legitimacy of medical interventions or asserting that practicing medicine was in line with the divine order.[45] At the same time these references emphasised that medical art and expertise could not be successful without the right faith and divine assistance.[46] The biblical story of Isaiah and King Hezekiah left no doubt that the miraculous recovery of Hezekiah was only made possible by his godliness, facilitating God's goodwill and the success of the medical treatment through Isaiah. Such (visual) links between the miraculous recovery of Hezekiah and medicine as a profession were still well-known in the eighteenth century.[47]

While the pictorial programme of the torquetum heavily features the mathematical sciences, the core message is clearly about the medical profession. The elaborate cycle of Isaiah and King Hezekiah highlights both, the healing power of physicians (with divine support) and the expertise any good and serious physician had to have in correctly interpreting astrological and cosmic phenomena. Thus, the prominent reference to Isaiah and King Hezekiah does not come as a surprise on an instrument that could both be practically used by a physician (at least potentially) and serve representative purposes as a patronage artefact in his rooms.

Schissler's Horologium Ahaz of 1578

The story of the miraculous recovery of Hezekiah is also featured *by* and *on* a refractive sundial, the Horologium Ahaz, made by Christoph Schissler (the Elder, ca. 1531–1608) in Augsburg in 1578. (Figure 4.1) The Horologium Ahaz is a sundial in the form of a drinking vessel which consists of two parts, an upper concave bowl and a circular base. It refers to the retrograding of the sun in two ways. On the one hand, the bowl serves as a sundial and, when filled with water, the refraction causes the shadow to jump back, thus imitating the miracle of the Horologium Ahaz. On the other hand, the story of Isaiah and King Hezekiah is depicted in two images on the reverse side of the base of the instrument, meaning that it cannot be seen when the instrument is properly

45 Cf. Peter Glaser, "Die XVI. Predigt uber die wort," in *Sechzig kurtze Leichpredigten Von dem grossen Landtsterben zur zeit des Königs Davids* [...] *Desgleichen von der Pestilentz des Königs Hiskiae* [...] (Dresden: Bergen, 1582).

46 Cf. Veit Dietrich, *Der gantz Prophet Esaias. Außgelegt durch Vitum Dietrich* (Nuremberg: vom Berg and Neuber, 1548).

47 See e.g., Johann Siegmund Stoy, *Bilder-Akademie für die Jugend*, vol. 1 (Nuremberg: Six & autor, 1784), 561–580, Plate 30. Cf. Anke te Heesen, *Der Weltkasten. Die Geschichte einer Bildenzyklopädie aus dem 18. Jahrhundert* (Göttingen: Wallstein-Verlag, 1997).

mounted. As Andrew Morrall discusses Schissler's Horologium Ahaz in detail in Chapter 4 in this volume and as many technical questions about the instrument have not yet been resolved, we will not examine the technical aspects in any detail but concentrate on the hidden presentation of the encounter of Isaiah and King Hezekiah.

The story of King Hezekiah's illness, the miracle of the dial of Ahaz, and the recovery of Hezekiah is told in two images on the lower side of a disk mounted into the reverse side of the base (Figure 5.6), which can only be seen when the instrument is picked up and turned around or, alternatively, when the base disk is taken out. Its upper side is decorated with carefully engraved ornaments. The limbus on the lower side is portioned in four sections of 90° each, which indicates that the disk must have had some function or practical use.[48] The designed space enclosed by the limbus is separated into five parts.

At the centre, a Latin inscription in the form of a helix winds out of a small, ornamented space. It reads (in translation): "This shell in the form of a hemicycle denotes the miracle of the 38th chapter of Isaiah. For if you fill it to the brim with water, the shadow of the sun is borne backwards by ten or twenty degrees. Besides this, it indicates the sign and degree of the sun, that is, the common hour of any day whatsoever, together with what they call the planetary hours."[49] The inscription begins just below a semi-circular grid of curves that in turn is captioned "Horolgii [!] Achaz hydrographica declinationes ad elevat: Poli · 47 · 48 · 49 · GRAD:" ("Hydrographical (= written in water) deviation of the Horologium Ahaz for the elevations of 47, 48, and 49 degrees"). The grid covers the lines of the common hours (two equal hour systems of twelve hours each) and the planetary hours (two unequal hour systems where day and night are each split into twelve equally long parts) for all months of the year. Accordingly, one would surmise that this could be used as a dial of the instrument, that could, for example, be installed on the upper concave bowl. However, this grid as well as the one on the opposite side of the circular base is

48 The limbus is mentioned in Maximilian Bobinger, *Christoph Schissler der Ältere und der Jüngere* (Augsburg: Verlag Die Brigg, 1954), 115. He does not, however, discuss its function and neither does the later literature.

49 "Notat concha isthac hemiciclea capitis 38 Esaia miraculum: nam hanc si aqua labrum usque impleveris: umbra solis · 10 · imo: 20 · gradibus retrorsum fertur: signum ac gradum solis: quin etiam horam diei vulgarem quamcuncque una cum planetarum quas vocant horis denuncians." We are grateful to Stefan Freund and Andrew Morrall for discussing this passage with us.

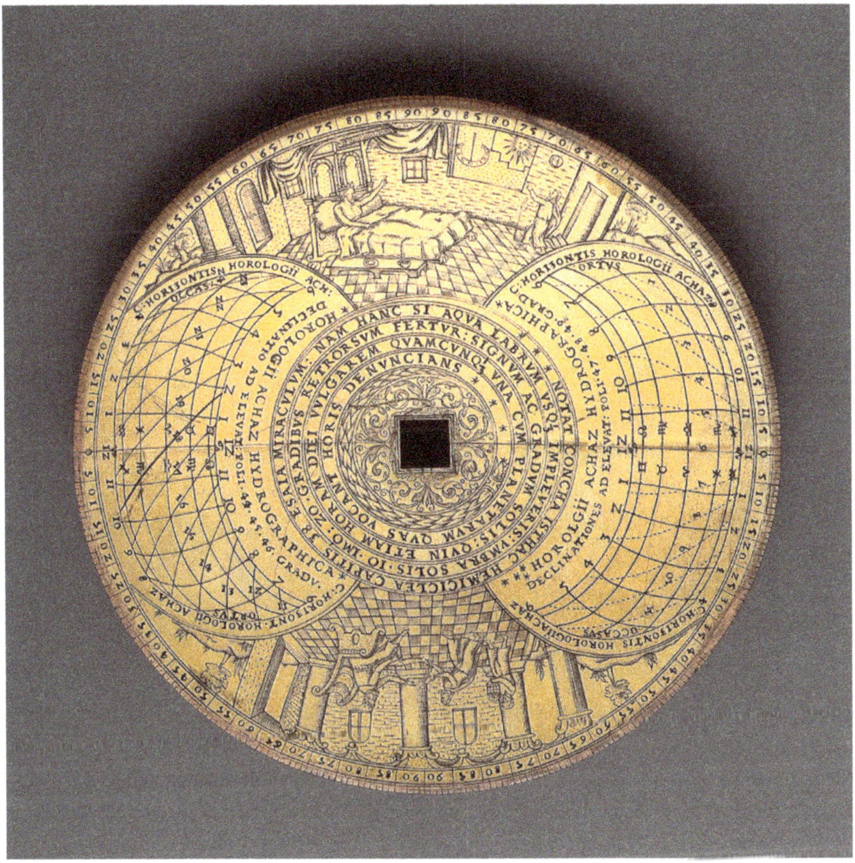

FIGURE 5.6 Christoph Schissler, *Refractive sundial: the reverse side of the base*, 1578
PHOTO: AMERICAN PHILOSOPHICAL SOCIETY, PHILADELPHIA, INV. NO. 58.66

usually interpreted as a conversion table.[50] This is not completely convincing as the dials for the upper concave bowl, that Schissler himself described as a horizontal dial, show identical data apart from having an enlarged range of latitudes. Thus a conversion would not have been necessary.

The fields between the two grids show two scenes from the story of Isaiah and King Hezekiah: (1) bedridden Hezekiah to the left and Isaiah to the right, both pointing towards a vertical sundial that can be seen through a window, and (2) the application of the fig poultice by Isaiah. The images are not of a

50 Cf. Bobinger, *Christoph Schissler*, 116; Robert P. Multhauf and David Davies, *A Catalogue of Instruments and Models in the Possession of the American Philosophical Society* (Philadelphia: American Philosophical Society, 1961), 52.

very high standard and their sources are unknown. While Jasmin Mersmann pointed to Hartmann's paper instrument of 1561 (see above, Figure 5.1) as a possible source, the images seem to be closer to the engravings on Praetorius's torquetum (see above, Figure 5.4).[51]

The programme of the instrument includes a turbaned and orientally dressed figurine mounted on the rotating rim of the upper concave bowl holding the string gnomon, as well as the base decorated with maritime imagery in relief such as fish, sea gods and sea creatures. The original stem, in the form of a supporting figure, has not been preserved.[52] Jasmin Mersmann has argued that the turbaned and orientally dressed figurine might be the prophet Isaiah. While this seems plausible at first glance, it has to be taken into account that there is no explicit iconographic hint in the appearance of the figurine that would characterise him as a prophet and that Schissler had used a similar figurine in a sundial without any reference to a prophet in 1562. While the base with its aquatic relief, too, had been used repeatedly by Schissler for other instruments,[53] its maritime imagery nicely fits a refractive dial.

All in all, the programme seems to be in close correspondence with the instrument and its technical aspects. That the engraved scenes from the story of Hezekiah and Isaiah are out of view in the lower side of the circular base is reminiscent of a parlour game: whoever did not immediately grasp the significance of the refractive dial, the shadow of the sun going back on the Horologium Ahaz, would find the solution of the puzzle down there. No allusion to the movement of sun or earth is intended by the reference to the biblical story.[54]

As we have seen there seems to be no hidden anti-Copernican message in any of the three instruments discussed above – Hartmann's paper instrument of 1561, Praetorius's torquetum of 1568, and Schissler's Horologium Ahaz of 1578.

51 Mersmann, "Moving Shadows," 116.

52 Bobinger, *Christoph Schissler*, 115, speculates that it might have been Hercules, similar to the figurine supporting Schissler's armillary sphere of 1569 (Munich, Bavarian National Museum, inv. no. Phys 27).

53 Cf. Mersmann, "Moving Shadows," 118, who also points to the sundial of 1562 (Dresden, Mathematisch-Physikalischer Salon, Staatliche Kunstsammlungen Dresden, inv. no. D I 37). An identical pedestal supports Schissler's armillary sphere of 1569 (Munich, Bavarian National Museum, inv. no. Phys 27).

54 This interpretation is in line with Mersmann, "Moving Shadows," 120, and Chapter 4 in this volume by Andrew Morall, "Turning Back the Sun. Christoph Schissler's 'Horologium Ahaz' as 'Kunststück.'"

Duo: Combinations of the Horologium Ahaz and the Sun Standing Still Over Gibeon

Schissler may well have been the first to have thought about combining the stories of the Horologium Ahaz and the sun standing still over Gibeon in the design or construction of an instrument or, in this case, a pair of related instruments. We know about this from a letter Schissler sent to August I, elector of Saxony in 1572 offering several instruments for sale. In particular he mentioned two drinking vessels based on the stories of the Horologium Ahaz and the sun standing still over Gibeon:

> also to make a curious contrivance, a drinking-vessel based on the text of chapter 20 of the 4th Book of Kings about the dial of King Ahaz: when the shadow moved backwards, etc. Also one based on the text of 10th chapter of Joshua: the sun stayed in mid-heaven.[55]

Schissler was probably referring to a pair of instruments, one based on each story.[56] It is important to notice that Schissler's description is not that of an instrument illustrated by images of the respective miracle, but rather an instrument re-enacting or replicating it (such as the Horologium Ahaz of 1578, discussed above). The matching drinking vessel would have been an analogical instrument replicating the miracle of the sun standing still over Gibeon. While we do not have any hint as to how such an instrument could have worked, it would have been quite a challenge to replicate both miracles in one and the same instrument. However, regardless of whether Schissler thought about a single instrument or a pair of instruments, he seems to have been the first among instrument makers considering intertwining both scenes.

55 Translation adapted from Peter Plaßmeyer, "Christoph Schissler: The Electors Dealer," in *European collections of Scientific Instruments, 1550–1750*, ed. Giorgio Strano et al. (Leiden: Brill, 2009), 20; German text in Bobinger, *Christoph Schissler*, 34–36: "item ein sonderlich kunststück ein trinkgeschür zu machen, auf den Text des 20. Cap. etc. im 4. buch der künigen von dem horologium Achaz: da der Schatten zuruckgelauffen etc. Item auf den Text deß 10. Capitels Josua, da die Sonnen ist stil gestanden."

56 This reading, favoured as well by Andrew Morrall and Peter Plaßmeyer, is based on the use of the German term "item" (also) each time a new instrument in the listing is introduced. We are grateful to Svetlana Petrova and Dominik Brückner for discussing this passage with us.

FIGURE 5.7 Wenzel Jamnitzer, *Measuring instrument: Sun-side* (for time-measuring), 1578
PHOTO: BIBLIOTHÈQUE DE L'OBSERVATOIRE, PARIS, INV. NO. 7 / DOMINIQUE MONSEIGNY

Jamnitzer's Measuring Instrument of 1578

The earliest instrument with a programme combining the stories of the Horologium Ahaz and the sun standing still over Gibeon is a measuring instrument made by Wenzel Jamnitzer (1508–1585) in 1578 (Figure 5.7 and 5.8).[57] The object is owned by the Observatoire de Paris and currently on view in the Bibliothèque Humaniste in Sélestat.

Today, the Nuremberg artist and entrepreneur Wenzel Jamnitzer is considered one of the most outstanding and prominent goldsmiths of sixteenth-century Germany.[58] Among contemporaries he was also highly esteemed as an

57 Brass, gilded, diameter 51,7 cm, Bibliothèque de l'Observatoire de Paris, inv. no. 7 (on long-term loan to the Bibliothèque Humaniste in Sélestat). The instrument has only rarely been discussed in the literature (e.g. Zinner, *Astronomische Instrumente*, 395); the first detailed analysis, especially of its uses as a sundial, has recently been published by Denis Savoie, "Le cadran solaire de hauteur de Wenzel Jamnitzer de l'Oberservatoire de Paris," *Cadran Info* 36 (October 2017): 136–151.

58 The literature on Jamnitzer is vast. For an orientation see Sven Hauschke, "Wenzel Jamnitzer," in *Allgemeines Künstlerlexikon*, ed. Günter Meißner and Andreas Beyer (Berlin: de Gruyter, 2013), 77:265–266.

FIGURE 5.8 Wenzel Jamnitzer, *Measuring instrument: Moon-side* (for surveying), 1578
PHOTO: BIBLIOTHÈQUE DE L'OBSERVATOIRE, PARIS, INV. NO. 7 / DOMINIQUE MONSEIGNY

instrument maker, mathematician, and aficionado of the art of perspective.[59] Many dignitaries bought or commissioned instruments from Jamnitzer: the Nuremberg City Council, the prince-electors of Saxony, various members of the high nobility in Germany, and four emperors, Charles V, Ferdinand I, Maximilian II, and Rudolf II.[60]

---

59 On Jamnitzer as instrument maker see Sven Hauschke, "Goldschmiede als Hersteller wissenschaftlicher Instrumente und Geräte," in *Nürnberger Goldschmiedekunst 1541–1868,* ed. Karin Tebbe, exh. cat., Germanisches Nationalmuseum Nuremberg (Nuremberg: Germanisches Nationalmuseum 2007), 2:219–223; Sven Hauschke, "Wenzel Jamnitzers Beschreibung seines 'kunstlichen und wolgetzierten Schreibtischs' und die Erfindung des Kunstschranks," *Jahrbuch der Coburger Landesstiftung* 58 (2014): 1–164; Sven Hauschke, "The Mathematical Instruments of Wenzel Jamnitzer (1508–1585)," in *European collections of Scientific Instruments, 1550–1750,* ed. Giorgio Strano et al. (Leiden: Brill, 2009), 1–13.

60 See Ulrike Berninger, "Im Auftrag von Kaisern, Fürsten und anderen hohen Herren, der Goldschmied Wenzel Jamnitzer," in *Von nah und fern, Zuwanderer in die Reichsstadt Nürnberg,* ed. Brigitte Korn, Michael Diefenbacher, and Steven M. Zahlaus (Petersberg: Imhof, 2014), 115–124.

The measuring instrument of 1578 had a technically identical twin instrument, albeit with a different programme. The prince-elector of Saxony paid for it in 1583 and by 1587 it had been incorporated into the Dresden Kunstkammer. It has been missing since the end of World War II.[61] We know from the 1587 inventory of the Kunstkammer that quite a bit of auxiliary equipment belonged to the measuring instrument such as a plumb bob, a telescopic rod to hang and adjust the instrument, etc.[62] The instrument of the Observatoire probably came with similar auxiliary equipment, but it seems to have been lost.

Jamnitzer did not engrave the decorations on the instrument himself. Already in 1568, in the preface to his *Perspectiva Corporum Regularium*, an impressive collection of engravings of regular polyhedra, Jamnitzer had explained that due to his age [and] his hand having become heavy he had commissioned this task to the multi-talented and well-educated Swiss artist Jost Amman (1539–1591).[63] Amman had settled in Nuremberg in 1561, first working with Virgil Solis and carrying on Solis's artistic and intellectual heritage after his untimely death in 1562.[64] Jamnitzer and Amman started working together in 1568. When Engelmann published his paper on Jamnitzer's mathematical instruments in 1914 he speculated about a possible collaboration of the two on the measuring instrument in Dresden.[65] While the Dresden instrument

61 Brass, gilded, diameter 51,8 cm, lost from the Staatliche Kunstsammlungen Dresden, Mathematisch-Physikalischer Salon, inv. no. III, 14. Cf. Richard Steche, "Astronomische Instrumente von W. Jamnitzer," *Anzeiger für Kunde der deutschen Vorzeit* 24, no. 2 (February 1877): 55. A description of the Dresden measuring instrument in Max Engelmann, "Mathematische Instrumente von Wenzel Jamnitzer," *Mitteilungen aus den sächsischen Kunstsammlungen* 5 (1914): 1–11. Cf. Wunderlich's reassessment of the instrument on the basis of photographs: Herbert Wunderlich, *Kursächsische Feldmeßkunst, artilleristische Richtverfahren und Ballistik im 16. und 17. Jahrhundert* […], (Berlin: Deutscher Verlag der Wissenschaften, 1977), 95–104. We follow Engelmann and Wunderlich in our description of the functions of the measuring instrument of the Observatoire.

62 Hauptstaatsarchiv Dresden, "Inventar über des Kurfürsten von Sachsen und Burggrafen zu Magdeburg Kunstkammer," 10009 Kunstkammer, Sammlungen und Galerien, No. 001, fol. 46v – 50r; see Dirk Syndram, Martina Minning, and Jochen Vötsch, eds., *Die kurfürstliche sächsische Kunstkammer in Dresden. Das Inventar der Kunstkammer 1587* (Dresden: Sandstein, 2010). Cf. Engelmann, "Mathematische Instrumente," 46–49; Wunderlich, *Kursächsische Feldmeßkunst*, 98.

63 Wenzel Jamnitzer, *Perspectiva Corporum Regularium. Das ist Ein fleyssige Fürweysung, Wie die Fünff Regulirten Cörper, darvon Plato inn Timaeo, Unnd Euclides inn sein Elementis schreibt, etc.* […] (Nuremberg: Heußler, 1568), see page 3 of the preface.

64 On Amman see Giulia Bartrum, ed., and Gero Seelig, comp., *The New Hollstein German Engravings, Etchings and Woodcuts 1400–1700: Jost Amman*, 2 vols. (Rotterdam: Sound & Vision Publishers, 2001), esp. vol. 1, 13–31, general introduction and earlier literature.

65 Engelmann, "Mathematische Instrumente," 50; Wunderlich, *Kursächsische Feldmeßkunst*, 98. Steche still claimed that Jamnitzer had been responsible for the engravings, Richard

apparently did not carry Amman's signature, the Paris instrument does. On the upper rim of the cartouche formed of scrollwork positioned beneath the coats of arms (and between the depictions of the Horologium Ahaz and the sun standing still over Gibeon) the sober initials "I" (left) and "A" (right) used by Amman can be seen.[66] This corresponds to the position of Amman's signature on a reversed print, that almost exactly matches the Observatoire's measuring instrument.[67] Many details (dots, scratches, discontinuous lines) suggest that the print was taken directly from the instrument itself, but it is puzzling that the diameter of the print and the original disk differ by nearly a centimetre.[68]

Description of the Observatoire's Instrument

The Observatoire's measuring instrument is a massive circular object made of gilded brass, with a diameter of nearly 52 centimetres, and weighing 5 kilograms. Virtually all surfaces apart from the functional scales, tables, and hour lines are richly decorated. The two sides serve different purposes:

(1) one is mainly geared at the measuring of time, including astronomical and astrological aspects,
(2) the other deals with observing and processing geodetic data, i.e. with surveying methods.

Steche, "Zwei Radierungen von Wenzel Jamitzer," *Mitteilungen aus dem Germanischen Nationalmuseum* (1891): 37.

66 Without recognising this monogram an exhibition catalogue already wrote in 1936: "grave de sujets dans le style de Jost Amman." Cf. *Exposition des Instruments et outils d'autrefois, Mars-Avril 1936,* exh. cat., Musée des arts décoratifs, Paris (Paris: La Musée, 1936), 46, cat. 327.

67 Etching, 50,8 cm (diameter), 58,3 x 51,5 cm (sheet), Bayerisches Nationalmuseum Munich, inv. no. NN 2611. A print like this with an attribution to Jost Amman (after Wenzel Jamnitzer) is described – without naming the owning institution – in Andreas Andresen, *Der deutsche Peintre-Graveur* [...], vol. 1 (Leipzig: Rudolph Weigel, 1864), 170, no. 213 together with a slightly smaller and true sided depiction of the other side of the instrument (no. 212). Kurt Pilz, "Jost Amman 1539–1591," *Mitteilungen des Vereins für Geschichte der Stadt Nürnberg* 37 (1940): 201–252, esp. 227–228, and Bartrum, ed., and Seelig, comp., *New Hollstein: Jost Amman,* 2:115–116, no. 294–295, could not trace these prints anywhere and considered them lost. They are, in fact, part of the Kupferstichkabinett in the Schlossmuseum Gotha, inv. no. 12,2 and 12,3: Justus Lange, Michael Eissenhauer, and Ludolf von Mackensen, eds., *Die Erfindung der Welt. Martin Schaffners bemalte Tischplatte von 1533,* exh. cat., Staatliche Museen Kassel (Kassel: Staatliche Museen, 2002), 97–98, cat. no. 108, 109.

68 All measurements have been checked on our request. We thank Emilie Kaftan, Paris, and her colleagues in Sélestat, and especially Raphael Beuing, Munich, for their help. The exemplars in Gotha, which came to our knowledge after writing this article, have not yet been examined.

(1) The first side is divided into an upper and a lower half (Figure 5.7). The compass that could be inserted in the centre of the upper half (or alternatively in the centre of the lower half of the reverse side), just as in the Dresden exemplar, is missing. Above the compass that would have been situated in the centre of a circle, a table of the planets ruling the hours at day and night can be seen. Below, a separate circle presents the planets ruling each day.[69] The free spaces on its left and right are filled by two depictions of men in oriental garb. The one on the left seems to be measuring a circle, divided into 14 unequal segments, while the one on the right is using a quadrant while a globe lies between his feet. Nude personifications of the four elements can be seen in the outer segments of the upper half of the instrument, earth and water on the left, air and fire on the right. The lower half comprises two vertical altitude dials. A frame of scrollwork between them includes Jamnitzers's signature as well as the date of production 1578. It also indicates that the sun-dials are suitable for latitudes between 49° and 51°, covering Nuremberg (49°), Prague (50°), and Dresden (51°).[70] The frame of scrollwork lies within a semi-circular scale, divided into two quadrants of 90°, used to determine the length of the day (between 16 and 8 hours) by measuring the height of the sun at noon with the help of a (lost) alidade and a plumb bob. The zodiac below encloses an anthropomorphic sun and is flanked by Hercules with the skin of a lion carrying a celestial globe (on the left) and Atlas with a crown (on the rock above his knee) carrying a terrestrial globe. The gnomon necessary to read the time was mounted at the centre of the circular instrument, but is now missing.

(2) The second side is concerned with observing and processing geodetic data (Figure 5.8). While the first side was usually used vertically and we can view the information and imagery on it from one perspective, two perspectives are necessary to marvel at the second side, which was used horizontally, as the information and imagery is split into two parts, each rotated by 180°.[71] When turned in such a way that the compass hole is in the upper side we can only read the lower part of the instrument. We will not go into the functional details of the double shadow square, a

69 With the inscriptions: "Der Sieben Planeten Regierung alle Stundt Bey dem Tag zu zelen Unnd bey iren Zaychen zu erkhennen"; "Der sieben Planeten Regierung alle Stundt Bey der Nacht zu zelen und bey iren zaichen zu erkennen." The days of the week are designated by the old names used in Southern Germany.

70 Inscription: "Diese zway Horologia sein gemacht worden, auff 49. 50. 51. Durch Wentzel Jamnitzer Goldschmid zu Nürmberg verfertiget M. D. LXXVIII."

71 Our sketch of the functions of this side follows Wunderlich, *Feldmeßkunst*, 98–104.

rectangular scaled frame inscribed with UMBRA RECTA, UMBRA VERSA, that could be used for surveying purposes. The semi-circular scales within are a conversion table for the most common units of length. They encompass half of a hexagon with three sides of equal length partitioned into 24 segments, possibly devised for astrological calculations.[72] The enclosed frame of scrollwork gives a short user instruction for the cropped circle below.[73] In analogy to the sun depicted at the very same position on the reverse side, the cropped circle can also be interpreted as a moon disk.[74] We could thus succinctly speak of a sun-side and a moon-side of the instrument. In the segments to the left and right of the cropped circle alias the moon disk personifications of the four seasons are depicted as nude men with the seasonal attributes. The segments to the left and right of the semi-circular scales show personifications of the sun (sceptre, wreath of laurel, sun) and the moon (sickle moon) or, possibly, of Day and Night (Dies and Nox).[75]

Rotating this side by 180° the empty field for the compass moves to the middle of the imagery in the lower part. In the frame of scrollwork just below, instructions for the deviation of the compass are given.[76] To the left and the right of the frame two putti hold various scientific instruments (square and compass on the left, astrolabe (?), measuring rod and compass on the right), each with a foot on a globe (celestial on the left, terrestrial on the right). On the segment of the annulus above them the coats of arms of the seven electorates are presented, the place of pride on top occupied by the coat of arms of Rudolf II, King of Bohemia since 1575 and German Emperor since 1576.

72 See Engelmann, "Mathematische Instrumente," 51; Wunderlich, *Feldmeßkunst*, 103.

73 Inscription: "In diesem Circkel mit A. B. C. D. betzeichent, findt man was vornen abgesehen ist, darnach zu verjünngen zu gebrauchen." For the use see Wunderlich, *Feldmeßkunst*, 102.

74 As can be seen by the inscription on the Dresden instrument, which is absent from the Observatoire instrument. On the left of the circle: "Diese thailung gehört gegen dem mon"; on the right: "In die Hoch unnd Dieffe zu messen."

75 The attributes of Day and Night can coincide with those of sun and moon as, for example, in designs attributed to Amman in the Herzog Anton Ulrich Museum in Braunschweig, inv. no. Z WB VII 40i (Dies and Nox) and 40j (Sol and Luna); see Christian Heusinger, *Herzog Anton Ulrich-Museum Braunschweig. Die Handzeichnungssammlung*, vol. 3, Tafelband 1 (Braunschweig: Herzog Anton Ulrich-Museum, 1992), fig. 91, and vol. 3, Textband (Braunschweig: Herzog Anton Ulrich-Museum, 1997), 292.

76 Inscription: "In diesem Circkel A. B. C. D. getzaichent, sihet man wie vil Grät die verruckung ist nach den Compast." For a description of the functions of this part see Wunderlich, *Feldmeßkunst*, 99–104.

The areas to the left and right of the coats of arms show the miracles of the sun standing still over Gibeon (left) and the Horologium Ahaz (right). Joshua, with a battle scene in the background, is shown in an antique military dress befitting his position and by far dominates the scene. His left arm is stretched out towards the sun as if he were commanding it to stand still, while a hail of stones showers down on the fleeing Amorites. The right side depicts Isaiah and Hezekiah and is no less impressive. The king, richly attired, sits in front of a magnificent canopy bed and looks up to Isaiah, who is pointing to a vertical sundial behind him. In his backdrop the view opens on an architectural prospect with obelisks and representative classical buildings.

The Pictorial Representation of the Two Miracles

There seem to be no printed templates for the monumental execution of these two scenes. While we have seen examples (Bible illustrations, illustration on instruments) for the miraculous recovery of Hezekiah, they cannot be considered templates for or compare to the mature and rich scene on Jamnitzer's measuring instrument. Amman's rendering of Hezekiah is particularly striking as he is not presented as naked, bedridden, and suffering. The majestic Hezekiah put centre stage here is not naked, but sumptuously attired, not bedridden, but sitting enthroned in front of his magnificent canopy bed, itself reminiscent of a throne, not suffering, but venerable. Moreover, in contrast to the presentations of the Hezekiah story on Praetorius's torquetum of 1568 (Figure 5.4) and on Schissler's Horologium Ahaz of 1578 (Figure 5.6), where the story had been split into two images, the miracle of the sun and the application of the fig poultice, the two scenes are fused into one here. The fig poultice is only alluded to by the arrangement at the lower left, reminiscent of a still life, of figs, grapes, and a little sealed jar (possibly containing soothing ointment).

A similar composition of the recovery of Hezekiah adorns the astronomical clock of 1583 by Hans Gruber (see below, Figure 5.11). The only precedent of the scene we know of is from a tripartite tapestry series that was woven around 1530 by the Weaver of the Hezekiah Series (Figure 5.9).[77] The tapestries are in the collection of the Bob Jones University in Greenville, South Carolina, and

77 *Tapestries on the Life of King Hezekiah: The Destruction of the Idols by Hezekiah, The Sacrifice of Hezekiah, and Hezekiah's Life Prolonged*, each tapestry ca. 396,3 x 350,5 cm, Greenville, South Carolina, Bob Jones University, Museum & Gallery; cf. Jean-Paul Asselberghs, *Les tapisseries flamandes aux États-Unis d'Amérique* (Brussels: Musées Royaux d'Art et d'Histoire, 1974), 39–40. There is a more complete set of this series, which also includes the Horologium Ahaz scene, in the Abbey of Sacromonte in Granada (Spain), cf. Margarita García Calvo, "Una serie de tapices de la historia del rey Ezequías en la Abadía del

their template is attributed to Jan Cornelisz. Vermeyen. The provenance of the tapestries can only be followed back to the seventeenth century when they hung in Somerset House in London, the residence of Queen Henrietta Maria, spouse of Charles I. The provisional name of the unknown weaver is based on a similar series of eight tapestries with the same weaver's mark in the Swedish Royal Collection in Stockholm, that King Eric XIV of Sweden had purchased on the occasion of his coronation in 1561.[78]

The impressive composition of Joshua stopping the sun, too, cannot be ascribed to printed templates. Comparing it to the design of the scene in previous printed material it is obvious that Amman's composition differs from the standard representations. Usually, the battle scene is centre stage and often Joshua cannot even be identified; sun and moon can be seen, but do not command particular attention.[79] The distinct and close connection between Joshua dominating the battle scene and the sun and the moon is singular. Joshua, extricated from the battle scene, is presented as an Old Testament military leader and his monumental pose is fitting for one of the Nine Worthies. Since the Middle Ages the Nine Worthies had represented a canon of exemplary individuals, three pagan (Hector, Alexander the Great, Julius Caesar), three Jewish (David, Joshua, Judah Maccabee), and three Christian (King Arthur, Charlemagne, Godfrey of Bouillon), that was widespread in the arts well into the sixteenth century.[80] However, the composition of Joshua watching and

Sacromonte (Granda)," *Boletín/Museo e Istituto Camón Aznar de Ibercaja* 106 (2010): 23–43, esp. 27, 38, fig. 5.

78 Iain Buchanan, "The Tapestries Purchased by King Eric XIV of Sweden in Antwerp in 1560–1561. New Documentation," in *Flemish Tapestry in European and American Collections. Studies in Honour of Guy Delmarcel*, ed. Koenraad Brosens (Turnhout: Brepols, 2003), 93–98, colour plates on pages 205–206. Cf. Johan Fredrik Böttiger, *Svenska statens samling af väfda tapeter: historik och beskrifvande förteckning, vol. 4 (Stockholm: Palmquists Aktiebolag, 1898)*, 39.

79 Cf. Germain Hoyau, *Joshua commanding the sun and moon to stand still*, 29,1 x 44,3 cm, woodcut, 1558–81, in Guillaume Guéroult, *Figures de la Bible, illustrees de huictains francoys pour l'interpretation et intelligence d'icelles* (Lyon: Guillaume Rouillé, 1564). The image was also used in the edition of 1565, and in various other French Bibles. Jost Amman, *Sol stetit immotus*, in Jost Amman, *Bibliorum utriusque Testamenti Icones* (Frankfurt: Feyerabend, 1571); identical in: Heinrich Peter Rebenstock, *Neuwe Biblische Figuren: Künstlich und artig gerissen durch den sinn und kunstreichen auch weitberühmte[n] Joß Amman* [...] (Frankfurt: Feyerabend, 1571); Johannes Henten, ed., *Biblia, ad vetvstissima exemplaria nunc recens castigata* (Venice: Bevilacqua, 1574), 203.

80 Cf. Robert Ludwig Wyss, "Die Neun Helden: eine ikonographische Studie," *Zeitschrift für schweizerische Archäologie und Kunstgeschichte* 17 (1957): 73–106; Horst Schroeder, *Der Topos der Nine Worthies in Literatur und bildender Kunst* (Göttingen: Vandenhoeck & Ruprecht, 1971).

FIGURE 5.9 Weaver of the Hezekiah series after Jan Cornelisz. Vermeyen, *Hezekiah's life prolonged*, ca. 1530
PHOTO: BOB JONES UNIVERSITY, MUSEUM & GALLERY, GREENVILLE

pointing up to the skies does not seem to derive from a print series of the Nine Worthies, where Joshua is usually characterised by a shield or a banner with a sun symbol.

The Pictorial Programme of Jamnitzer's Measuring Instrument

To properly understand the programme and the specific role of the representations of the two astronomical miracles it is helpful to compare the Observatoire instrument to the Dresden instrument. Both were manufactured in 1578, but as corrections were made in the moon side's limbus of the Observatoire instrument it can be surmised that the Dresden instrument preceded it.

Almost all of the imagery on the Dresden instrument is closely intertwined with its functions. On the side dealing with time, personifications of summer and winter flank the zodiac. The planetary gods Sol (characterised by a sun and the sign of Leo as attributes) and Luna (characterised by a moon and the sign of Cancer as attributes) stand to the left and right of the table of the planets ruling the hours. This also points to the duality of the instrument's sides – a sun-side and a moon-side, as we previously referred to them. The putti seem to demonstrate the astronomical (on the left) and geodetical (on the right) uses of the instrument, as the disks they are handling resemble the sun- and the moon-side of the instrument itself. Indeed, we know from a manuscript that Jamnitzer thought of visual elements on his instruments as parts of a user manual for his instruments.[81]

The reverse side (of the moon) deals with surveying and is in line with the ongoing theme of duality. The scholar in the left upper corner of the shadow square is characterised as a cosmographer or surveyor, surrounded by maps, while Chronos fills the opposite corner amid various sundials. To the left and right of the coats of arms, where the miracles of the sun standing still over Gibeon and the Horologium Ahaz are presented on the Observatoire instrument, are personifications of geometry (with a polyhedron on her lap and a compass and a ring in her hands) and astronomy (taking measurements with a compass on a sphere).[82] As a whole the programme is coherent and finely tuned to the instrument and its uses.

Obviously, the programme of the Observatoire instrument differs significantly. Its mission is not restricted to a clear and sober illustration of the functions and uses of the instrument, where only imagery and allegories pertaining to the mathematical sciences would be featured and appropriately placed, as with the Dresden instrument. Instead, it abounds with an opulent gamut of imagery, combining biblical and mythological stories with cosmic allegories and more down to earth aspects of the mathematical sciences. The programme is more complex and at the same time less concerned with exactness, in that the imagery is not systematically linked to the functions of the instrument and not always appropriately placed on the instrument.

While the imagery of the Dresden instrument covers astronomy and surveying equally, the imagery of the Observatoire instrument almost exclusively highlights astronomy. On the side of the instrument designated for time measurement Atlas and Hercules represent a long line of tradition in astronomy, referring to the old story that Atlas had been its first teacher and Hercules his

81 National Art Library of the Victoria & Albert Museum, "Ein gar Kunstlicher und wolgetzierter Schreibtisch sampt allerhant kunstlichen silbern und vergulten newerfunden Instrumenten so darin zufinden. Zum gebrauch der Geometrischen und Astronomischen, auch andern schönen und nützlichen Kunsten," Wenzel Jamnitzer, MSL/1893/1600 and MSL/1893/1601, 2 vols. (1585). Vol. 1, fol. 37r. printed in Hauschke, "Beschreibung," 109.

82 Engelmann, "Mathematische Instrumente," 49.

first disciple.[83] Among the scholars on the upper half the one on the right is clearly an astronomer. The four elements as well as the four seasons on the other side of the instrument symbolise the entanglement of humans in cosmic contexts. The planetary gods, Sol and Luna, are of a less programmatic character than on the Dresden instrument. However, they correspond to the two biblical astronomical miracles, the sun above the miracle of the retrograding shadow of the Horologium Ahaz and the moon above Joshua stopping sun and moon over Gibeon. Seen as personifications of Dies and Nox instead of Sol and Luna, they equally point to aspects of the passage of time: the biblical miracles deal with hours, Dies and Nox with day-time and night-time, and the four seasons see months and years elapsing. The story of Hezekiah points to the oldest dial mentioned in the Bible and its creator – King Ahaz – and thus highlights the long tradition of clockmaking as Andrew Morrall indicates.[84]

But the programme goes far beyond the praise of astronomy at large. The specific and unprecedented representations of a charismatic Joshua in the commanding and powerful pose of a military leader and a pious Hezekiah in his royal grandeur and dignity (his recovery being imminent) indicate another iconographic layer addressed to a royal or even imperial patron.

To present Hezekiah as the ideal of a pious leader is firmly rooted in early modern iconography. Scripture stresses "that after him was none like him among all the kings of Judah, nor *any* that were before him" (2 Kings 18:5), and only King Josiah was equally distinguished as a pious leader. Hezekiah and Isaiah, together with David, symbolising justice, and Solomon, symbolising piety and wisdom, are featured on the four enamel plaques on the imperial crown. Isaiah holds a banderole with the prophecy that Hezekiah will be granted another 15 years of life (ADICIAM SVPER DIES TVOS XV ANNOS), thus emphasising the privileges granted to pious and virtuous rulers such as Hezekiah.[85] Naturally, the emperors of the Holy Roman Empire connected this prophecy to themselves and saw themselves reflected in Hezekiah as pious and model rulers. This aspect is stressed on various occasions, such as in the hymn on the coronation of Maximilian II as King of Bohemia in 1562 by Lorenz

83 Volker Remmert, "Tycho Brahes Nase, oder: Atlas und Herkules und die visuelle Legitimierung der neuen Astronomie im 17. Jahrhundert," *Zeitschrift für Kunstgeschichte* 66 (2003): 177–206; cf. Remmert, *Picturing the Scientific Revolution*, 127–165.

84 See Chapter 4 in this volume by Andrew Morrall, "Turning Back the Sun. Christoph Schissler's 'Horologium Ahaz' as 'Kunststück.'"

85 Cf. Georg Johannes Kugler, *Die Reichskrone*, rev. ed. (Munich: Verlag Herold Wien, 1986), 42 on the symbolism of the enamel plaques.

Wessel von Essen,[86] or in the funeral sermon (1740) on Charles VI by the Jesuit Ignatius Pittermann, who compares Charles to Hezekiah.[87]

Joshua, too, as one of the Nine Worthies, was counted among the model rulers. In the large print "Apotheosis of Emperor Maximilian II" of 1571, an allegory on earthly rule designed by Jamnitzer and executed by Amman, Joshua is included in the line of tradition of Old Testament paragons along with David, Moses, Samson, Ezekiel, and Solomon. Together they adorn the alcoves of the rotunda on which Maximilian is kneeling.[88]

On the other side of the instrument the images of Atlas, as the first teacher of astronomy, and Hercules, as his first disciple, too, were excellently suited to feed into the political layer of the programme. The role of Atlas here reaches far beyond the fallen Titan of Greek mythology and founder of astronomy, but clearly refers to the figure of Atlas that had emerged by the ninth century, namely that of King Atlas, as indicated by the crown on the rock above his knee.[89] Regardless of the location of Atlas's kingdom (Libya, Mauretania or Spain), the crucial aspect here is that in early modern iconography he embodied wisdom, readiness to sacrifice the self, and power. Hercules, too, is not just a docile disciple of astronomy. Rather, given his image as a virtuous and resourceful hero (*Tugendheld*) he was seen as a princely ideal and the perfect patron of many a dynasty, and especially of Charles V and the Habsburgs from the mid-sixteenth century onwards.[90]

Thus, the programme of Jamnitzer's measuring instrument presented various role models for active rulers based on Greek mythology as well as biblical legends. Atlas and Hercules beyond being revered as virtuous, wise, and

86 Lorenz Wessel von Essen, *Ein schöner Lobspruch zu Ehren Dem Allerdurchleuchtigtsten [...] Maximiliano Römischen, Hungarischen vnd Behaimischem Künig etc. Auff irer Kü: May: Titel unnd Namen gestellet. Auch wirdt darinnen angezaigt vom leben und ende der Aller Gottsförchtigisten, unnd frumbsten dreyen Künigen, so in Juda regieret haben, als nemblich David, Ezechias vnnd Joseias etc.* (Vienna: Michael Zimmermann, 1563).

87 Ignatius Pittermann, *Leich- und Lob-Red Der Römisch-Kayserlich, und Königlich-Katholischen Majestät, Ertz-Herzogen von Öesterreich, u.u. Karl des Sechsten Da Sr. Kayserl. Majetsät Leich-Besingnuß Den 16. 17. und 18.ten Novembris 1740 In der Hof-Kirchen bey herzlichen Trauer-Gerüst gehalten worden* [...] (Vienna: Heyinger, 1740), 33.

88 Bartrum, ed., and Seelig, comp., *The New Hollstein German* [...] *Jost Amman, Part 1*, 23–25, no. 15. For the identification of Charles V with Joshua in a frontispiece, see Figure 3.5 in Chapter 3 by Brent Purkaple, "Appealing to Ahaz. The Jesuits and the Sundial of Ahaz in the Era of Copernican Astronomy."

89 The first image of Atlas with a crown may have been a woodcut in Johannes Schöner's *Opera mathematica* (Nuremberg: J. Montanus and U. Neuber, 1551). On this and Atlas as a king see Remmert, *Picturing the Scientific Revolution*, 130–144.

90 See Remmert, *Picturing the Scientific Revolution*, 131 and 139, with further references.

powerful, were knowledgeable in astrology, a discipline highly relevant to early modern rulers. Joshua stood for the power that God could confer on a virtuous leader, and Hezekiah for the rewards that godliness and piety would ensure.

Provenance

Not much is known about the provenance of the measuring instrument. It was found in a cabinet at the Observatoire together with other discarded instruments in the late nineteenth century.[91] It has been claimed that both, the Dresden and the Observatoire instrument, must have been commissioned by August I, elector of Saxony.[92] This does not seem plausible for various reasons. For once, there is no indication in the Dresden inventories and archives of a second instrument and no counterpart would have been needed to use the instrument. Moreover, as we have seen, the programmes are not finely tuned to one another. Finally, it cannot even be taken for granted that the Dresden exemplar was indeed commissioned by August I, because, if so, a much stronger emphasis on the Saxon coat of arms would have seemed appropriate.

On the other hand, an entry in the inventory of the Kunstkammer of Rudolf II in Prague (1607–1611) possibly refers to the Observatoire instrument: "a large gilded disk with all equipment [...] made by the old Jamnitzer in order to measure all kinds of things."[93] In fact we know that in 1581 – three years after the measuring instrument had been made – Jamnitzer's son Abraham delivered precious silver work and "all kinds of strange instruments" ("allerlei fremde Instrumente") at the Rudolfine court in Prague.[94] The gilded disk ("vergulte Scheiben") cannot be traced in later, less detailed inventories.[95] There

91 We thank Emilie Kaftan at the Observatoire for this information.

92 Savoie, "Le cadran solaire," 138. He argues that the precious and complex disk must have been made for a German ruler who understood mathematics and gnomonics very well.

93 Wolfgang Eckhardt, "Erasmus und Josua Habermel – kunstgeschichtliche Anmerkungen zu den Werken der beiden Instrumentenmacher," *Jahrbuch der Hamburger Kunstsammlungen* 22 (1977): 13–74, esp. 18; Klaus Pechstein, "Der Goldschmied Wenzel Jamnitzer," in *Wenzel Jamnitzer und die Nürnberger Goldschmiedekunst 1500–1700. Goldschmiedearbeiten, Entwürfe, Modelle, Medaillen, Ornamentstiche, Schmuck, Porträts*, ed. Gerhard Bott, exh. cat., Germanisches Nationalmuseum Nürnberg (Munich: Klinkhardt & Biermann, 1985), 64 and 69: "Ein gefierte nidere grosse schwartze laden, darin ein grosse vergulte scheiben sambt aller zugehör mit compaß, laitter klein und groß, ein seulen inmitten, ist vom alten Gamitzer gemacht, zu allerley abmessen." Cf. Rotraut Bauer and Herbert Haupt, eds., *Das Kunstkammerinventar Kaiser Rudolfs II., 1607–1611* (Vienna: Schroll, 1976), 118.

94 Ulrich Thieme and Felix Becker, eds., *Allgemeines Lexikon der bildenden Künstler von der Antike bis zur Gegenwart*, vol. 18 (Leipzig: E.A. Seemann, 1925), 371.

95 Cf. the 1621 inventory in Heinrich Zimmermann, ed., "Das Inventar der Prager Schatz- und Kunstkammer vom 6. Dezember 1621. Nach Akten des k. und k. Reichsfinanzarchivs in Wien," *Jahrbuch der Kunsthistorischen Sammlungen des Allerhöchsten Kaiserhauses* 25

is currently no way to retrace the trajectory of the instrument after Rudolf's death, when the best pieces of his collection were brought to Vienna, where the collection was finally dissolved. Denis Savoie offered two speculations as to how the instrument may have ended up in the Observatoire. Either the Swedish army may have taken it from Prague in 1648 and Queen Christina could later have handed it on as a gift to astronomer Jean-Dominique Cassini (1625–1712), first director of the Observatoire, or it may have been among the booty taken from Bohemia by Louis XIV or Napoleon Bonaparte.[96]

Be that as it may, a programme seemingly made to measure for the emperor of the Holy Roman Empire, as outlined above, supports the hypothesis that the Observatoire instrument came from the imperial collection in Prague.

Gruber's Astronomical Table Clock of 1583

The hypothesis that the representations of Joshua stopping the sun and the recovery of Hezekiah on the Observatoire instrument may have been addressed to an imperial patron is supported by a similar pictorial arrangement, explained by inscriptions, on Hans Gruber's astronomical table clock of 1583 (Figures 5.10 and 5.11). The clock was probably commissioned by the Danish King Frederik II, patron of Tycho Brahe, as we know that he ordered a travel clock from the Nuremberg clockmaker Hans Gruber (ca. 1530–1597) in 1583.[97] The decoration of the clock may seem a bit too extravagant for practical use as a travel clock. But its size corresponds to Frederik's wishes and the sundial on the right inner side technically covers the latitudes from 40° to 56°, i.e. a travel range from Madrid to Helsingør (north of Copenhagen). A second sundial in the left inner side is calibrated to the latitude of 57° only.[98] The clock is in the collection of the Landesmuseum Württemberg in Stuttgart.[99]

(1905): XIII–LXXV. We also consulted the inventories of 1619, 1635, 1646/47, 1648 and the inventory of Christina of Sweden of 1652.

96 Savoie, "Le cadran solaire," 137.

97 Bering Liisberg, *Urmagere og ure i Danmark* (Copenhagen: Martius Truelsen, 1908), 138.

98 In the sixteenth and seventeenth centuries latitudes between 55° and 60° were current for Copenhagen, cf. Sven Hauschke, "Globen und Wissenschaftliche Instrumente. Die europäischen Höfe als Kunden Nürnberger Mathematiker," in *Quasi Centrum Europae. Europa kauft in Nürnberg 1400–1600*, ed. Hermann Maué et al., exh. cat., Germanisches Nationalmuseum Nürnberg (Nuremberg: Germanisches Nationalmuseum, 2002), 385–389, f.n. 53.

99 Its measurements are 23 x 23 cm, with a height of 37 cm; inv. no. 1973-021. Cf. Volker Himmelein and John H. Leopold, *Prunkuhren des 16. Jahrhunderts. Sammlung Joseph Fremersdorf* (Stuttgart: Württembergisches Landesmuseum, 1974), 59–61, cat. no. 9; Andrea Schaller, *Prunkuhren der Renaissance* (Stuttgart: Württembergisches Landesmuseum, 2001), 33–39, cat. no. 6.

FIGURE 5.10 Hans Gruber, *Astronomical clock*, 1583, *Joshua stopping the sun*
PHOTO: LANDESMUSEUM WÜRTTEMBERG, STUTTGART, INV. NO. 1973-021 / P. FRANKENSTEIN, H. ZWIETASCH

The clock's case was designed by an unknown artist. The quadratic base is supported by four lions and embellished with allegories of the four seasons. The middle part, in form of a richly decorated tower, bears the various dials. The bell on top is a later addition. The main, silver dial on the front side, shows the time in various systems, along with the calendar, the zodiac, date and the position of sun and moon in the zodiac, as well as a disk with the phases of the moon and an astrological aspectarium. The two smaller silver dials below and the dial on the back side have been engraved with busts of scholars with

FIGURE 5.11 Hans Gruber, *Astronomical clock*, 1583, *Hezekiah's life prolonged*
PHOTO: LANDESMUSEUM WÜRTTEMBERG, STUTTGART, INV. NO. 1973-021 / P. FRANKENSTEIN, H. ZWIETASCH

instruments, copied from prints by Jost Amman and Heinrich Vogtherr the Elder.[100]

100 One astronomer is from Jost Amman, *Eygentliche Beschreibung aller Stände* (Frankfurt, 1568); cf. Schaller, *Prunkuhren der Renaissance*, 34; the other goes back to Hermetus (Hermes Trismegistos?) in Amman's etching *Globus Terrestris* of 1564 (but the quadrant has been replaced by a celestial globe); the other two busts are based on Azarchel (2nd sheet) und Solinus (3rd sheet) of the three broadsheets engraved by Heinrich Vogtherr and printed by Eustachius Froschauer with 48 portrait medallions of famous astronomers of 1545/46.

In the relief to the left and right of the back dial, personifications of Virtus (Bravery, her foot posed on a turtle) and Honos (Honour, her foot posed on a helmet) point to military accomplishment and virtue in general.[101] As both served as devices for various noble families, their appearance in this arrangement is insufficient for the identification of a specific patron. Below, looking up to the two personifications, two putti with a quadrant (on the left) and a plumb (on the right) sit on the scrollwork frame of a cartouche. In the cartouche a female figure is sitting on the ground surrounded by mathematical instruments, including an arithmetical table leaning on an irregular polyhedron, a table clock (similar to the one it adorns), a ruler, and an (astrological) disk with a dial. As a personification of *mathematica*, encompassing arithmetic, geometry, and astronomy, she is measuring a celestial or terrestrial globe with a compass. Astronomy and geometry are also the main topics on the engravings of the inner, detachable sides of the clock, where the sundials necessary to set the clock are mounted. Two putti each holding a compass and a celestial (left) and terrestrial (right) globe are positioned next to the sundial on the left inner side. On the right inner side the planetary gods Sol and Luna together with the signs of Leo and Cancer stand on columns laden with mathematical instruments.

On the outer sides of the clock the two biblical scenes are of particular interest here. On the left Joshua stops the sun (Figure 5.10), on the right the Horologium Ahaz is shown in between Isaiah and Hezekiah (Figure 5.11). Both scenes, monumental and full of detail, are displayed in an architectural frame enclosed by a semi-circular alcove. Joshua clearly dominates the scene on the left, placed prominently in front of the battle scene and almost touching the sun he is looking at. The moon and the stars can be seen on the left of the sun while on the right the hail of stones is coming down on the fleeing Amorites. On the pendant on the right side of the clock, Hezekiah sits on his throne and Isaiah faces him pointing to and almost touching the sundial on the tower in the background between the two. On a chest between them a bowl, a pitcher, a carafe, and a round object, possibly a fig, can be seen.

The iconographic similarities between the expositions of the two miracles on Gruber's clock and Jamnitzer's measuring instrument are striking. Even though Gruber had already made visual reference to Hezekiah in 1573

101 Partly misidentified in Himmelein and Leopold, *Prunkuhren des 16. Jahrhunderts*, 61, and Schaller, *Prunkuhren der Renaissance*, 34. On the representations of Honos und Virtus in Roman antiquity see Margarete Bieber, "Honos and Virtus," in *American Journal of Archaeology* 49, no. 1 (1945): 25–34.

on a table clock,[102] the prominent presentation of Joshua and the design of Hezekiah in his chair facing Isaiah raise the question of whether the artist of the 1583 programme could have seen either Jamnitzer's measuring instrument of 1578 or a print of it.[103]

Inscriptions below the two images give hints on how to properly decode the images. The story of Joshua stopping the sun emphasises the spectacular power that God can bestow on an exemplary and virtuous ruler: "It is not in the power of humans to stop the sun and the moon, and to escape the venerable course of nature. This is only in the power of the lord of the heavens. How great you were, Joshua, to be able to inhibit the stars of the world."[104]

The miraculous recovery of Hezekiah serves as a reminder to the rulers of the world to be pious and godly, which in turn will please God: "The fate of Hezekiah ends in three times five years. Even the heavens move backwards. It is easy for God to be omnipotent. Do you see how great his power is, who moves backwards the mobile shadow by a step?"[105]

These allegories of the virtuous ruler, whose life and glory still are perishable and ephemeral, have been immersed in an overall programme by way of the two astronomical miracles under the wings of the concept of vanitas. The aspect of time flying is reflected in the programme in a similar way as on the Jamnitzer instrument by the scholars measuring time, by Sol and Luna ruling day and night, and by the four seasons. But different from the latter, where the vanitas aspect might only be inferred, the inscription running around the top of the clock in this case explicitly testifies to it and warns: "A happy day

---

102 Cf. the inscription "VOM KVNG HISKA" and the information for the auction at Lempertz from 2017, "Tischuhr von Hans Gruber aus Nürnberg. Auction 1096, Lot 1607," Lempertz, accessed February 6 2021, https://www.lempertz.com/de/kataloge/lot/1096-3/1607-tischuhr-von-hans-gruber-aus-nuernberg.html.

103 Susanne Thürigen presented this thesis in her presentation "Pointing it out: Miracles of the Sun in Hans Gruber's Astronomical Table Clock of 1583" during the "Manipulating the Sun" workshop held at the Bergische Universität Wuppertal, from 21–23 August 2019. She also deals with the clock in *Turm, Spiegel, Buch. Astronomische Tischuhren in Süddeutschland zwischen 1450 und 1650* (Berlin/Boston: De Gruyter, 2022), 67–73 and passim.

104 Our own, rather free translation of: "NON OPVS HVMANVM EST CVM LVNA SISTERE SOLEM / NATVRAEQVE ALTAS EVARIARE VICES. / SOLA HAEC IN CAELI DOMINVM VIRTVS CADIT OQVI / SIDERA MVNDI INHIBES, IOSVA, QVANTVS ERAS!" For a German translation cf. Himmelein and Leopold, *Prunkuhren des 16. Jahrhunderts*, 60.

105 Our own, rather free translation of: "EZECHIAE FATVM TER QVINOS EXIT IN ANNOS: / PRAETER NATVRAM VERTITVR IPSE POLVS. / OMNIA POSSE, DEO FACILE EST. VIS QVANTA EA, DENOS / QVOD RECLINATVR MOBILIS VMBRA GRADVS?" For a German translation cf. Schaller, *Prunkuhren der Renaissance*, 34.

runs rapidly with vanishing hours. The sad day flows slowly under heavy fate. However much time you may have: pass it in a way, that you think of death as long as you live, whenever the bell tolls!"[106]

Conclusion

The analysis of the role of the two biblical miracles of the retrograding sun (Horologium Ahaz) and the sun standing still over Gibeon in the pictorial programmes of five instruments has shown a variety of messages conveyed in their specific contexts. As we have seen, the imagery of the miraculous recovery of Hezekiah on Hartmann's paper sundial of 1561 (Figure 5.1) and on Schissler's refractive sundial of 1578 (Figure 5.6) is each finely tuned to its respective instrument. Hartmann draws an analogy between the shadow going backwards on the Horologium Ahaz and the similar effect on his instrument, where he presents a natural way to achieve the miracle.[107] Thus his imagery basically restricts itself to the Horologium Ahaz itself. Schissler also shows an image illustrating the effect that the instrument shows or re-enacts, but adds one referring to the actual recovery of Hezekiah. In contrast to these two instruments, the story of the miraculous recovery of Hezekiah on Praetorius's torquetum of 1568 (Figure 5.4) was presented in four parts. In this case, the miracle itself was not centre stage; rather, the Isaiah and Hezekiah cycle was embedded within a programme both illustrating the uses of the torquetum and emphasising the role of the prophet Isaiah as a physician. Thus the iconography was closely adapted to the professional identity of its owner, the physician Melchior Ayrer, and the instrument could serve as a high-end patronage artefact.

The story of Isaiah and Hezekiah carries a third type of message on the measuring instrument of 1578 designed by Jamnitzer and etched by Amman (Figure 5.8) and the astronomical table clock of 1583 by Gruber (Figure 5.11). Here Hezekiah and Joshua stand as role models and ideals for virtuous rulers.

---

106 Our own, rather free translation of: "LAETA VOLVBILIBVS RAPIDE LVX AVOLAT HORIS. / MAESTA TRAHIT TARDAM SORTE GRAVANTE MORAM / QVOD TE CVMQVE MANET TEMPVS, SIC TRANSIGE, LETHI / VT VIVAS, AVRES CVM FERIT HORA, MEMOR." For a German translation cf. Schaller, *Prunkuhren der Renaissance*, 33.

107 Cf. Chapter 2 by Henrique Leitão, "Mathematical Certainty and Biblical Inerrancy. Pedro Nunes and the Retrogradation of Shadows at the Dial of Ahaz" and Chapter 3 by Brent Purkaple, "Appealing to Ahaz. The Jesuits and the Sundial of Ahaz in the Era of Copernican Astronomy" in this volume.

In the programme the two of them impressively and unequivocally exemplify the powers (God changes the course of nature on Joshua's demand) and the divine grace (God gives 15 further years of governance to Hezekiah) that God grants the virtuous and pious ruler.

The instruments that we discussed in this chapter were selected on the grounds that their imagery included either the Horologium Ahaz or both the Horologium Ahaz and the miracle of Joshua stopping the sun. Apart from systematically analysing and comparing the role of the two miracles in their respective programmes, the chapter had the goal of addressing the question of whether the two images would have carried any anti-Copernican undertones. The answer is clear: there are no traces of the Copernican debate on the instruments discussed. The association of visual representations of the two astronomical miracles with the Copernican debate, which seems so plausible, must be handled with care.

PART 3

The Horologium Ahaz and Joshua Stopping the Sun in Art and Architecture in Specific Locations

CHAPTER 6

Stairway to Heaven

The Marvellous Staircases of the Medici Villa at Pratolino

Denis Ribouillault

For Hervé Brunon, friend and mentor

Abstract

This article deals with a neglected aspect of the famous Villa Medici of Pratolino, built for the Grand Duke Francesco I de' Medici at the end of the sixteenth century, namely the stairs on the south façade of the villa, which served as sundials. These stairs, which were also equipped with small water jets, are described and interpreted by the philosopher Francesco de' Vieri, professor at the University of Pisa, in his *Discorsi delle maravigliose opere di Pratolino e d'Amore*, published in 1586 and dedicated to Grand Duke Francesco. In his book, Vieri links this rare and complex staircase to notions relating to meteorological science and the destiny of the soul directly borrowed from Aristotle and Plato. Other examples of sundials of this type are then examined and linked to various expressions of divine wonder and spiritual ascent to God, from Francesco Colonna's *Hypnerotomachia Poliphili* (1499) to the Jantar Mantars of Maharajah Jai Singh II (1724–1727) in Mughal India.

Aside from the garden of the Villa d'Este at Tivoli, one of the most extraordinary Renaissance gardens in view of its hydraulic mechanisms and its Mannerist inventions is the garden of the Villa Medici at Pratolino, north of Florence (Figure 6.1). It was created for the Grand Duke Francesco I de' Medici from 1568 to 1586 by a team led by the architect and engineer Bernardo Buontalenti (1531–1608). This famous villa garden, now largely a ruin, has been admirably studied by Luigi Zangheri and Hervé Brunon, who dedicated to it his monumental doctoral dissertation.[1] Historian of science Matteo Valleriani

1 I want to thank Anatole Tchikine, Denis Savoie, Filippo Camerota, Matteo Valleriani, Bruce D. Chilton as well as the editors for discussing with me various aspects of this research, initiated in the fall of 2017 whilst a Fellow in Garden and Landscape Studies at Dumbarton Oaks Research Library and Collection. The bibliography on the Villa Medici at Pratolino

 | DOI:10.1163/9789004677654_008

has dealt more recently with the many technological innovations developed for the garden fountains, wondrous machines and automata, locating their origin in the impulse given to Renaissance science by the rediscovery of Hero of Alexandria's ancient treatise on *pneumatics*, a translation of which was dedicated to Bernardo Buontalenti as early as 1582.[2] Despite this wealth of scholarship however, one of Pratolino's wonders seems to have escaped scrutiny entirely, namely the main staircases of the villa's southern façade, which functioned as sundials.[3] This is the subject of the present paper.

The presence of sundials at Pratolino should not come as a surprise. Although they have been systematically forgotten, such instruments, often spectacular, are documented in the most important European gardens of the early modern period.[4] Already in the thirteenth century, in the park of Hesdin, in northern France, a monumental sundial stood alongside other marvellous buildings, automata and fountains. Called "Le miedi," it was carried by statues of six wild men and six lions and leopards.[5] Documents also mention several clocks and a mechanical organ, instruments which would be found three centuries later in the gardens of Tivoli and Pratolino.[6]

---

until 2008 is presented and analysed in Hervé Brunon, "Pratolino: art des jardins et imaginaire de la nature dans l'Italie de la seconde moitié du XVIe siècle" (PhD diss., Université Paris 1 Panthéon-Sorbonne, 2001); revised edition accessible online: TEL Multidisciplinary Theses Server, last modified March 2, 2020, https://tel.archives-ouvertes.fr/tel-00349346.

2 Biblioteca Nazionale Centrale, "Della natura del voto di Erone Alessandrino," Bernardo Davanzati, MS Rari 223, formerly MS Palat. 1166 (1582); reedited as Bernardo Davanzati, *Della natura del voto di Erone Alessandrino. Volgarizzamento inedito di Bernardo Davanzati* (Florence: Stamperia del Monitore, 1862). The same year, Oreste Vannoci Biringucci dedicated to Buontalenti a complete translation of Hero's treatise: Biblioteca degli Intronati, "Libro degli artifizii spiritali over di fiato d'Herone Alessandrino," Oreste Vannoci Biringucci, MS L. VI. 44 (1582). See further the work of Matteo Valleriani, especially Matteo Valleriani, "Ancient Pneumatics Transformed during the Early Modern Period," *Nuncius* 29, no. 1 (January 2014): 127–173 and the website on Pratolino he created and edited: Matteo Valleriani, "Pratolino: The History of Science in a Garden," accessed July 24, 2020, http://pratolino.mpiwg-berlin.mpg.de/.

3 To my knowledge, the only mention of Pratolino's sundials (with no further explanation) is in Marco Dezzi Bardeschi, "Le fonti degli automi di Pratolino," in *La fonte delle fonti: iconologia degli artifizi d'acqua*, ed. Alessandro Vezzosi and Cristina Acidini Luchinat (Florence: Aliena Editrice, 1985), 13–24.

4 Denis Ribouillault, *Gardens of the Heavens*, forthcoming.

5 Jules-Marie Richard, *Inventaire-sommaire des archives départementales antérieures à 1790. Pas-de-Calais*, vol. 1 (Arras: Impr. de la Société du Pas-de-Calais, 1878), 229–230. On the Garden Park at Hesdin, see Sharon Farmer, "Aristocratic Power and the 'Natural' Landscape: The Garden Park at Hesdin, ca. 1291–1302," *Speculum* 88, no. 3 (July 2013): 644–680.

6 Elly Rachel Truitt, *Medieval Robots: Mechanism, Magic, Nature, and Art* (Philadelphia: University of Pennsylvania Press, 2015), 124; Fabienne Pomel, ed., *Cloches*

FIGURE 6.1 Giusto Utens, *Villa Medici at Pratolino*, ca. 1599, tempera on canvas, 145 x 245 cm, Florence, formerly at the Museo Storico-Topografico, now at Villa Petraia
PHOTO: HERVÉ BRUNON

Gnomonics, i.e. the study of the design, construction and use of sundials, was, like pneumatics and mechanics, a branch of mathematics. Ancient authors interested in hydraulics also frequently dealt with problems concerning sun and light rays and the course of the stars and planets. For instance, Hero of Alexandria, the author of the *Pneumatica*, wrote a treatise on Catoptrics.[7] Likewise, if Book X of Vitruvius's *De architectura* is dedicated to Mechanics, book IX is a sophisticated new treatise on sundials. Arab science too, heir to the knowledge of the ancient Greeks, shows the intricate character of the various mathematical branches. As in the Park of Hesdin, Arab science may have played an important part in making the great Renaissance gardens of Europe a marvellous showcase for technology and science.[8]

---

et horloges dans les textes médiévaux. Mesurer et maîtriser le temps (Rennes: Presses Universitaires de Rennes, 2012), 23; and Simone Kaiser and Matteo Valleriani, "The Organ of the Villa d'Este in Tivoli and the Standards of Pneumatic Engineering in the Renaissance," in *Gardens, Knowledge and the Sciences in the Early Modern Period*, ed. H. Fischer, Volker Remmert, and Joachim Wolschke-Bulmahn, 77–96. Basel: Birkhäuser, 2016.

7 Alain Boutot, "Modernité de la catoptrique de Héron d'Alexandrie," *Philosophie antique* 12 (2012): 157–196.

8 See further Michel Baridon, "L'islam, les jardins et la migration des savoirs en Méditerranée," *Horizons Maghrébins – Le droit à la mémoire* 45 (2001): 7–13. Comparisons between Greek and Arab sources, the Park at Hesdin and the gardens at Pratolino have been made recently

A careful review of the evidence for the shape and possible meaning of the Pratolino sundials is proposed in the first part of the paper. In the second part, we turn to other examples of sundial-staircases in order to assess the exceptional nature of Buontalenti's invention for Pratolino.

The Description and Exegesis of the Staircases in Francesco de Vieri's *Discorsi delle maravigliose opere di Pratolino*

The staircases of Pratolino's villa are described and interpreted by the philosopher Francesco de' Vieri (1524–1591), professor at the University of Pisa, in his *Discorsi delle maravigliose opere di Pratolino e d'Amore*, published in 1586 and dedicated to Grand Duke Francesco I.[9] This work is a complex allegorical and philosophical reading of the garden that distinguishes it from a more classical *ekphrasis*. For Vieri, all the works of Pratolino belong to the category of the fable [*favola*] which he defines as "a fiction or invention of divine or human actions, which has something marvellous and delightful whilst containing in an obscure way some useful truth."[10] Thus, for the visitor-philosopher, statues, fountains and other marvels visible in the garden must be deciphered via a close exegetical reading, at both a philosophical and moral level. This does not mean however that his description is unreliable. It is, after all, an "official" description and a woodcut representing the entire garden appended to the second edition of 1587 is indicated as having been done under the guidance of Bernardo Buontalenti himself (Figure 6.2).[11] Accordingly, modern scholars have made extensive use of the descriptive part of the text and found it generally reliable. Nevertheless, Vieri warns his reader that his description is

---

in Lily Filson, "Magical and Mechanical Evidence: The Late-Renaissance Automata of Francesco I de' Medici," in *Evidence in the Age of the New Sciences*, ed. James A.T. Lancaster and Richard Raiswell (Cham: Springer International Publishing, 2018), 177–206.

9 Francesco de' Vieri, *Discorsi delle maravigliose opere di Pratolino e d'Amore* (Florence: Giovanni Marescotti, 1587). Beside Hervé Brunon's dissertation, many sources are now available online: See "Historical Sources Related to Design and Construction of the Garden of Pratolino," European Cultural Heritage Online, last modified June 2015, http://echo.mpiwg-berlin.mpg.de/content/pratolino/sources. On Vieri's *Discorsi*, Brunon's work is essential, especially Brunon, "Pratolino," 475–485. See also Luigi Zangheri, "Pratolino. La grande macchina del cosmo," in *Il giardino e la memoria del mondo*, ed. Giuliana Baldan Zenoni-Politeo and Antonella Pietrogrande (Florence: Leo S. Olschki, 2002), 43–52.

10 "la favola è una fintione, o un trovato dell'attioni divine, o humane, che ha del mirabile, & del dilettevole contenente oscuramente qualche utili verità." Vieri, *Discorsi*, 67. See also Brunon, "Pratolino," 481.

11 Brunon, "Pratolino," 50.

not perfect. This is the best he could come up with, he writes apologetically, given the fact that he saw the whole garden in the space of a day and that Buontalenti and his son gave him only an hour of their time to help him in his endeavour. Furthermore, he insists that his mind is constantly distracted by "gentle speculation."[12] In sum, Vieri's text should not be seen as a "programme" for the garden and its architecture, but rather as an example of the type of reflection it could occasion in the mind of an intellectual well-introduced at the court of Francesco I de' Medici. With this warning in mind, let us see what he has to say regarding the villa's staircases (Figure 6.3):

> If ancient authors considered the instrument called the clock a marvellous thing, because with it, one knows the hour of the day and of the night, it is also great marvel that one knows the same thanks to lines and numbers exposed to the sun that are [traced] on the edges of the stairs of the palace [*sponde delle scale del palazzo*], without having to rewind the clock so many times.[13]

Here, Vieri indicates at the outset one of the main functions sundials had in the early modern period, that is giving the exact (solar) time in order to set mechanical clocks, whose precision was notoriously problematic. Next, Vieri integrates the sundial staircase into his great exegetical reading of the garden:

> Those lines to know the hours being traced on the steps of the palace signify that all things are subjected to Time. Located underneath the celestial palace, all things [the souls] go down into this world when, produced by divine art they are driven from non-being into being, then ascend again as rational creatures and finally leave with divine grace. Such a passage is happy and desired even by the very wise Socrates, who was a pagan.

12 "Di quì è, che io innanzi à tutti gli altri modi di scrivere terrò primieramente l'historico, narrando à parte à parte tutte le piu maravigliose opere di Pratolino, così appunto, & con ordine: Come per me sarà possible, & secondo, che io in un sol giorno, che l'hò vedute, ne hò potuto tenere à memoria; & secondo, che in brevissimo tempo di un'hora l'Eccellente Architetto, & Ingegniere, Messer Bernardo Buontalenti, & Messer Francesco suo figliuolo m'hanno aiutato mettere in carta; oltre che come molti sanno l'animo mio è molto distrahatto da ogni gentile specolatione, per tante, & tante mie sventure." Vieri, *Discorsi*, 22–23.

13 "Quinto, se gl'antichi hebbero per cosa mirabile lo strumento, detto oriuolo, perche con esso si sà, che hora corre del giorno, & della notte, è ancora di gran maraviglia, che questo stesso si conosca per via di alcune linee, & numeri opposti al sole, che sono su per le sponde delle scale del palazzo, senza havere tante, & tante volte à ritemperare l'oriuolo." Vieri, *Discorsi*, 63.

AL SERENISSIMO SIGNOR DON FRANCESCO
DE' MEDICI SECONDO GRAN DVCA DI TOSCANA
SIGNORE, ET PADRON MIO OSSERVANDISSIMO.

HAVENDO à giorni passati, Serenissimo Gran Duca, stampato la descrizione di Pratolino, composta dall Eccellentiss. Filosofo M. Francesco de Vieri cognominato il Verino secondo, mi parue poi esser molto a proposito lo stampare ancora il disegno del Palazzo, con vna parte delle marauigliose bellezze, e delle opere, che l'accompagnano: e tanto più, che da molti Signori, e personaggi forestieri, mi era con instanza richiesto. Onde io per compiacer loro, e dare in oltre qualche contento à chi non hauesse veduto in fatto la sua grandezza, e bellezza; e per mostrarlo di nuouo sotto altra forma à quelli i quali veduto lo hauessero, l'ho fatto intagliare co'l mezo dell'Eccellente Pittore Architettore, & ingegnere di V. S. Alt. M. Bernardo Buontalenti, ilqual, doppo di lei, è stato il principale inuentore di così bella, e stupenda, rara, e ben intesa fabbrica, e delle altre tante ingegnose, ricche, e marauigliose opere, che dentro, e fuori, e d'intorno a quella si veggono: allequali hà ancora non poco giouato la diligente cura hautane dal molto Magnifico M. Benedetto Vguccioni, sopraintendente in tutte le fabbriche, che del continouo si fanno nelli stati di V. A. S. e trà le altre di quella della Gran Machina di S. Maria del Fiore, procurando con ogni diligenza, & sollecitudine, che secondo l'ordine dato da V. A. S. e conforme al disegno di detto M. Bernardo Buontalenti, alla grandezza, e nobiltà di cotale fabbrica la facciata di essa Chiesa, Duomo di Fiorenza si cominci, e fornisca. Ho stampato adunque il presente disegno nel modo migliore, che per mè si è potuto, e alfine di già detto: e per mostrarmi in qualche parte ancora, che minima, grato de tanti fauori, e delle tante grazie riceuute dalla benignità di V. A. S. e rappresentarle non come io vorrei; ma come mi permettono le mie picciole, e pouere forze, la humile, e diuota seruitù mia verso di quella, glie l'ho dedicato, anzi come cosa sua ritornato gliel'ho; Accettilo con lieta fronte, perche è cosa sua, e degnisi di hauere in protezione, e per raccomandato Giorgio Marescotti suo humilissimo fedele, e diuottissimo seruidore, il quale inchinandosele, prega Dio N. Signore per la sua felicità.

Riscontro di quelle sole cose, che appariscono nel Disegno di Pratolino.

1 LA statua di Gioue di marmo, di mano del Caualier Bartolomeo Bandinelli.
2 Laberinto di allori, ilquale ne contiene dentro di se altri quattro.
3 Colosso, e statua di grandezza di braccia 60. fatta di pezzi di pietra, finta per lo Apennino, ilquale ha per lo dosso diuerse sorte di piante, e per tutto versa acqua con molta marauiglia de riguardanti.
4 Peschiera in guisa di Teatro: nellaqual casca l'acqua che versa dall'Apēnino
5 Prato dinanzi al Palazzo, cō molte statue intorno tutte di marmo, e antiche lequali sono tramezate da Piramidi.
6 Tempio bellissimo doue si dice la santissima Messa.
7 Grotta nella quale è vna Orsa con gli suoi orsacchini, che getta acqua.
8 Palazzo per abitazione del Ser. Gran Duca con numero di piu di 100. camere, e con vn salone sotto, e sopra di lunghezza di 50. passi, e largo 25. E tutto il Palazzo ricco, bello, e vago al possibile, e luogo la state deliziosissimo, e sopra modo fresco quantunque non habbia cortile. La scala cammina per tutto il Palazzo, senza occupare o guastare stanza alcuna. Le stanze sono tutte libere, e hanno i lumi viui: E quelle, che sono fatte per comodo, e vso delli vfiziali di casa, sono appartate dalle altre con giudizioso diuisamento, e ordine.
9 Prato dinanzi alla Grotta: dentro alla quale sono piu fontane, e vn'altra Grotta piena di fessure, e di spaccature, e in modo aperta che pare che sempre stia per cadere a dosso ad altrui, in questa seconda si vede Galatea Ninfa marauigliosamente vscire di vn'antro, e andare attorno per vna fontana e [illegible], e nel medesimo tempo altre figure fare il medesimo.
10 Strada lunga 300 passi laquale quanto piu si discosta dal palazzo, piu pende all'ingiù, ma tanto dolcemente, che a pena l'huomo se ne accorge: & ha 30. fontane da ogni suo lato, le quali con grande, e dolce mormorio, versano l'acqua l'vna nell'altra.
11 Statua di marmo finta per vna lauandaia di panni, che gli sprema, e di quei faccia vscire l'acqua, questa ha a i pie di vn putto marmoreo, che piscia.
12 Monte Parnaso, con le Muse, gli stromenti delle quali, con il moto, e per forza di acqua, sono fatti sonare Madriali, e canzonette in Musica: cosa piaceuole in vero, e di stupore.
13 Peschiera, nellaquale sono varie, e distinte sorte, e qualità di pesci, soliti di viuere, e nutrirsi nell'acque dolci.
14 Fontana con molte statue, di mano di M. bartolomeo Ammannati scultore eccellente.
15 Vccelliera grande coperta di rete di ferro, nella quale sono Grotte, fontane, e [illegible], con grandissima quantità d'vccelli.
16 Giardino segreto con fontane di pietre fine, e colonne di pastiglio, cosa molto vaga, e piaceuole.
17 Casa doue abita il Casiere, e guardaroba del Palazzo.
18 Fontana che da l'acqua à tutti li Tonfani piena di Trote.
19 Fontana chiamata la Rossa. Questa è differente da tutte le altre, e ad essa possono bere i grandi, i mezzani, e i piccoli di statura, di maniera, è accomodata.
20 Molti Tonfani, i quali sgorgano l'acqua loro l'vno nell'altro, e dentro a essi si mantiene viua inestimabile quantità di pesce.
21 Fontane che sono dalle latora dello stradone.
22 Tōfani, che si partono dalla Vccelliera.
23 Grotta di Cupido, ilquale si vede piangere, perche la sua facella è spenta, Getta acqua, e dal'acqua è fatto muouere quasi che in giro.

Altre cose vi sono, e degne d'esser vedute; delle quali per breuità non si ragiona: e di continouo se ne va mettendo in ordine delle nuoue, non punto manco marauigliose delle sopradette.

E stata condotta così ricca bella, e stupenda fabbrica al termine che l'è oggi, con spesa infinita nello spazio di dodici anni: Il che pare impossibile a credersi, eziandio per coloro, che si ricordono quale fosse quel luogo prima che S. A. S. mettesse le mani a cotale impresa, veramente degna, e propria della sua grandezza.

IN FIRENZE, Appresso di Giorgio Marescotti. MDLXXXVII.

FIGURE 6.2 Workshop of Bernardo Buontalenti, *Pratolino*, 1587, woodcut inserted in Francesco de' Vieri, *Discorsi delle maravigliose opere di Pratolino e d'Amore* (Firenze: Giovanni Marescotti, 1587), private collection
PHOTO: HERVÉ BRUNON

> Plato in the *Phaedo* writes of this passage of the soul from this world to the other world with religious piety. See, see, gentle and judicious minds how these instruments of Pratolino to know the hour are located on each side of the staircase [*da ogni banda*] and how they are marvellous and amazing.[14]

In this passage, Vieri mentions Plato's *Phaedo* on the immortality of the soul, set in the last hour before the death of Socrates, and compares the ascent and descent of the staircase with the platonic doctrine of the movements of the soul (for instance *Phaedros*, 248c–249c), later developed by Porphyry, Hermes Trismegistus and Augustine, among others. As Brian Dobell explains, "The soul is capable of two types of movement: *progressus* and *regressus*. The *progressus animae* is the soul's 'fall' into the transient world of sensible objects, and the *regressus animae* is the soul's 'flight' back to the unchanging world of reason."[15] This world of ideas and pure thought is here metaphorically represented by the *celeste palazzo del sole*, the celestial palace of the sun, that is, the villa itself.[16] Note that, for Plato, *progressus* and *regressus animae* were directly connected to the stars, as souls originate from them and return to them, as recalled for instance in *Timaeus* (41d).[17] The platonic metaphor is here remarkably well adapted to the gnomonic dimension of the staircase and reinforced by the presence of four monumental obelisks symbolising sun rays (Figures 6.3 and 6.4).

14 "Et poi che sendo queste linee da sapere l'hore disegnate su per le scale del palazzo ci è significato, che al Tempo son soggette le cose tutte, le quali sono sotto al celeste palazzo, le quali scendono in questo mondo, quando procedendo dall'arte divina son condotte dal non essere allo essere, salgono poi di nuovo all'insù le creature rationali, se di quà si partono come la divina gratia, il qual passaggio in tal caso è felice, & tale desiderava fino al sapientissimo Socrate, che fu gentile: come di esso, & della sua passata all'altro mondo con religiosa pietà è scritto da Platone nel Fedone, ò vero dell'anima. Vedete, vedete gentili, & giudistiosi spiriti, come questi stromenti di Pratolino da conoscere l'hore siano da ogni banda, che eglino si considerano maravigliosi, & stupendi." Vieri, *Discorsi*, 63.

15 Brian Dobbel, *Augustine's Intellectual Conversion. The Journey from Platonism to Christianity* (Cambridge: Cambridge University Press, 2009), 126.

16 The expression is used three times by Vieri on pages 34, 38 and 63.

17 An interesting example of the interpretation of the journey of the Platonic soul in a Renaissance mathematical-astronomical context is in Philip Melanchthon, *Orations on Philosophy and Education*, ed. Sachiko Kusukawa (Cambridge: Cambridge University Press, 1999), 93. Melanchthon refers to *Phaedrus* 246a–e, formulating for the first time the popular image of arithmetic and geometry as the two wings of astronomy, for which see Volker Remmert, "On Picturing the Past: Arithmetic and Geometry as Wings of the Mind," *The Mathematical Intelligencer* 31, no. 3 (June 2009): 44.

FIGURE 6.3 Giusto Utens, *Villa Medici at Pratolino, Detail of the façade and the staircases to the South*, ca. 1599, Villa Petraia, Florence
PHOTO: HERVÉ BRUNON

Furthermore, this analogy between the gnomonic architecture of the staircase and the cycles of the descent and ascent of the soul described by Plato is complicated by a second meteorological reading, this time of Aristotelian inspiration. Besides the hour lines traced on the steps, the ramps of the stairs were equipped with small jets of water.[18] The effect of the sun rays on the water drops created marvellous effects, such as rainbows, which enchanted visitors.

18 Besides Vieri, *Discorsi*, 33–34 (see note 22 below), see the anonymous description preserved at the Biblioteca Riccardiana in Florence, MS Riccardiana 2312, fol. 108v (accessible online: http://echo.mpiwg-berlin.mpg.de/MPIWG:9MHXWTWK): "Questo è grande quanto la facciata dinanzi, che guarda verso levante e rigira altresi quella parte che guarda verso mezzodi, onde non è meno di 150 braccia, e largo almeno 20 da qual ricontro alla porta si scende per due scale di nuova architettura, che fanno un ovato perfetto, li cui scaglioni sono tutti pieni di cannelli, che quand'altri vuole, tutti schizzano acqua."

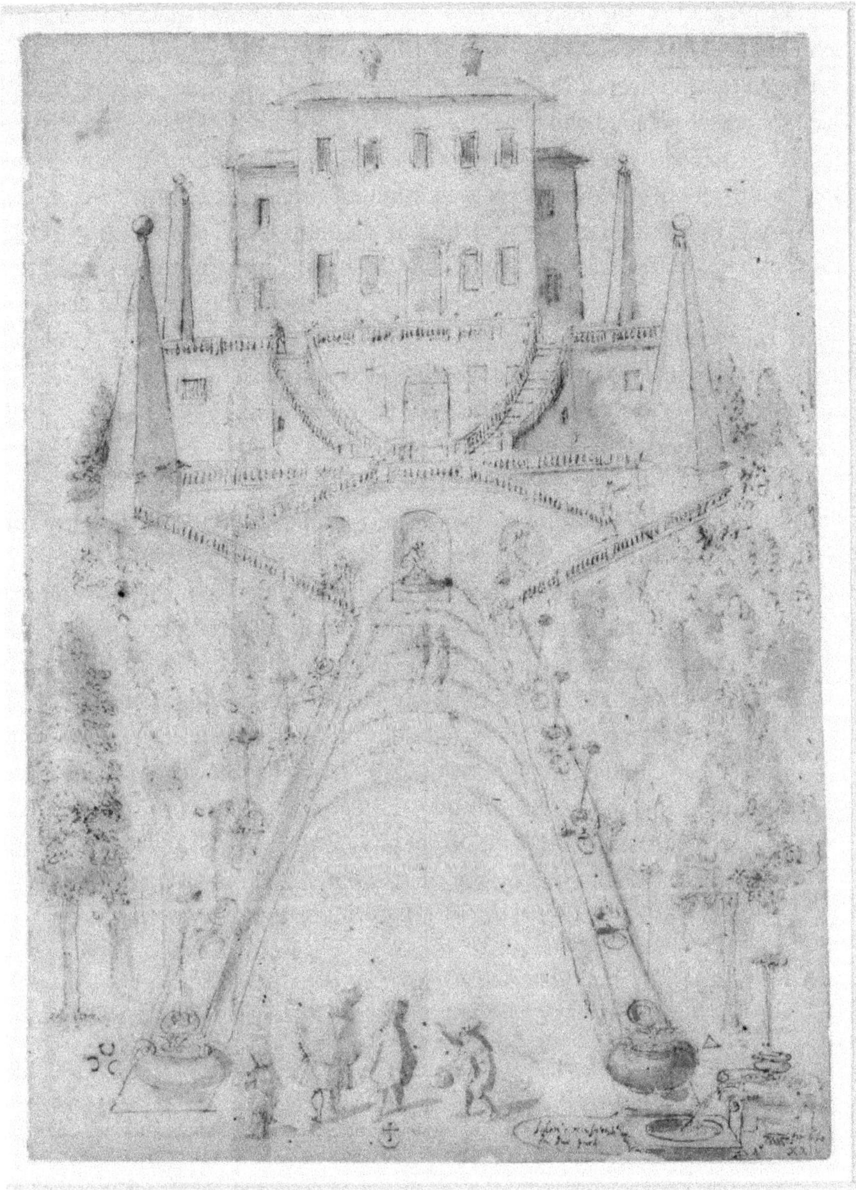

FIGURE 6.4 Giovanni Guerra, *Southern façade of the Villa Medici at Pratolino*, 1598, drawing, Albertina, Vienna, Graphische Sammlung
PHOTO: MAX PLANCK INSTITUTE FOR THE HISTORY OF SCIENCE, BERLIN

This is attested to, for instance, by the notes the German traveller Heinrich Schickhardt left regarding his visit to Pratolino on 11 January 1600.[19]

For Vieri, the effect of the sun on the water, theatricalised as it were by the oval staircase, recalled the theory exposed in Aristotle's *Meteorologica* (I, 9 and II, 2, 354b 23–33). It is also from this work that the idea of creating rainbows by projecting water droplets in sunlight originates (III, 4, 374a 35–b). Vieri knew this text especially well. In 1582, he dedicated a commentary of the *Meteorologica* to Francesco I de' Medici.[20] The staircases thus represented the course of the sun and its action on the "movement of the waters of the universe" [*movimenti dell'acque dell'Universo*], to use Vieri's own words. The oval shape of the stairs, in this sense, would derive from the circularity of such a movement, which follows the course of the sun according to Aristotle.[21] This constitutes a central metaphor for the garden as a whole.

> Is it not wondrous that the water of so many statues, set in so many places, proceeding with force in so many ways water us in such delicious surprises? [These waters] imitate the different modes of the movements of the waters of the universe. Some take their material principle here in the earth, since from [this earth], which is humid, vapours ascend by the power of the sun up to the middle region of the air, where the cold and the end of the dry and hot exhalation transform them into clouds, and from these clouds they fall down again as raindrops. Other [waters] are generated inside the summit and the caves of mountains, from which springs originate, and these springs give birth to rivers, which are waters flowing continually, and finally these waters reach the sea, as the lowest

19 Württembergische Landesbibliothek Stuttgart, [Description of Pratolino, 1600], Heinrich Schickhardt, MS Cod. hist. qt. 148b, fol. 49r–53v, quoted in Luigi Zangheri, *Pratolino. Il giardino delle meraviglie*, vol. 1 (Florence: Gonelli, 1979) 221–225.

20 Francesco de' Vieri, *Trattato nel quale si contengono i tre primi libri delle Metheore, nuovamente ristampati, et da lui coretti con l'aggiunta del quarto Libro* (Florence: Giorgio Marescotti, 1582). See Brunon, "Pratolino," 500 et seq., "Les mouvements des eaux de l'univers."

21 "So we get a circular process that follows the course of the sun. For according as the sun moves to this side or that, the moisture in this process rises or falls. We must think of it as a river flowing up and down in a circle and made up partly of air, partly of water. When the sun is near, the stream of vapour flows upwards; when it recedes, the stream of water flows down: and the order of sequence, at all events, in this process always remains the same. So if 'Oceanus' had some secret meaning in early writers, perhaps they may have meant this river that flows in a circle about the earth." Aristotle, *Meteorology* I.9, trans. E. W. Webster (Oxford: Clarendon Press, 1908; Internet Classics Archive, n.d), http://classics.mit.edu/Aristotle/meteorology.1.i.html.

> place and totality of the waters, as I have explained in vulgar tongue in my *Metheora*, with order, principle and clarity.[22]

Vieri's interpretation is astute. Indeed, the great grotto (*Grotta grande*, later designated as *Grotta del diluvio*) set underneath the monumental staircase could rapidly fill with water thus evoking the subterranean origin of water, although when it comes to interpreting it, the philosopher abandons Aristotle altogether to opt instead for a moral reading (Figure 6.5).[23] Vieri's meteorological and geological metaphor of the staircase could also be linked to the geography of the place, for a statue of the local river Mugnone by Giambologna was located in the central niche beneath the great double ramp of the lower level (Figure 6.6).

The analogy between the oval-shaped staircase and the sun as marker of days, seasons and weather is already apparent in an earlier passage of Vieri's text, which is worth quoting fully even if the sundials are not mentioned:

> coming out of the palace towards the south, one finds a magnificent terrace with balustrades and two oval staircases for ascent and descent. And underneath the staircase is a very great and magnificent grotto, as we will see below [*grotta grande*]. From these stairs come out, after the goodwill of who controls them, jets of water in great abundance which wittily water those that ascend and descend the stairs. Here we can note three very important points. The first: the stairs with the terrace to the south.

22 "Come non è cosa stupenda, che l'acqua di tante, & tante statue in tanti luoghi, & in tante maniere con furia procedenti ci bagnino con piacevolissimi inganni ? imitanto le maniere de' movimenti dell'acque dell'Universo: delle quali alcune prendono il principio materiale di quaggiù di Terra, in quanto da essa, sendo humida, per virtù del Sole se ne levano vapori, & sono condotti fino alla mezza regione dell'aria; & dalla freddezza di essa, & abandonati dallo halito secco, & caldo, si condensano in nugole [sic.], & di nugole ne vengon giù in forma di gocciole d'acqua. Altre acque si generano dentro alla sommità, & cavità de' monti; Et quivi si fanno le fonti, & da fonti hanno il lor nascimento i fiumi, che sono acque, che all'ingiù vengono con velocità scorrendo del continuo, & finalmente si conducono al mare, come a luogo bassissimo, & come al suo tutto; si come da me è stato dichiarato nella mia Metheora in questa lingua, con fondamento, facilità, & con ordine." Vieri, *Discorsi*, 61–62.

23 See Vieri, *Discorsi*, 35–36. Montaigne's description is eloquent in this regard: "À un seul mouvement toute la grotte est pleine d'eau, tous les sièges vous rejaillissent l'eau aux fesses; et fuyant de la grotte, montant contremont les escaliers du château, il sort de deux en deux degré de ce escalier, qui veut se donner ce plaisir, mille filets d'eau qui vous vont baignant jusqu'au haut du logis." Michel de Montaigne, *Journal de voyage*, ed. Fausta Garavini (Paris: Gallimard, 1983), 175–176.

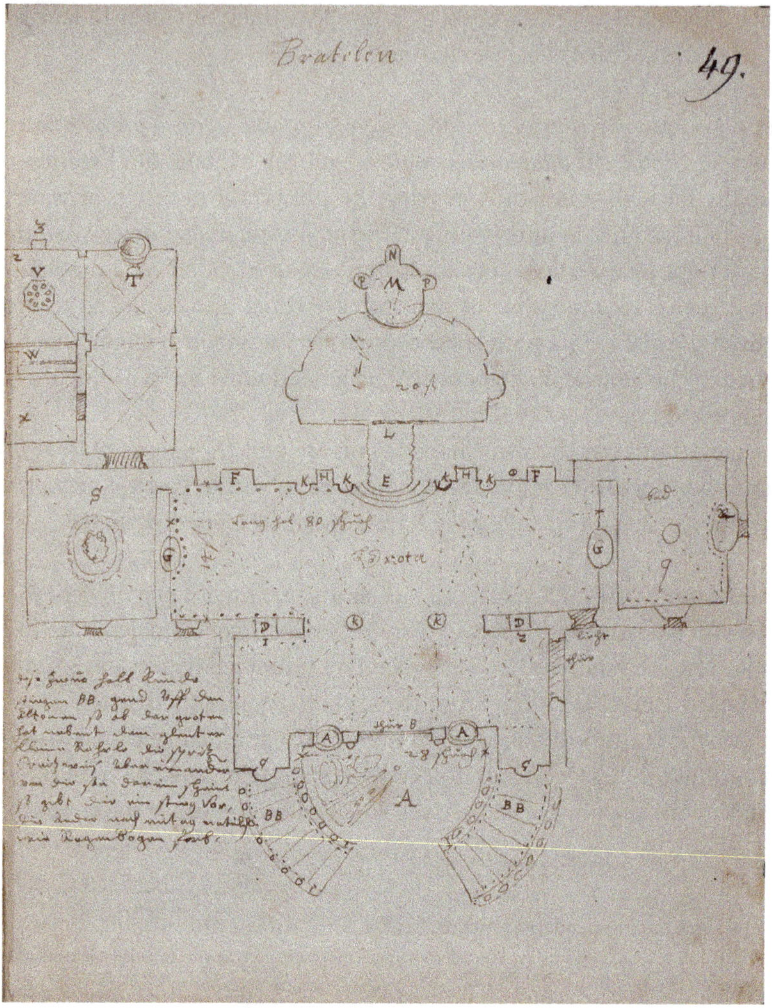

FIGURE 6.5 *'Bratelen'* (*Pratolino*) in Heinrich Schickhardt, *Travel journal* [*Tagebuch der Romreise*], 1599–1600
PHOTO: WÜRTTEMBERGISCHE LANDESBIBLIOTHEK, STUTTGART, MS COD. HIST. QT. 148,B, FOL. 49R

The second, the oval shape of those stairs. The third, that from these stairs water shoots surreptitiously. All of this can make reference to the celestial palace of the sun, which ascend and descend [in the sky], at times coming near us and then going away from us, giving birth to the various seasons of the year. And within the same day, it rises up to the meridian

FIGURE 6.6 Stefano della Bella, *Southern façade with the great stairs*, ca. 1653, etching, 25,2 x 38,5 cm, plate III from the series *View of the Pratolino villa*
PHOTO: THE METROPOLITAN MUSEUM OF ART, BEQUEST OF PHYLLIS MASSAR, 2011, INV. 1986.1180.670

> circle then sets to the west and from the virtue of the sun, waters are produced, rising in the air as vapours and falling down again as rain.[24]

This astronomical and meteorological figure of the stairs based on Aristotle is again compared with the course of the platonic soul as described by Plato in

24 "uscendo del palazzo di verso Mezzo giorno, vi è un magnifico ballatoio con balaustri, & con due scale aovate [sic.], per le quali si sale, & scende: & sotto alle dette scale v'è una grandissima, & stupenda grotta, come di sotto si dirà. Delle dette scale escono, à beneplacito di chi ne ha cura, zampilli d'acqua in gran copia, che per ischerzo immollano chi ne sale, & scende. Quisi possono considerare tre importantissimi punti. Il primo le dette scale col ballatoio verso Mezzo giorno. Il secondo è la figura ouata delle scale. Il terzo, che di dette scale eschino zampilli d'acqua à certo tempo occultamente. [...] Tutti questi si possonno intendere per il celeste palazzo del Sole, per il quale e'sale, et scende, hora accostandosi à noi, et poi discostandosi, onde nascono diuerse stagioni dell'anno; Et nel medesimo giorno si alza fino al circolo meridiano, poi va calando verso Ponente, et per virtù del Sole si producono quaggiù le acque, le quali si alzano su all'aria, ma in vapori, et poi ripiovano in giù " Vieri, *Discorsi*, 33. The staircases are also described in Bernardo Sansone Sgrilli, *Descrizione della regia villa, fontane e fabbriche di Pratolino*, Florence, Tartini et Franchi, 1742, p. 12.

the final myth of the *Phaedo* (107a–108c) on the destiny of the souls judged in the Underworld:

> One can say furthermore that our soul comes out by God's creation, which is the invisible sun and goes down into our body; and when, with this body, the soul comes out of the mother's womb, it cries, not only because it feels that the air is not well tempered, but even more because it foresees and guesses future misery: thus, at the moment of the corruption of its body, when it needs to get rid of it, [the soul] laments, fearing the future judgement. So, see, judicious and sweet minds, how the marvellous works of Pratolino can be explained through the elevated and important means of speculation and morality, besides being magnificent works of art, linked to a royal soul full of magnificence.[25]

The fact that Vieri interprets the staircase using both Plato and Aristotle is typical of the philosopher's syncretism, well-illustrated by the publication in 1576 of his *Compendio della dottrina di Platone, in quello che ella è conforme con la fede nostra* and another opuscule published in 1590, *Vere conclusioni di Platone conformi alla dottrina christina et a quella d'Aristotile*.[26] An important part of Vieri's work is the attempt to reconcile Platonism, Aristotelianism and Christianism in a search for virtue. For Vieri, this is what Pratolino was meant to teach visitors. In the *Discorsi*, he states that the marvels of Pratolino were "useful to recall good things and a virtuous conduct."[27] A visit to Pratolino's garden is thus associated with a quest for virtue, knowledge and wisdom, culminating in the villa itself, the place of the apotheosis of the soul. A similar meaning existed at the villa d'Este at Tivoli where the journey through the gardens was metaphorically presented as that of the hero Hercules choosing

25 "Puolsi ancora dire, che l'anima nostra escie per creazione da Dio, che è l'invisibile Sole, et discende in questo corpo; Et quando con esso esce del ventre della madre, piange, non solamente perche sente l'aria manco temperata: ma più oltre perche è presaga, & indovina delle future miserie: così al tempo della corruzione del suo corpo quando di qua dee far partita si addolora temendo il futuro giudizio. Per tanto vedete giuditiosi, et gentili spiriti: come le maravigliose opere di Pratolino si possono esporre con altissimi, & importantissimi sensi di specolatione, & di moralità: oltre allo essere come artifiziose opere stupendissime, et come dependenti da animo regale, et magnificentissimo." Vieri, *Discorsi*, 33.

26 On Vieri's career and thoughts, see Brunon, "Pratolino," 475–485.

27 "Dovendo dunque io ora discorrere dell'opere di Pratolino così meravigliose e così magnifiche e dilettevoli et utili al ricordarci il bene e virtuosamente adoperare." Vieri, *Discorsi*, 8.

FIGURE 6.7 Giuseppe Vasi, *Palazzo Farnese at Caprarola*, 1746–1748, engraving, 47 x 71,1 cm
PHOTO: METROPOLITAN MUSEUM OF ART, NEW YORK, INV. NO. 53.601.323

between virtue and vice and, upon reaching the palace, joining the Olympian Gods.[28]

Stairways as Figures of Time: Villa Medici at Poggio a Caiano and Villa Farnese at Caprarola

Vieri insists that an important aspect of Pratolino's architecture is the oval shape of the stairs, which was indeed innovative at the time. The anonymous description preserved at the Biblioteca Riccardiana mentions "due scale di nova architettura che fanno un ovato perfetto."[29] Amelio Fara suggested that

28 See Denis Ribouillault, "Le Salone de la Villa d'Este à Tivoli: un théâtre des jardins et du territoire," *Studiolo* 3 (2005): 65–94.

29 Biblioteca Riccardiana, MS Riccardiana 2312, fol. 108v. See also Biblioteca Apostolica Vaticana, Barb. 5341, fol. 207v, quoted in Zangheri, *Pratolino*, 1:174; Brunon, "Pratolino," 204. On the use of stairs in Medicean villas, see Sabine Frommel, "Medici Villas as an Expression of a New Way of Life and Perception," in *The Early Modern Villa*, ed. Barbara Arciszewska (Warsaw: Museum of King Jan III's Palace at Wilanów, 2017), 17–34. On the typology of staircases in Italian Renaissance Gardens, see Günter Mader and Laila

FIGURE 6.8 Giovanni Battista di Bianchi, *One of four statues representing the hours, holding a polyhedral sundial*, ca. 1572–1577, Caprarola, Gardens of the Palazzo Farnese
PHOTO: DENIS RIBOUILLAULT

the staircase was inspired by Baldassare Peruzzi's project for the Villa Medici at Poggio a Caiano and by the solution adopted by Vignola for the monumental ramps of Villa Farnese at Caprarola, a building that Buontalenti could have seen in 1557.[30] Michel de Montaigne, who visited Pratolino in 1580, also recalls that Poggio a Caiano is the "model" for Pratolino.[31] Such comparisons are not only interesting from an architectural and formal point of view, they are also highly

Neubert-Mader, *Jardins italiens*, trans. Aude Virey-Wallon (Fribourg: Office du Livre, 1987), 42–47.

30 Amelio Fara, *Bernardo Buontalenti* (Milan: Electa, 1995), 57; Marcello Fagiolo, "Le due anime nelle ville della Tuscia," in *Il giardino d'Europa. Pratolino come modello nella cultura europea*, ed. Marco Dezzi Bardeschi, Alessandro Vezzosi and Luigi Zangheri (Milan: Mazzotta, 1986), 68–81 and 74–75; and Brunon, "Pratolino," 204 and 296.

31 "la forme de ce bastiment est le modele de Pratolino." Montaigne, *Journal*, 264.

suggestive on an iconographical level since the façades of both these buildings were closely connected with the passage of time, the cycle of the seasons and the course of the soul. At Caprarola, the grand oval and rectilinear ramps should be read in conjunction with the monumental sundials and mechanical clock located on the bastions of the pentagonal villa-fortress, itself a perfect example of cosmic architecture (Figure 6.7).[32] On the garden side, pairs of caryatids representing the Hours flank the entrance of the two drawbridges leading to the summer and winter apartments respectively. Each of them carries on its head a polyhedral stone block that once bore three sundials – for a total of twelve sundials (the hour lines are no longer visible) (Figure 6.8).[33] At Poggio a Caiano, the precious terra cotta frieze on the villa's temple-like façade, dated around 1490, is also dedicated to the cycles of Time, in accordance with an interpretation of texts by Ovid and Claudian (Figure 6.9).[34] The frieze shows Eternity, the birth of the Age of Jupiter, the birth of the year, the seasons, the months and the days. Taken as a whole, it represents the allegorical travel of the virtuous soul and its choice between good and evil. Whilst the sundial on the villa's façade was added later in the seventeenth or eighteenth century, the mechanical clock that crowns the façade's pediment is documented since at least the middle of the sixteenth century (Figure 6.10).[35]

At Poggio a Caiano, the theme of the circularity of time also alludes, on a more political level, to the famous *motto* of the Medici family, *Le temps revient*, heralding Lorenzo il Magnifico as harbinger of a new Golden Age.[36] This meaning is also well suited to the gardens and villa of Pratolino, which not only represented an image of the universe, "the great machine of the Cosmos," to use Luigi Zangheri's expression, but also an image of the Good Government

32 Paolo Marconi, "Una chiave per l'interpretazione dell'urbanistica rinascimentale: la cittadella come microcosmo," in *Quaderni dell'Istituto di Storia dell'Architettura. Serie XV. Fascicoli 85–90* (Rome: Facoltà Di Architettura Università Di Roma, 1968), 53–94.

33 Ribouillault, *Gardens of the Heavens*, forthcoming. The Caprarola sundials can be dated around 1572–1577.

34 See especially Janet Cox-Rearick, "Themes of Time and Rule at Poggio a Caiano: The Portico Frieze of Lorenzo il Magnifico," *Mitteilungen des Kunsthistorischen Institutes in Florenz* 26, no. 2 (1982): 167–210; Fabrizia Landi, *Le temps revient: il fregio di Poggio a Caiano* (Florence: Landi, 1986).

35 The clock is visible on Utens's view painted around 1600. See also Ugo Pancani, "L'orologio di Villa Medicea di Poggio a Caiano," AVRVM, accessed July 25, 2020, http://www.avrvm.it/lorologio-di-villa-medicea-di-poggio-a-caiano/.

36 Cox-Rearick, "Themes of Time and Rule," 183–184. See also Janet Cox-Rearick, *Dynasty and Destiny in Medici Art: Pontormo, Leo X and the two Cosimos* (Princeton: Princeton University Press, 1984) and Paola Ventrone, ed., *Le Temps Revient, 'L Tempo si Rinuova; Feste e Spettacoli nella Firenze di Lorenzo il Magnifico*, exh. cat. (Milan: Silvana, 1992).

FIGURE 6.9 Giusto Utens, *Poggio a Caiano: Detail of the façade*, ca. 1599, tempera on canvas, 145 x 245 cm, Florence, formerly at the Museo Storico-Topografico, now Villa Petraia
PHOTO: DENIS RIBOUILLAULT

established by the Medici. In the words of Vieri, "Pratolino is the portrait of a republic well governed here on earth and then of the one that is above heaven and that is divine."[37] Furthermore, the idea of the cyclicality of time contained in the Medicean *motto* was associated very early on with two crucial elements of the great oval staircase of Pratolino. Early poems describe Andrea del Verrocchio's banner for Lorenzo il Magnifico on the occasion of the great Joust of 1469: Lorenzo's *motto Le temps revient*, written in golden letters, was associated with the sun and a rainbow.[38] It thus appears that Pratolino's façade and

37 "Et à fine, che di qui ancora apparisca, come Pratolino sia un'ritratto prima di una Republica ben governata quaggiù in terra, & poi della sopraceleste et divina." Vieri, *Discorsi*, 8.

38 See Luigi Pulci, *La Giostra fatta in Fiorenza dal Magnifico Lorenzo de' Medici il Vecchio l'anno MCCCC–LXVIII* (Florence, 1518) stanz. 64: "E' mi parea sentir Miseno, / quando in sul campo Lorenzo giugnea / sopra un caval che tremar fa il terreno; / e nel suo bel vexillo si vedea / di sopra un sole e poi l'arcobaleno, / dove a lettere d'oro si leggea: / 'Le tems revient,' che può interpretarsi / tornare il tempo e'l secol rinnovarsi." See also Cox-Rearick, "Themes of Time and Rule," 183.

FIGURE 6.10 Poggio a Caiano, *Façade*
PHOTO: NICCOLO RIGACCI, WIKIMEDIA COMMONS, CC BY 2.5

its hydraulic-gnomonic oval staircase were meant as a grandiose translation of this early family device.

Finally, some scholars have affirmed that in the court of Lorenzo de' Medici, where the work of Marsile Ficino was held in great esteem, the path of the soul depicted on the façade of Poggio a Caiano also derives from astral magic, whose ultimate goal is the union of the soul with the divine.[39] Without taking a side here on this complex debate, it should be noted that the field of natural magic has also been considered by several important scholars as a suitable field of investigation for Pratolino and the court of Francesco I de' Medici, as he was a fervent believer in magic and alchemy.[40] Vieri himself knew of his

39 Patricia Aakhus, "Astral Magic in the Renaissance. Gems, Poetry, and Patronage of Lorenzo de' Medici," *Magic, Ritual and Witchcraft* 3, no. 2 (Winter 2008): 200–202.

40 See Suzie Butters, "Natural Magic, Artificial Music, and Birds at Francesco de' Medici's Pratolino," in *Sense and the Senses in Early Modern Art and Cultural Practice*, ed. Alice E. Sanger and Siv Tove Kulbrandstad Walker (London: Routledge, 2016), 31–63; and Filson, "Magical and Mechanical Evidence," 177–206. See also the more nuanced approach of Zangheri, "Pratolino," 43–52. Brunon also remained cautious on the issue, conceding,

interests and dedicated around 1579 a “Breve discorso intorno all'arte dell'alchimia” to him, noting however that he knew very little on the subject.[41]

A Very Rare Type of Sundial: Some Reflections on its Use, Sources and Precedents

If one considers that Vieri's basic information regarding the sundial-staircase is accurate, how then did it work? Given the orientation of the villa (slightly south-west), the two obelisks of the upper terrace could not have served as styles, for the shadow would have fallen on the stairs only for a brief moment at sunrise and sunset. According to the French astronomer Denis Savoie, one of today's foremost experts on sundials and their history, one must conclude that it was the shadow of the parapets on each side of the curved ramps that served to indicate the time (Figure 6.11). Examples of such sundials have been realised recently, showing not only the hours but also the date, although the mathematics behind the theory of this complex dial type was not yet available to Buontalenti.[42] However, it is perfectly possible that the sundial was built using an empirical method. Many early modern sundials were built in this way. The procedure is quite simple: with a well-adjusted reference sundial (or a clock set to solar time), the shadow of the parapet is recorded on the stairs when it is such and such a time on the sundial; it is enough to do this operation periodically (between two solstices) and finally to join the points obtained by a curve. This is done in theory for each hour, at least in part, because one cannot have many hours (the shadow of the parapets falls quickly outside the stairs in the morning and afternoon and the time range also depends on the width of the steps and the height of the parapet); but one could have a stair-meridian without any problem. Furthermore, it is not impossible that the globe on the top of the eastern obelisk served as a gnomon to project a meridian shadow on

---

however, that “serious investigation on the side of alchemy would be required.” Brunon, “Pratolino,” 28.

41 Biblioteca Nazionale Centrale, “Breve discorso intorno all'arte dell'alchimia,” Francesco de' Vieri, MS. Magl. XVI 78 (sixteenth century).

42 Denis Savoie, Private correspondence with author, January 8, 2020. I am much indebted to Denis Savoie for taking the time to study this case for me. See especially his *Recherches sur les cadrans solaires* (Turnhout: Brepols, 2014). More specifically, for the mathematical theory for sundials using a cornice or parapet as style, see Denis Savoie, “Les cadrans solaires à corniche,” *Cadran Info* 19 (May 2009): 71–80. For an example of such a sundial built by the author, see “Le cadran solaire du barrage de Castillon,” *L'astronomie* 124 (February 2010): 12–19.

the stairs, but without precise information regarding its height and distance, only a conjectural reconstruction can be proposed here.

Pratolino's sundial staircase is thus highly original and extremely rare. Yet, a very famous occurrence concerning a similar sundial is found in the Bible itself. We must now ask ourselves if this had any bearing on Buontalenti's invention for Pratolino.

Two closely related passages in the Hebrew Bible refer to a sundial constructed under King Ahaz. It features in a story about his son and successor King Hezekiah (2 Kings 20:8–11 and Isaiah 38:7–8). Here the prophet Isaiah calls on God to reverse the shadow cast on the sundial by ten 'steps' (*ma'aloth*). The usage of that term has led to the inference, common among interpreters, that Ahaz, at the time he engaged in building work – which involved devotional objects dedicated to the sun and other deities later demolished as idolatrous (2 Kings 23:11–12) – used the steps of an extension to mark time by means of a shadow cast by a pole or other device for casting the shadow. Attempts at a

FIGURE 6.11 Giusto Utens, *Villa Medici at Pratolino, Detail of the façade and the staircases to the South*, ca. 1599, Villa Petraia, Florence, with hypothetical reconstruction of the sundial-staircase by Denis Savoie
PHOTO: HERVÉ BRUNON/DENIS SAVOIE

rationalist explanation have not been successful; the story appears to reflect a conception similar to that expressed in Joshua 10:12–13. The translation of this passage of the Bible varies from version to version, although the underlying conception seems fairly clear. The word *ma'alah* might be translated as 'step,' 'degree,' or 'interval' in this context, on the surmise that the Horologium Ahaz was shaped like a staircase whose steps enabled measurement of the passage of time.[43]

Various Egyptian models of such sundials in the form of stairs are known, dated to the sixth century B.C., but archaeologists do not know if they were already known in Judea in the eighth century B.C.[44] In his famous book on the history of sundials, René Rohr links the biblical episode to another one in 2 Kings 16, which concerns the construction of a famous Assyrian altar commissioned by King Ahaz in the temple of Jerusalem. This altar, equipped with steps, could have functioned as a kind of sundial.[45] It is interesting to note that a reference to a staircase sundial reappears in fourteenth century Egypt and Syria in the illustrated treatise of the Egyptian astronomer Najm al-Dīn al-Miṣrī, calling to mind the ancient Egyptian 'zikkurat' time-telling device or 'stepped sundial' now preserved at the Egyptian Museum in Cairo (CG33401) (Figure 6.12). According to François Charrette, "It consists of a series of stairs which are laid out in such a manner that whenever one hour passes before noon, the shadow of the gnomon jumps one step, and inversely in the afternoon. [...] The procedure may have been empirical rather than mathematical."[46]

For us, two things matter here. First, the fact that the biblical text mobilised the energy of many mathematicians and engineers. And second, the fact that some visual ideas of the sundial of Ahaz, quite similar to the configuration of the palatial complex of Pratolino, existed. An interesting example is in the splendid byzantine manuscript from the tenth century called Psalter of Paris, at

43 I thank Professor Bruce D. Chilton, Director of the Institute of Advanced Theology, and Bernard Iddings Bell Professor of Philosophy and Religion at Bard College, for summarising for me the state of the research on this much discussed topic.

44 On the first Egyptian sundials, see Ludwig Borchardt, *Die Geschichte der Zeitmessung und der Uhren*, ed. Ernst von Bassermann-Jordan, vol. 1 (Berlin: De Gruyter, 1920); R. W. Sloley, "Primitive Methods of Measuring Time: With Special Reference to Egypt," *The Journal of Egyptian Archaeology*, 17, no. 3–4 (Nov. 1931): 166–178; Anne-Claire Salmas, "La mesure du temps de la journée (II). Modules et fonctionnement des horloges à ombre tardives et des cadrans solaires," *Bulletin de l'Institut français d'archéologie orientale* 114 (2014): 419–446.

45 René R. J. Rohr, *Sundials. History, Theory and Practice*, rev. ed. (New York: Dover Publication, 1996), 7–9.

46 François Charette, *Mathematical Instrumentation in Fourteenth-Century Egypt and Syria: The Illustrated Treatise of Najm al-Dīn al-Mißrī* (Leiden: Brill, 2003), 208.

FIGURE 6.12 Drawing by René J. Rohr after the *Model of a stepped sundial*, 35 x 13 cm, date unknown, Cairo, Egyptian Museum, inv. no. CG33401 = SR3/2463
PHOTO: ROHR, *SUNDIALS*, FIG. 5

the Bibliothèque Nationale de France (MS. gr. 139, fol. 446v), where the shadow is clearly depicted falling on the stairs of the Royal palace (Figure 6.13).[47] In the sixteenth century, great astronomers such as Christopher Clavius and Galileo pondered this passage. Engineers and makers of scientific instruments like Georg Hartmann and Christoph Schissler invented ingenious refractive dials to replicate the miracle described in the Bible.[48] The Portuguese Pedro Nunes or Nonius discussed the problem from a scientific point of view in a treatise published in 1537, opening the possibility of a "natural" explanation for the phenomenon described in the Bible.[49]

So far, there is no indication that Buontalenti's invention had anything to do with the biblical Horologium, but it is certainly not impossible given the fame of this biblical prototype. Buontalenti's knowledge in the field of astronomy and gnomonics is, at least, attested to by several sources. According to Gaetano Milanesi, he owned drawings of machines and clocks, including sundials, which had belonged to the famous Florentine family of clock makers,[50]

47 Bibliothèque Nationale de France, "Psautier de Paris," MS. gr. 139, fol. 446v, available online since March 9, 2015, https://gallica.bnf.fr/ark:/12148/btv1b10515446x/f916.item.

48 See Anthony J. Turner, "A Biblical Miracle in a Renaissance Sundial," *Bulletin of the Scientific Instrument Society* 61 (1999): 11–14, and Jasmin Mersmann, "Moving Shadows, Moving Sun. Early Modern Sundials Restaging Miracles," *Nuncius* 30, no. 1 (January 2015): 96–123.

49 Henrique Leitão, "A brief note on the power of mathematics. Pedro Nunes and the retrogradation of shadows," in *Proceedings of Recreational Mathematics Colloquium V – G4G (Europe)*, ed. Jorge Nuno Silva (Lisbon: Ludus, 2017), 45–52. See also Chapter 2 in this volume: Henrique Leitão, "Mathematical Certainty and Biblical Inerrancy. Pedro Nunes and the Retrogradation of Shadows at the Dial of Ahaz."

50 Gaetano Milanesi, Editor's note to *Le vite de' più eccellenti pittori, scultori ed architettori*, by Giorgio Vasari, ed. Gaetano Milanesi, vol. 2 (Florence: Sansoni, 1878), 593.

FIGURE 6.13 *Hezekiah and the miracle of the Horologium Ahaz*, byzantine manuscript, ca. 940–960, Bibliothèque Nationale de France, Paris, MS Grec139, fol. 446v
PHOTO: BNF/ GALLICA

the Della Volpaia ("Libro di machine diverse," MS. 5363, now at the Biblioteca Marciana in Venice).[51] His knowledge, typical of an engineer of his time, served incidentally for the practice of astrological prognostication which was very

51 Biblioteca Marciana, "Libro di macchine diverse," Benvenuto della Volpaia, MS. 5363, formerly cod. It. IV.41, 96 fols. (ca. 1520). Among the folios dealing for the most part with clocks, astronomy, astrology and mechanics, there are several meridian sundials (46r and 66v) and a sundial (46v). See Carlo Pedretti, "Il Codice di Benvenuto di Lorenzo Golpaja," in *Studi Vinciani. Documenti, Analisi, e Inediti leonardeschi* (Geneva: Librairie E. Droz, 1957), 23–41.

popular at the time, especially in Medici circles. For instance, he determined the opportune moment for the foundation of the new city of Livorno on 28 March 1577 using *horiuoli et astrolabi*.[52] More importantly, he was the inventor of the famous Tribuna degli Uffizi built between 1581 and 1583, which included a great meridian sundial. It postdates only by a few years the grand staircase of Pratolino, finished in 1577 according to documents.[53] The meridian line of the Tribuna was inspired by the gnomon of the cathedral of Florence built by Paolo dal Pozzo Toscanelli in 1475 and by the famous Tower of the Winds at the Vatican. The gnomon was installed around 1590 under Francesco's successor Ferdinando de' Medici, and was coupled with an anemometer similar to the one the mathematician Egnazio Danti had installed on the Vatican tower a decade earlier. According to the description given by Francesco Bocchi in his 1591 guide of Florence, *Le bellezze della città di Firenze*, the cupola was decorated with a pattern of mother-of-pearl with the signs of the winds and the zodiac. The four elements were represented by statues of Zephyr (air), Apollo (sun), Juno (earth) and Venus and Galatea (water). The decoration as a whole expressed the mastery of the cosmic order and the good government of the Medici.[54] As in Pratolino, the programme was organised around the theory of the elements, their mutation and the cosmic influence to which they are subjected. These were popular themes in Medicean decorations. They can be found in the *Studiolo* of Francesco I de' Medici at the Palazzo Vecchio, in the Sala degli Elementi of Palazzo Firenze in Rome and in several *intermezzi* such

52 "con astrolabi e orivoli credé di esplorare nel cielo il momento più felice [...] e trovato alle ore sedici e due terzi, il prelato obbedì esattamente alla sua intimazione." Quoted in Fara, *Bernardo Buontalenti*, 18.

53 In February 1576, work was progressing on the façades' staircases. A payment from April 1577 documents the construction of the staircases and the completion of the subterranean grottoes. See Brunon, "Pratolino," 53 and 903, and Webster Smith, "Pratolino," *Journal of the Society of Architectural Historians* 20, no. 4 (December 1961): 161–162.

54 "nella Lanterna poscia sopra la Cupola si vede per certo segno il vento, che regna; perche quando soffia per l'aria, e domina in sua parte, agitando di fuori una banderuola, tosto dentro si vede certa lancetta, che senza errore, onde viene questo vento, dimostra acconciamente l'equinozio di Primavera nell'Ariete, quello di Autunno nella Libra: il Solstizio nel Granchio, quello di Capricorno à tempo assignato quando viene il sole à questi punti, passando il lume Solare per certo luogo forato, con tanta certezza si conoscono, che ancora, che altro sia poco pratico di corso di pianeti, del moto del cielo, e delle stelle nel contemplare così sovrano artifizio si fa intendere tuttavia." Francesco Bocchi, *Le bellezze della città di Firenze* [...], rev. ed., ed. M. Giovanni Cinelli (Florence: G. Gugliantini, 1677), 109–110. See also Antonio Natali, Alessandro Nova, and Massimiliano Rossi, eds., *La Tribuna del principe* (Florence: Giunti, 2014). On the Tribuna's astronomical function, see Filippo Camerota, *La linea del sole. Le grandi meridiane fiorentine* (Florence: Edizioni della Meridiana, 2007), 93–98.

as the fifth *intermezzo* of l'*Amico Fido* performed in 1586 at the theatre of the Uffizi, transformed for the occasion by Bernardo Buontalenti in a magnificent garden.[55]

There is at least one theory that explicitly links the biblical sundial of Ahaz with the Aristotelian sources of Pratolino's garden. To explain the miracle of the retrogradation of the sun, one scholar in particular formulated the hypothesis that it could have been provoked by the incidence of sunrays on clouds or water vapour, thus mobilising Aristotle's meteorological theory staged at Pratolino, according to Vieri's reading. This was Levi ben Gerson (1288–1344) also called Gersonides or Leo Hebraeus, one of the greatest French Jewish scholars and exegetes of the late Middle-ages, author of various ground-breaking works on mathematics and astronomy.[56] In his *Wars of the Lord*, completed in 1329, which contains his great astronomical treatise, he explains several biblical miracles as meteorological phenomena using two Arab commentaries on Aristotle's *Meteorologica*, including the famous one by Averroes.[57] Here is the passage on the Horologium Ahaz translated into English by Bernard Goldstein:

> You already know that clouds that move under the Sun sometimes cause the ray (*nisos*) of the Sun to be displaced, and this is clearly perceptible. Hezekiah saw that the motion of the wind at that time, and some vapours (*edim*) under the Sun which move easily, caused the shadow to return 10 steps, for their motion was in the same direction (*sad*). Therefore, he requested from Isaiah that the opposite happen in order to recognize a miracle from God, and the shadow returned backwards 10 steps as Hezekiah had asked. The miracle thus resulted from God's miraculous creation (*hiddush*) of vapours moving in that direction, and the ray of the Sun was impressed on them such that with their motion they brought it with wondrous speed to that place.[58]

55 See further Hervé Brunon, "'Les mouvements des eaux de l'Univers': Pratolino, jardin météorologique," in *Les Éléments et les métamorphoses de la nature: imaginaire et symbolique des arts dans la culture européenne du XVIe au XVIIIe siècle*, ed. Hervé Brunon, Monique Mosser, and Daniel Rabreau (Bordeaux: Blake, 2004), 35.

56 Bernard R. Goldstein, *The Astronomy of Levi ben Gerson (1288–1344)* (New York: Springer, 1985).

57 Sara Kelin-Braslavy, "Gersonides' Use of the *Meteorology* in his Accounts of some Biblical Miracles," *Historical Studies in Science and Judaism* 10, no. 2 (July 2010): 240–313. The *Wars of the Lord* (*Milhamot ha-Shem*) was published in northern Italy at Riva di Trento in 1560. Gersonides's commentaries on Aristotle appeared in compilations on Aristotle in the middle of the sixteenth century. See especially Bernard R. Goldstein, "The Physical Astronomy of Levi ben Gerson," *Perspectives on Science* 5 (1997): 1–30.

58 Bernard R. Goldstein, "Galileo's Account of Astronomical Miracles in the Bible: A Confusion of Sources," *Nuncius* 5, no. 1 (January 1990): 8.

Furthermore, Gersonides does not mention a sundial but the steps of the staircase of the royal palace of King Ahaz. The connection between Gersonides's interpretation and the Pratolino staircase is thus plausible, if entirely conjectural for the moment. Gersonides's work was known in Italy. Pico della Mirandola calls him *insignis et celeber mathematicus*. Federico Commandino had access to the manuscript of his *Astronomy* (Ms. Vat. lat. 3380), which was then in the library of Fulvio Orsini. Finally, a very interesting idea proposed by Bernard Goldstein is that Levi's *Astronomy* and especially his pioneering chapter on the *camera obscura* had a bearing on the building of the great meridian sundials in Italian churches, starting with the one devised by Paolo dal Pozzo Toscanelli for the cathedral of Florence in 1475.[59]

Obviously, further research is needed. One thing, however, is certain. Buontalenti's inventions were, as Vieri's opuscule demonstrates, remarkably suited to syncretic exegesis simultaneously philosophical, biblical and scientific. For erudite visitors and natural philosophers, who were numerous at the Medicean court, such marvellous *invenzioni* enabled the display of their *ingegno*, i.e. their cultivation and their capacity to understand complex and sometimes veiled allusions, a quality that was deemed central in Renaissance society. For those who did not master Plato, Aristotle and the science of astronomy and gnomonics, Pratolino's staircase could be more prosaically compared with the numerous staircases representing spiritual ascent, making the ducal villa akin to a divine abode: for instance, the Virgin ascending the staircase of the Temple of Solomon, *Philosophy* in Cesare Ripa's *Iconologia*, *Intellectus* ascending the ladder of Creation as in Raymond Lull or the devout soul reaching towards the Virgin Mary as in Michelangelo's *Madonna della Scala*, to give only a handful of examples.[60]

Other inventions by Buontalenti definitely possessed such hermeneutic richness. For instance, the effects of the sunrays on water is staged again in his *Grotta grande* in the Boboli gardens. In the first room, the central oculus of the vault was originally obturated with a circular fish bowl made of crystal

---

59 On Gersonides's influence on Renaissance astronomers, see Goldstein, *The Astronomy of Levi ben Gerson*, 1–17. On the Latin translations of his *Astronomy*, see José Luis Mancha, "The Latin Translation of Levi Ben Gerson's Astronomy," in *Studies in Medieval Astronomy and Optics* (Burlington: Ashgate, 2006), 1–25. Bernard Goldstein suggests that Toscanelli used Gersonides's work to build the gnomon of Santa Maria del Fiore in "Before the Sun in the Church," *Journal for the History of Astronomy* 32, no. 1 (2001), 73–77.

60 For these examples, see Linda Pellecchia, "Stepping up: Observations on the Renaissance Staircase in Florence," *Opus Incertum* 2, no. 4 (2008): 43, 46. On the Madonna della Scala, see especially Howard Hibbard, *Michelangelo*, 2nd ed. (Cambridge, MA: Harper & Row, 1985), 28.

letting light enter the dark grotto. This invention – Baldinucci praised it as Buontalenti's *nuovo concetto* – was inspired by the Pantheon in Rome and created marvellous light effects on the wet walls of the grotto. It has also been connected to a famous passage of Plato's *Phaedo* (109–110) where man is located in the hollows of the earth, "seeing the sun and the stars through the water" (109c).[61] Such marvellous apparatus – *dispositif* in the sense Michel Foucault gave to this term[62] – could simultaneously have fuelled *conversazioni* among natural philosophers, whose interest for the refraction of light rays in water is well known to historians of Renaissance science.[63] In short, one can see here again how Buontalenti's inventions were purposefully located at the intersection of art and science.

Astronomical Architecture from the *Hypnerotomachia Poliphili* of Francesco Colonna (1499) to the *Jantar Mantars* of Maharajah Jai Singh II (1724–1727)

Besides the famous Horologium Ahaz and some very old Egyptian models, there are very few gnomonic staircases one can compare to Pratolino's and they are generally of a later date. Of course, there are numerous examples of cosmic

61 See Brunon, "Pratolino," 272 and 496. Baldinucci describes the oculus as follows: "Ha questa grotta una grande apertura nel bel mezzo della volta, a simiglianza della Rotonda di Roma; di tale apertura servissi il Buontalenti per effetturare un suo nuovo concetto, e fu d'accomodare in essa alcuni cristalli, che coprendola tutta, formavano un grosso vaso, dove tenevasi acqua e pesci, i quali da chi era sotto vedeansu entro il medesimo andar vagando, senza togliere alla grotta la necessaria luce." Filippo Baldinucci, *Notizie dei professori del disegno da Cimabue in qua*, ed. Ferdinando Ranalli, Paola Barocchi, vol. 2 (Florence: Studio per Edizioni Scelte, 1974), 499.

62 "What I'm trying to pick out with this term is, firstly, a thoroughly heterogeneous ensemble consisting of discourses, institutions, architectural forms, regulatory decisions, laws, administrative measures, scientific statements, philosophical and philanthropic propositions – in short, the said as much as the unsaid. Such are the elements of the apparatus. The apparatus itself is the system of relations that can be established between these elements. Secondly, what I am trying to identify in this apparatus is precisely the nature of the connection that can exist between these heterogeneous elements." Michel Foucault, "The Confession of the Flesh," in *Power/Knowledge: Selected Interviews and Other Writings, 1972–1977*, ed. Colin Gordon (New York: Pantheon Books, 1980), 194.

63 On this topic, see Sven Dupré, "Galileo, the Telescope and the Science of Optics in the Sixteenth Century, A Case Study of Instrumental Practice in Art and Science" (PhD diss., University of Ghent, 2002), especially chap. 3, "Optics and Instrument Design in the Sixteenth Century: Breaking Rays and Shadows Running Backwards," which deals very thoroughly with refractive sundials.

architecture throughout history, which are all deeply connected with the belief in astral influences.[64] The Roman Pantheon for instance, which served as a model for Buontalenti's *grotta grande*, was a temple of the sun and a colossal sundial.[65] One of the oldest Renaissance villas organised in relation with the sun is the villa of Belriguardo near Ferrara (1435). The central axis around which the architecture and the gardens are organised corresponds with that of the rising sun at the summer solstice, which appears framed by the arched entrance of the main gate.[66] Likewise, a precise disposition was calculated for the oculus of the great reception room of villa Medici at Poggio a Caiano, which was later incorporated within the famous fresco of *Vertumnus and Pomona* by Jacopo Pontormo (ca. 1519–1521). According to Claudia Rousseau, each year, on the birthday of Lorenzo de' Medici, the longitude of the rising sun coincides precisely with the oculus. In the fresco, the oculus is associated with the Medicean *motto* GLOVIS which, read backward, means "*si volge*," meaning "it turns," in reference to the cyclicality of time, destiny and eternal return so dear to the Medici family.[67] In Vincenzo Scamozzi's treatise *Dell'idea dell'architettura universale*, published in 1615, villa Bardellini at Monfumo near Treviso (ca. 1594) is presented as true sundial-architecture. The architect used this example to establish several typologies of light effects on architecture, indicating its source and radius, the amplitude of the light cones and shadows. In the same treatise, a rare discussion of elaborate sundials shows that Scamozzi's interest for light intersects precisely with the art and science of gnomonics based on geometry and optics.[68]

64 Yaël Nazé, *L'astronomie des anciens* (Paris: Belin, 2009).

65 On cosmological architecture and the Pantheon in Rome, see Marcello Fagiolo, "Creare un'architettura cosmologica: l'invenzione del Pantheon e le infinite variazioni sul tema," in *Creatività, biologia, psicologia, struttura del processo creativo, Atti dei Convegni Lincei* (Rome: Accademia Nazionale dei Lincei, 2020), 245–267.

66 Maria Teresa Sambin de Norcen, "*Ut apud Plinium*: Giardino e paesaggio a Belriguardo nel Quattrocento," in *Delizie in villa. Il giardino rinscimentale e i suoi committenti*, ed. Francesco Ceccarelli and Gianni Venturi (Florence: Olschki, 2008), 70–73.

67 "Very simply, each year on Lorenzo's birthday, and at his relative birth time, the longitude of the rising Sun comes into precise alignment with the oculus in the south-east wall of the Salone. This fact cannot be accidental, and it dramatically accents the iconography of the Villa under Lorenzo as a "temple of time" and time's returning cycles which are in part illustrated in the portico frieze." Claudia Rousseau, "Cosimo I de' Medici and Astrology: the Symbolism of Prophecy" (PhD diss., Columbia University, 1983), 152.

68 Vincenzo Scamozzi, *Dell'idea dell'architettura universale*, vol. 2 (Venice, 1615), 136–152 (151–152 on sundials). See also Charles Davis, "Vincenzo Scamozzi architetto della luce," in *Vincenzo Scamozzi, 1548–1616*, ed. Franco Barbieri and Guido Beltramini (Venice: Marsilio, 2003), 33–45.

Turning more specifically to staircases, an interesting prototype is described inside the pyramid dedicated to the sun in the famous *Hypnerotomachia Poliphili* by Francesco Colonna, published in 1499 and first translated into English, albeit partially and imperfectly, in 1592:

> And the aforesayde entrie cut out of the firme stone, led to the scale and compassing passage in the center, with winding steps tending to the highest parte of the stately Pyramides, and opening vpon the outside of the catill or cube: vpon the which the shining obeliske was founded. And among the rest of such notable partes that I beheld, me thought that this deuise was woorth the noting, because the artifitious and most cunning architect with an exquisite and perspicuous inuention, had made to the stayres certaine loopes or small windowes, imbracing the bountifull beames of the sunne correspondently on three parts, the lower, the middle, and supreame: The lower taking light from the higher, and the higher from the catabasse or lower with their opposite reflexions shewing a maruellous faire light, they were so fitly disposed by the calculate rule of the artificious Mathematrician, to the Orientall Meridionall and Occidentall partes of the ayre, that euery houre of the day the sunne shined in, and gaue light to the whole scale, the same loopes or windolets in diuerse places symmetrially and definitely dispersed and set.[69]

Mino Gabriele considers the ascent of this truly "catoptric" staircase by Polifilo a "typical allegory of the *ascensio animae* (that is the ascent of the Neoplatonic soul towards illumination)." For him, the shape of the stairs, rectilinear first then helicoidal is a reference to the first degrees of contemplative life, according to the most probable sources used for this passage, i.e. Denys the Aeropagyte and Dante.[70] The first known catoptric sundial built inside a staircase however dates from 1632. It was devised by Athanasius Kircher in the staircase of the observatory tower at the Jesuit college in Avignon. Its mathematical principles are given in his *Primitiae gnomonicae catoptricae*, published in the same town in 1635.[71] Other examples are the sundial built by the Jesuit Father Jean Bonfa

69 I quote from Francesco Colonna, *Hypnerotomachia. Strife of Loue in a Dreame*, trans. Robert Dallington (London: Simon Waterson, 1592), fol. 8v–9r.

70 Mino Gabriele, "Voyage de l'âme et architecture symbolique dans le *Songe de Poliphile*," *Annuaire de l'École pratique des hautes études* (EPHE). *Section des sciences religieuses* 121 (2014): 309.

71 Athanasius Kircher, *Primitiae gnomonicae catoptricae* (Avignon: Ioannis Piot, 1635). Note that one of the rare studies of this early work, R. P. Marcel Chossat, *Les jésuites et leurs œuvres à Avignon: 1553–1768* (Avignon: F. Seguin, 1896), 234–237, shows a completely

in the staircase of the Lycée Stendhal in Grenoble, France, dated 1673,[72] and, a century later, the sundial in the staircase of the Château of Denainvilliers, attributed to François Bedos de Celles. This important astronomer dedicates in his 1774 treatise *La gnomonique pratique* a rare passage on the tracing of hour lines on stair steps.[73]

After Pratolino, the first project of an astronomical villa with an external staircase-sundial was the now famous project for Villa Pamphilj as a "mathematical villa," devised by the architect Francesco Borromini and the French Mathematician Emmanuel Maignan. This project, documented in several descriptions and drawings, should have been realised in 1644, displaying the most advanced technology of the time embedded within the architecture (Figure 6.14). Point 9 of the programme, which contains 21 in total, describes the staircase leading to the villa's entrance as follows:

> The stairs that ascend from the garden to the southern loggia on two sides can be laid out in such a way, according to the principles of gnomonics, that the corners of each step represents lines of the hours or the half hours, and a statue placed in both the corners casts a shadow with its head or index finger. In this way, there will be two clocks on the stairs that reveal on one side the Italic hours, and on the other the Babylonian hours.[74]

erroneous understanding of the sundial, as was kindly communicated to me by Denis Savoie.

72 J. de Rey Pailhade, A. Rome, and Aug. Favot, "Le Cadran solaire du Lycée de Jeunes Filles de Grenoble," *Bulletin de la Société de Statistique, des sciences naturelles et des arts industriels du département de l'Isère*, 4th series, no. 15 (1920): 213–310.

73 Denis Savoie, "Les cadrans solaires du Château de Denainvilliers," *Bulletin* ANCAHA 97 (Summer 2003): 79–82. François Bedos de Celles, *La gnomonique pratique ou l'art de tracer les cadrans solaires avec la plus grande précision* … (Paris, 1774), 262–263: "Si le plan étoit, par exemple, un escalier, l'on pourroit également y tracer une Méridienne par la même méthode. En supposant qu'on ne peut marquer qu'un point à l'instant de midi, & que cependant il en faudroit deux ou trois à chaque marche, l'on peut marquer tous ceux qui manqueront au moyen d'une lanterne, lorsqu'il sera bien nuit."

74 "Li scalini per li quali si salirà da due parti dal Giardino alla loggia posta à mezzogiorno si potranno disporre in modo tale, con l'arte dell'horologio à sole, che i principij di ciaschedun scalino ci mostrino una liena ò serie d'hore intiere, overo di mezz'hore, et una statua collocata in tutti due gl'angoli, faccia l'ombra col capo, ò con l'indice: et in questa maniera nelle scale si faranno due horioli da una parte all'Italiana, e dall'altra alla Babilonica." Filippo Camerota, "Architecture and Science in Baroque Rome. The Mathematical Ornaments of Villa Pamphilj," *Nuncius* 15, no. 2 (2000): 624–625.

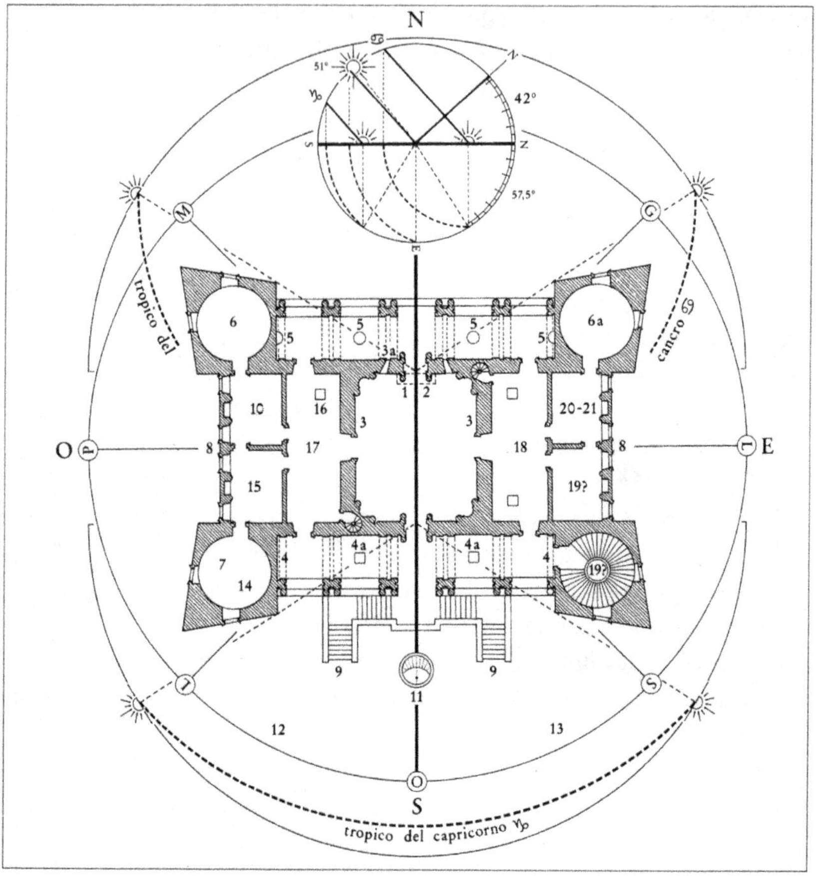

FIGURE 6.14 Filippo Camerota, *Reconstruction of the Villa Pamphilj in Rome*
1) Statue of Innocent x; 2) Mechanical dove; 3) Optical anamorphosis; 3a) Viewfinders; 4) Dioptric anamorphosis; 4a) Telescopes; 5) Catoptric anamorphosis; 6–6a) Catoptric sundial and geographical hemisphere; 7) Planetarium; 8) Windows – wind rose; 9) Sundials; 10) Catoptric and dioptric prodigies; 11) Dioptric-gnomonic sundial; 12) Topiary sundial; 13) Hedge in the shape of a fortress; 14) Galilean observatory; 15) Catoptric prodigies; 16) Talking statues; 17) Acoustical prodigies; 18) Talking statues; 19) Nero's celestial globe; 20–21) Magnetic prodigies
PHOTO: CAMEROTA, "ARCHITECTURE AND SCIENCE," FIG. 4

The dispositive is also described by Maignan in his treatise on sundials *Perspectiva horaria* (I, VII, p. 21), published in 1648.[75] It is possible that, like

75 Camerota, "Architecture and Science," 624–625: "ipso autem rostro vel capite umbram facientis posset et figura Mercurij indices vel Caduceo demonstrantis."

Maignan's invention, the staircase of Pratolino indicated different kinds of hours. In one letter connected with the project, Maignan indicates that the stairs also served as a calendar. The similarity between the villa Pamphilj project and Pratolino is interesting. It may be accounted for by the knowledge Borromini had of Buontalenti's inventions. As noted by Filippo Camerota, a manuscript note in one of his drawings mentions Francesco Bocchi's guide of Florence, which contains, as we have seen, the description of several of Buontalenti's works in Florence.[76]

Another project can be found in the treatise of mathematical architecture *Architectura civil recta y obliqua* by Juan Caramuel y Lobkowitz (1606–1682), published in 1678 (Figure 6.15). Parts of the treatise are dedicated to architectural ornament, painting, sculpture, as well as music and astronomy. Thus, one chapter deals with astronomical architecture, which Caramuel defines as "a science on its own which deals solely with the edification of palaces where one can observe the stars."[77] Its main inspiration is Tycho Brahe's famous observatory of Uraniborg, located on the island of Hven near Copenhagen and surrounded by a magnificent geometrical garden. Caramuel wants to surpass Tycho's observatory, which was carefully described in Brahe's *Epistolarum astronomicum libri*, published in 1596.[78] However, rather than containing mathematical instruments, Caramuel wants the building itself to serve as an instrument to observe and measure celestial movements. The façade he imagines is oriented east and the stairs are not used to indicate the hour lines as in Pratolino or the Villa Pamphilj project but enable astronomers to read the immense curved dials located on the façade, on each side of the great double ramp. According to Caramuel, the sundials could be used to observe the movements of the sun, the moon, the planets and the stars. The villa was also to be equipped with several catoptric sundials.

76 Camerota, "Architecture and Science," 623–624.

77 "Una Ciencia a parte, que se occupa solamente en fabricar Palacios, en que se puedan observar las Estrellas." Juan Caramuel y Lobkowitz, *Architectura civil recta y obliqua* ..., vol. 3 (Vigevano: C. Corrado, 1678), 65–66. On this astronomical villa, see Maria Elisa Navarro Morales, "The Astronomer's Palace: an Instrument for the Observation of the Sky," *Rupkatha Journal on Interdisciplinary Studies in Humanities* 3, no. 1 (2011), available online at https://rupkatha.com/V3/n1/02Astronomer%27s-Palace.pdf, accessed July 29, 2020.

78 Tycho Brahe, *Epistolarum astronomicarum*, 2nd ed. (Nuremberg: Apud Levinum Hulsium, 1601), 231–264, especially 261–264. On the gardens, see also Margherita Azzi Visentini, "Il giardino in Scandinavia tra Rinascimento e Romanticismo," in *Scandinavia. Luoghi, figure, gesti di una civiltà del paesaggio*, ed. Domenico Luciano and Luigi Latini (Treviso: Fondazione Benetton Studi Ricerche, 1998), 133–135.

FIGURE 6.15 *Project for an astronomical palace*, in Juan Caramuel y Lobkowitz, *Architectura civil recta y obliqua* ..., vol. 3 (Vigevano: C. Corrado, 1678), 65–66, plate XXXVIII
PHOTO: DENIS RIBOUILLAULT

Caramuel's invention anticipates the immense measuring instruments or *yantra* of the Jantar Mantar, i.e. astronomical gardens, built between 1724 and 1727 for the Maharaja Jai Singh II, in Delhi, Jaipur and other locations in Moghul India (Figure 6.16). According to Jai Singh II's own words, these observatories were to permit a more precise reading of the positions and movements of the planets, fixed stars, the sun and the moon, in order to build an almanac at the service of religion and the state. As for the other astronomical gardens and villas examined in this paper, these constructions had an important social and political function. This is testified to, for instance, by the frequent use of solar allegories to define the Maharaja's rule.[79]

79 Literature on the topic includes G. R. Kaye, *The Astronomical Observatories of Jai Singh*, New Imperial Series vol. 40 (Calcutta: Superintendent Government Printing, 1918); Bonnie G. MacDougall, "Jantar Mantar: Architecture, Astronomy, and Solar Kingship in Princely India," *The Cornell Journal of Architecture* 5 (1996): 16–33; Andreas Volwahsen, *Cosmic Architecture in India. The Astronomical Monuments of Maharaja Jai Singh II* (Munich: Prestel, 2001); Susan N. Johnson-Roehr, "The Spatialization of Knowledge and

FIGURE 6.16 *Misra Yantra* of the Jantar Mantar, ca. 1750–1754, New Delhi
PHOTO: AKS.9955, WIKIMEDIA COMMONS, CC BY-SA 4.0

To conclude, the staircase of the Villa Medici at Pratolino plays an important role in the understanding of this famous garden, which not only theatricalises nature in its territorial, geographical, telluric and meteorological dimensions, but situates it also within the cosmos as a whole. A walk in the garden becomes an initiation into the secrets of nature and of the stars to which everything is subjected. Gilles Clément, the French landscape and garden architect and philosopher, is one of the very rare authors who have dealt, albeit briefly, with this fundamental aspect of garden and villa culture in a small chapter of his *Brief History of the Garden* entitled "the garden of the stars." Instead of recalling again the commonplace, according to which the early modern garden is an expression of the domination of man over nature, he stresses how much the cosmic and astronomical dimension of the garden situates "man in front of the universe, where the cosmos always interacts with living systems and place humanity in a state of dependence. To pay attention to the sky is to accept to

---

Power at the Astronomical Observatories of Sawai Jai Singh II, c. 1721–1743 CE" (Ph.D. diss., University of Illinois at Urbana-Champaign, 2011) (with extensive bibliography); Barry Perlus, *Celestial Mirror. The Astronomical Observatories of Jai Singh II* (New Haven: Yale University Press, 2020). On the Misra Yantra in New Delhi, thought to have been added by Jai Singh's son, Madho Sing, between 1750 and 1754, and shown in the photograph Figure. 6.16, see Virendra Nath Sharma, "Míshra Yantra of the Delhi Observatory," *Indian Journal of History of Science* 29, no. 3 (1994): 477–487. Other useful references can be found on the following website: http://www.jantarmantar.org.

collaborate with the stars and to abandon any project of domination by man over nature."[80]

80 "l'Homme face à l'univers où le cosmos interagit toujours avec le système vivant et place l'humanité dans la dépendance. Être attentif au ciel c'est accepter de collaborer avec les astres, abandonner le projet d'une quelconque domination de l'Homme sur la nature." Gilles Clément, "Le jardin des astres," in *Une brève histoire du jardin* (Paris: J.C. Béhar, 2011), 67–72.

CHAPTER 7

Joshua Stopping the Sun in the Gesù

A Hypothesis

Evonne Levy

Abstract

This essay reconsiders the unexecuted plan dating to the 1670s to paint Joshua Stopping the Sun as a monumental fresco in the most prominent spot in the mother church of the Jesuit order, il Gesù in Rome. The Joshua scene had numerous equally important valences in Jesuit iconography, as the figure of Joshua himself was an Old Testament type for both Saint Ignatius, founder of the Society of Jesus, and for the name of Jesus after which the Society was named. While various explanations have been offered by art historians as to why this plan was ultimately discarded and replaced with the more universal theme of the Adoration of the Lamb, here I examine in detail the heretofore rejected idea that the episode from the book of Joshua was also of interest to the Jesuits because of its centrality to the Copernican debates and Galileo's trial. What makes the argument even more specific to this planned location for the fresco was the presence right below the spot of Cardinal Robert Bellarmine's tomb.

Visitors to the Roman church of Il Gesù today who approach the apse will see the damaged but still resplendent fresco, painted in the semi-dome between 1680 and 1683 by Giovanni Battista Gaulli (1629–1709), called il Baciccio, of the *Adoration of the Mystic Lamb* (Figure 7.1).[1] Painted over a century after the church's consecration, the scene from the book of Revelations (7:11–10:7) projects to the end of time, to the day of judgment, when the work of spreading the name of Jesus by religious like the Jesuits, will have been accomplished. The choice of this subject late in the seventeenth century and for the mother

1 The apse fresco and Gaulli's others in the vaults of the Gesù and dispersed documents published previously by Pio Pecchiai, *Il Gesù di Roma* (Rome: Società Grafica Romana, 1952) and Robert Enggass, *The Painting of Baciccio: Giovanni Battista Gaulli 1639–1709* (University Park, PA: Penn State University Press, 1964), have been comprehensively re-edited and supplemented in Jacopo Curzietti, *La decorazione della chiesa del SS. Nome di Gesù* (Rome: Gangemi Editore, 2011).

 | DOI:10.1163/9789004677654_009

church of the Society of Jesus has always been puzzling. A frequent subject of the mosaic decorations of Rome's Medieval churches,[2] the Apocalyptic Lamb is not an obviously Jesuit subject and by the 1670s the Jesuits were clearly devising decorations for their mother church that were explicit about their own history, spiritual practices, and theological perspectives.[3] We know from documents of two schemes contemplated for the globe-like space of the Gesù apse in the 1670s that would have better accomplished those goals. The first was the Old Testament scene *Joshua Stopping the Sun*, based on Joshua (10:12–14):

> Then spake Joshua to the LORD in the day when the LORD delivered up the Amorites before the children of Israel, and he said in the sight of Israel, Sun, stand thou still upon Gibeon; and thou, Moon, in the valley of Ajalon.
>
> And the sun stood still, and the moon stayed, until the people had avenged themselves upon their enemies. Is not this written in the book of Jasher? So the sun stood still in the midst of heaven, and hasted not to go down about a whole day.
>
> And there was no day like that before it or after it, that the LORD hearkened unto the voice of a man: for the LORD fought for Israel.

It was this miracle and battle scene that the Jesuit General Giovanni Paolo Oliva (1600–1681) planned to have painted by the Jesuit battle specialist Jacques Courtois (1621–1676), referred to in his long years in Italy as Giacomo Cortesi or il Borgognone. The artist's death in 1676, three years before the patron's support for the scene and scheme would be secured, has served as an enduring alibi for the change of programme.[4] But as Jesuit officials were well aware, *Joshua Stopping the Sun* was also the biblical passage at the vortex of the

2 See Erik Thunø, *The Apse Mosaic in Early Medieval Rome: Time, Network, and Repetition* (Cambridge: Cambridge University Press, 2015). Prominent Medieval examples include the apsidal arches of S. Prassede (ninth century) and SS. Cosmas and Damian (sixth century). The Apocalyptic lamb already appeared in Agostino Ciampelli's dome fresco in the Chapel of the Martyrs (1st on the right) of the Gesù dating to the 1590s where the sacrifices of the church's martyrs populated the heavens before the end of time. See Gauvin A. Bailey, "The Church of the Gesù in Rome and the First Phase of Decorations," in *The Holy Name: Art of the Gesù*, ed. Linda Wolk-Simon (Philadelphia: St. Joseph's University Press, 2018), fig. 10.

3 Evonne Levy, "The Honor of the altar, the fate of cults," in *The Holy Name: Art of the Gesù* (see f.n. 2), 145–165.

4 Francis Haskell, *Patrons and Painters. A Study of the Relations between Italian Art and Society in the Age of the Baroque* (London: Chatoo & Windus, 1963), 78–79; Enggass, *The Painting of Baciccio*.

FIGURE 7.1 Giovanni Battista Gaulli, called il Baciccio, *Adoration of the Mystic Lamb*, 1680–1683, fresco, Il Gesù, Rome
PHOTO: STEVEN ZUCKER

conflict between the Catholic Church and Galileo over the Copernican world system. The plans for the Joshua scene and several of the meanings the Jesuits invested in it have been considered, though the prominent, even emblematic role the passage played in Galileo's dispute with the Catholic Church over whether the new science contradicted the literal understanding of scripture has only recently been acknowledged and the connection dismissed.[5] The Gesù was not the only Jesuit church for which the Joshua scene was planned, but this location – the mother church, not far from the Collegio Romano where the Jesuit fathers grappled with how to teach the new science – absolutely commands our attention. Here the question as to whether the notoriety of the subject in the great Catholic conflict with the new science had a role in the

5 In his recent monograph on Gaulli's Gesù frescoes Jacopo Curzietti noted the connection. Most recently, Louise Rice pondered the possibility and also dismissed it: "It might be tempting to view the planned apse fresco in a similar light, especially given the presence of the memorial to Cardinal Bellarmine, Galileo's nemesis, on the left wall of the tribune directly below the semi-dome. But I find it hard to imagine what possible reason Oliva could have had for turning the apse of the mother church into a billboard for geocentrism, especially given that by the second half of the century the flaws in the theory were becoming increasingly obvious, and I agree with Curzietti that the explanation for the Joshua story lies elsewhere." Louise Rice, "Joshua and the Jesuits: The Battle for the Apse of the Gesù," in *The Holy Name: Art of the Gesù* (see f.n. 2), 243 n.3.

decision both for and against its representation in the most prominent space of the mother church of the Society is placed in the foreground. Although the debates over the Joshua scene are not preserved, the enormous controversy attached to the Old Testament passage and its continuous presence in Biblical exegesis and mathematical and astronomical tracts produced at a steady clip by leading Jesuit intellectuals, must have weighed heavily in both the proposal and ultimate rejection of the scene. The theological-scientific historical context for such a visible artistic programme cannot be ignored by us just as it could not be ignored by the Jesuits responsible for such choices, who were usually counted amongst the most respected intellectuals of the Society.

The apse fresco came to fruition more than a century after the church was consecrated and what came before offers up potentially tantalising clues as to the Jesuits' later intentions. When Cardinal Alessandro Farnese first built Il Gesù, the apse was to be clad with marbles and its semi-dome to be decorated with mosaics, presumably a material evocation of Early Christian churches that supports the later decision to paint the Lamb of the Apocalypse.[6] During Farnese's lifetime all he managed to pay for was the two-story aedicula, an elaborate tabernacle of the sacrament, and the altarpiece by Girolamo Muziano (ca. 1532–1592) of the *Circumcision of Christ* (1589) (Figure 7.2). A modest tomb for the founder of the Society, Ignatius of Loyola, placed to the left of the altar, was moved to the left transept once Ignatius was canonised (in 1622) and he earned the honour of the altar. The site of his vacant tomb was almost immediately filled with a cenotaph for the Jesuit Cardinal Robert Bellarmine (1542–1621), defender of church dogma against Galileo's heresy (Figure 7.2). Given the Jesuit aspirations for Bellarmine's canonisation, the site for his tomb, which remains on the spot, with the original portrait by young Gianlorenzo Bernini (Figure 7.3), partially rearranged in the nineteenth century, could not have been more auspicious.[7]

6 Pecchiai, *Il Gesù*, 86.

7 The original architectural wall tomb of 1623–24 with allegorical figures of Religion and Faith carved by Pietro Bernini and the portrait of Bellarmine by Gianlorenzo Bernini, was partially preserved in the nineteenth-century renovation of the apse. For a thorough study of the tomb, about which questions remain open, see now Xavier F. Solomon, "The Lost Monument to Cardinal Robert Bellarmine," in *The Holy Name: Art of the Gesù* (see f.n. 2), 125–144.

FIGURE 7.2 Andrea Sacchi, *Celebration of the centenary of the Society of Jesus at the Gesù* (Detail of the high Altar with *Circumcision* by Girolamo Muziano), ca. 1640, Galleria Nazionale d'Arte Antica, Rome
PHOTO: EVONNE LEVY

Three Plans for the Gesù Apse

To get at the layers of meaning the Jesuits attributed to the unrealised Joshua scene, we will need to place it in the overall scheme for the Gesù redecoration, the plans for which were initiated by the General Oliva around 1670. At that time the large open public spaces of the church (nave, transept chapels and dome, apse) were to be unified with a fresco programme for the vaults of all the public spaces, which was completed, and new apse and transept altars, which were not.[8] Considered seriously at this time was the renewed proposal

8 See the most detailed account of the entire scheme in Evonne Levy, "The Institutional Memory of the Roman Gesù: Plans for Renovation in the 1670s by Carlo Fontana, Pietro

FIGURE 7.3 Gian Lorenzo Bernini, *Cardinal Robert Bellarmine*, ca. 1622–1623, Il Gesù, Rome
PHOTO: EVONNE LEVY

to move St. Ignatius's relics from a chapel in the left transept back to the apse.[9] Envisioned was a newly designed crypt chapel below the high altar. Oliva's comprehensive plan for the apse would have intensified the iconography of Jesus's Name. The Old Testament figure of Joshua provided a genealogy for the Name of Jesus (Joshua is a transliteration in Greek of Jesus). And the Jesuits understood the genealogy Joshua-Jesus as supporting their evangelising mission, as explained by the Jesuit exegete Cornelius a Lapide:

da Cortona and Luca Berrettini," *Römisches Jahrbuch der Bibliotheca Hertziana* 33 (1999/2000): 406–407 (for the architectural plans) and Curzietti, *Giovan Battista Gaulli*, 27–37 and 175–218 (for the vault frescoes).

9 For these plans see Levy, "The Institutional Memory," esp. 402–409. For an ambitious open confession plan dated to the 1640s by the Jesuit architect Orazio Grassi, see Richard Bösel, *Orazio Grassi architetto e matematico gesuita: un album conservato nell'Archivio della Pontificia Università Gregoriana a Roma* (Rome: Argos, 2004).

> The name of Jesus signifies the function of savior in its greatest fulness, inasmuch as He not only saved men Himself, but gave to His apostles and to those like them, the power of saving. This is what is implied by the word Josue, or, as the Hebrews say, *Jehosua*. Let the faithful then remember that they are children of Jesus, and that they ought therefore to imitate Him in bringing about the salvation of souls.[10]

Below the proposed Joshua fresco in the semi-dome would have been Muziano's *Circumcision* on the altar, the scene that marks the giving of the Name to Jesus on his eighth day of life in accordance with Jewish law. The Circumcision was frequently represented on the high altars of Jesuit churches because it makes explicit the Jesuit investment in the Name of Jesus and their mission to diffuse Jesus's name worldwide.[11] All of those monograms of Jesus (IHS) surrounded by sunbursts that dot the surfaces of Jesuit establishments, every sunburst on a coat of arms (including the massive one on the Gesù façade), keystone, or organ loft is a reminder of the centrality of Jesus's Name to the Jesuit order. The third part of the proposed scheme was constituted by the proposal to move the tomb of Ignatius below the high altar. The tomb was to show the scene of Ignatius's Vision at La Storta, as a result of which Ignatius gave the Name of Jesus to the Society in Ignatius's foundational act. Christ appeared to Ignatius at this crossroad, en route to the papal city to seek approval to found the Society of Jesus, and said that he would be propitious to him in Rome. It has always been emphasised that following this vision, Ignatius chose for his company, on the eve of its approval by Paul III, the Name of Jesus, as several representations of the scene make clear.[12]

With Joshua above, the Circumcision on the altar and the relics of Ignatius, giver of the name, below the altar, the Jesuits thus posed a tripartite scheme – Old Order, New Order, Jesuit Order – around the Name of Jesus. The programme spatialised along a vertical axis three epochs in the enunciation, of the spreading farther and wider in the world, of Jesus's name. There is visual evidence that the Jesuits thought in these terms: an engraving put two of these scenes

10 Cornelius a Lapide, *The Great Commentary*, trans. Thomas W. Mossman (Edinburgh: John Grant, 1908), 103.

11 Christine Goettler, "Nomen Mirificum: Rubens' Beschneidung Jesu für den Hochaltar der Jesuitenkirche in Genua," in "Aspekte der Gegenreformation," ed. Victoria von Flemming, special issue, *Zeitsprünge: Forschungen zur Frühen Neuzeit* 1 (1997): 796–844.

12 The IHS monogram was not by any means always included in representations of the Vision at La Storta. Examples in which it does appear include an engraving in one of the most important hagiographical accounts engraved by Jean Baptiste Barbé, in *Vita beati P. Ignatii Loiolae, Societatis Iesu fundatoris* (Rome, 1609), plate 53.

and their outcome – the giving of the name of Jesus – together (Figure 7.4).[13] But more to the point, the preaching of, the enunciation of The Name of Jesus is at the very centre of the Jesuit purpose in the world and it is, correspondingly, the focus of the entire scheme of decoration for the Gesù of the 1670s, the centrepiece of which was the adoration of the Name of Jesus on the nave vault, painted by Gaulli.

A valuable addition to our understanding of the Jesuit investment in the figure of Joshua was made recently by Reshma Nayyar, who brought back into view the proclamation of Ignatius's canonisation in 1622 in which the new saint was cast as an alter Joshua:

> [I]t appears to us that the eulogy the Sacred Scriptures apply to Joshua may well be applied to Blessed Ignatius: 'He became, as his name implies, a great savior of God's elect, to take vengeance on the enemies that rose against them, so that he might give Israel its inheritance.'[14]

In Nayyar's view, the Joshua scene for the Gesù was motivated by the Ignatian connection and thus depended on the presence of the founder's tomb under the high altar – the plan for which was very much alive in 1672 when Gaulli signed the contract to paint all of the frescoes in the church except the apse, and at least until 1675. Nayyar also detected a remnant of the Joshua plan for the apse on the nave vault: a golden stucco relief barely visible amongst the agitated limbs of angels showing the *Fall of Jericho* and the *Transport of the Ark*, both episodes from the book of Joshua which were included in a 40 hours decoration (Quarant'Ore) erected in the Gesù in 1671, an occasion during which the church was fitted with a special, ephemeral decoration featuring a dramatic installation of the eucharist illuminated continuously for 40 hours (see below for more on the 40 hours *apparato*).[15] The inscription on the vault

---

13 Ignatius appears with the IHS emblazoned on his chest alongside St. Ignatius of Antioch (who carried the IHS monogram in his heart) below the Circumcision with a blazing IHS in glory above in an altarpiece by Juan de Roelas (1604–1606) in the Jesuit church of the Annunciation in Seville. In a *Vision at La Storta* painted by Alonso Vázquez (Sacristia mayor, Cathedral of Seville, ca. 1600) the IHS monogram which is seen transferred to him by God and the Holy Trinity as a result of the vision of Christ on the Cross appears also emblazoned on Ignatius's heart.

14 Reshma Nayyar, "*Joshua Stopping the Sun* and Ignatius of Loyola at Il Gesù in Rome," *Journal of Jesuit Studies* 3 (2016): 211–237.

15 The expanded subject of the 40 hours decoration and its connection to the Gesù vault was made by Nayyar, "*Joshua Stopping the Sun*," 226–227.

FIGURE 7.4
Jean Valdor, *Circumcision and Vision at La Storta,* early 17th century, engraving
PHOTO: COMPAGNIE DE JÉSUS – ARCHIVES JÉSUITES, VANVES

ceiling quotes Paul's letter to the Philippians, exhorting "In the Name of Jesus, every knee shall bend, in Heaven, on Earth, and under the Earth" (Philippians 2:10). This passage had been chosen at the time of Ignatius's canonisation and became part of his official mass, entered into the Roman missal, building a story around for the name chosen for the Society by its founder.[16] Nayyar is persuasive in her argument that the passage, wrapped around scenes taken from the book of Joshua, reinforced the Jesuit investment in the Name as well as Joshua as the Old Testament type for Ignatius.[17]

16 I connected that passage to Ignatius's mass many years ago – when I argued that the ceiling fresco celebrated Ignatius's role in founding the society as much as it celebrated the Society itself. Evonne Levy, "Giovanni Battista Gaulli, *Adoration of the Name of Jesus* [Rome, Galleria Spada]," in *Saint, Site and Sacred Strategy: Ignatius, Rome and Jesuit Urbanism,* ed. Thomas Lucas (Rome: Biblioteca Vaticana, 1991), 195 (cat. no. 122).

17 Nayyar, "*Joshua Stopping the Sun,*" 225–226.

The Joshua scene had yet another meaning for the Jesuits that is relevant to the Gesù programme that has not been emphasised. Ignatius himself had written in a letter of 1547 about the passage in Joshua 10 as an analogy for obedience, the virtue of virtues that stood at the centre of Jesuit life. Ignatius took Gregory the Great as a guide to his view of obedience in this passage:

> And before him the Bible, speaking of Joshua, who was a model of obedience towards Moses, his superior, [Gregory] not only says that the sun obeyed him, stopping in its course at his voice ('Sun, stop thou still at Gabaon' [Joshua 10:12]), but even that God almighty, who made the sun and all creatures, obeyed him ('The Lord hearkened to a human voice' [Joshua 10:14]). Thus great benefits for the increase in virtue accrue to those under a superior, as the One who is the source of virtues is obedient to their prayer, and also, to quote a sage "What you take away from your own will, you will add to your virtue."[18]

Ignatius invoked Joshua here in a letter in which he was working out the power that should be accorded the Jesuit superior. It is a sympathetic interpretation of the positive effects of giving up one's will to power, which is assigned and accepted in obedience; even God, the most powerful, will obey a superior to whom power is assigned. The point here is that Ignatius, and the Jesuits in turn, had understood the Joshua scene for over a century as an Old Testament manifestation of obedience. For the Jesuits, who took four vows of obedience, one more than the usual three, obedience was absolutely central to religious life.[19]

The Joshua Stopping the Sun plan fell as the tripartite programme that linked Ignatius to the Name of Jesus around the tomb crumbled. A primary impetus for the proposal to move Ignatius's relics from transept to high altar was to make the founder's reliquary chapel more prominent and thereby to encourage his lackluster popular cult.[20] Amidst arguments that the cult would be further undermined by its movement into an underground crypt, the

---

18 Ignatius of Loyola, "Letter to the Brothers in Gandia, 29 July 1547," in *Personal Writings*, ed. and trans. Joseph A. Munitiz and Philip Endean (London: Penguin, 1996), 184.

19 For a discussion of the role of obedience in compelling strict adherence to a body of thought in the Jesuit order, with specific regard to the Galileo controversies, see Richard J. Blackwell, *Galileo, Bellarmine, and the Bible: including a translation of Foscarini's Letter on the motion of the earth* (Notre Dame, IN: University of Notre Dame Press, 1991), Chapter 6.

20 See Evonne Levy, *Propaganda and the Jesuit Baroque* (Berkeley: University of California Press, 2004).

architect Carlo Fontana made a drawing for a hybrid crypt confessio, partially open.[21] But the plans did not convince: Ignatius's relics would stay in the left transept, a space that was dignified with a new chapel design by Andrea Pozzo (realised 1695–1699) that far surpassed the previous chapel in the coherence of its programme and splendor of its materials.[22] With the abandonment of the Ignatius crypt tomb, one argument for the Joshua scene was weakened. For although *Joshua Stopping the Sun* was not a name-giving scene like the Circumcision, Ignatius was an alter Joshua and Joshua's name signalled the Name of Jesus which Ignatius had been inspired through a vision to give to the Society. With the *Adoration of the Lamb* the *Ignatian* content of the church decorations became more subtle. The overall message of the mother church would have to stay focused on the larger place in history of the Society that Ignatius founded, rather than on the founder's role in the history of the Society.

The Jesuits did though propose yet a third subject for the apse, also focusing on the Name of Jesus, that signals motivations for the Joshua scene as well. In an unsigned document (presumably penned by a Jesuit) that was clearly composed in the context of the Gesù programme, it was argued that the *Adoration of the Lamb* was not specific enough to the Jesuits.[23] As an alternative the author suggested an invented scene based on an oft-illustrated passage in the Book of Revelations (3:7–13) in which Christ promises to the angel of the church of Philadelphia his "name given anew." The attraction of a passage that focuses on the giving of Christ's name is clear. For the Gesù our author thus proposes the following (invented) scene:

> Christ [...] in the act of giving his new name to the Angel of Philadelphia who [...] appear[s] alone, with his candlestick, and [...] designate[s] the first ten fathers of the Society of Jesus with [...] 10 lamps[,] and around

---

21 See the illustration in Levy, *Propaganda and the Jesuit Baroque*, fig. 20.

22 Levy, *Propaganda and the Jesuit Baroque.*

23 Archivum Romanum Societatis Iesu (ARSI), Rom. 143/II, fols. 359r–v (correspondence, 1675). The letter, which I found three decades ago and have discussed in unpublished lectures, was recently transcribed with little commentary in Curzietti, *La decorazione della chiesa del SS. Nome di Gesù*, 209. It is briefly discussed in Nayyar, "*Joshua Stopping the Sun*," 231–232. She interprets the letter as an affirmation that the Jesuits were looking for an iconography about the Society. I agree on this point but not on her larger argument that the Joshua scene was rejected as a reaction to Innocent XI's possible displeasure at such a self-aggrandising scene. I believe she incorrectly conflates Innocent's pecuniary conservatism with the Society's choice of imagery. If the propriety around the lavish decoration of the Gesù had been an issue, I see no reason why this would not apply equally to another scene in the same place by the same artist.

> them ... the 24 elders adoring Christ and praising him for such a great feat and for the succor sent to the holy church [in this order].

The author backed up his idea with an authoritative interpretation by the medieval commentator Joachim of Fiore, who understood Revelations 3:7–10 as prophesying the coming of a new religious order, an order of "felices homines" which would help to usher man into the third age. These men would live life halfway between the active and the contemplative life, engaged in earthly affairs and not wholly withdrawn in a contemplative life. As Marjorie Reeves noted, "as a piece of real prophecy, there could hardly be a better delineation of the Society of Jesus and the role it conceived for itself."[24] By the 1670s, when this programme was proposed, the Jesuits, as missionaries amongst heretics and heathen, had already made great strides in implanting the Church amongst peoples in many parts of the earth, an effort that must be made before the end of time. But our anonymous Jesuit did not convince his fellow Jesuits and so the Joachimite interpretation of the coming of the Jesuit order, to my knowledge, never took on visual form, although the text from Revelations was much discussed in printed sources and sermons. It is implicit nonetheless in the choice of the *Adoration of the Lamb*, for the latter encapsulates the eschatological goal that the Society believes is their role in history. Two of the three proposed subjects for the apse – Joshua Stopping the Sun, the Bishop of Philadelphia – show a deep investment in a Jesuit iconography around the Name of Jesus. The Adoration of the Mystic Lamb did so as well, though far less explicitly.

But to turn now to the Joshua scene and why General Oliva was fixed on it. Oliva had been circling around the Joshua scene in the years leading up to the Gesù renovation plan. The scene featured prominently in the new programme of decoration of the Jesuit church in Perugia in 1666–1667, an apparent trial run for the decoration of the Gesù. There *Joshua Stopping the Sun* was one of four Old Testament Battle scenes (Figure 7.5). This cycle, as Louise Rice has pointed out, aligns in key aspects with the plans for the Roman Gesù. With a Circumcision on the high altar and references to the Jesuit saints in the vault, Joshua was being cast as a useful Old Testament precedent for the Jesuits (Figure 7.6).[25] Might such a controversial subject have passed unnoticed in the

24 Marjorie Reeves, *Joachim de Fiore and the Prophetic Future* (New York: Harper and Row, 1977), 116–121.

25 The central and most prominent scene, it too aligns with the scene of the Circumcision. Oliva was certainly behind the scenes of the Perugia frescoes, which were executed by the painter Giovanni Andrea Carlone, who like Oliva, was Genoese. Rice, "Joshua and the Jesuits," 234–235.

FIGURE 7.5 Giovanni Andrea Carlone, *Joshua Stopping the Sun*, 1666–1667, Il Gesù, Perugia
PHOTO: SANDRO BELLU

provincial city Perugia? A visitor to Perugia, the Neapolitan Francesco d'Andrea (1625–1698), a leader in science, lamented in 1671 as Oliva forged ahead with his plans in Rome: "You realize that news of his [Galileo's] works is about as good here as one could have in the unknown land of the southern hemisphere."[26]

The Joshua scene makes its first known appearance in the Gesù in Rome as the subject of a 40 hours *apparato* in 1671, precisely at the time that the church renovation plans were under discussion (Figure 7.7).[27] The Joshua scene bended towards many meanings – in this case the perpetual light of a

26 Francesco d'Andrea, "Letter to Lucantonio Porzio, 17 October 1671," as cited in Paula Findlen, "Living in the Shadow of Galileo: Antonio Baldigiani (1647–1711), a Jesuit Scientist in Late 17th-Century Rome," in *Conflicting Duties: Science, Medicine and Religion in Rome, 1550–1750*, ed. Maria Pia Donato and Jill Kraye (London: The Warburg Institute, 2009), 211–254.

27 The 40 hours Joshua scene followed one year after the publication of Giovanni Domenico Roccamora, *Le cifre dell'Eucharistia*, vol. 2 (Rome: Giacomo Fei, 1670), which included an engraving showing the scene. The two were first connected by Mark Weil, "The Devotion of the Forty Hours and Roman Baroque Illusions," *Journal of the Warburg and Courtauld Institutes* 37 (1974): 239–240, suggesting that it was a tryout for the apse fresco. The fact that in 1675 the Adoration of the Lamb was the 40 hours decoration reinforces the connection between ephemeral decoration and permanent decorations.

FIGURE 7.6 Federico Barocci, *Circumcision*, 1575, high altar, Il Gesù, Perugia
PHOTO: SANDRO BELLU

stopped sun was made to support the continuous veneration of the Eucharist. And associations continued to accrue to Joshua Stopping the Sun. As Louise Rice has explained, the Jesuits had helped to build their argument for a Joshua scene by encouraging a thesis print around the Joshua subject for the doctorate ceremony of Stanislav Andrei Lubienski, a Polish nobleman and canon at the Cracow Cathedral. The print (Figure 7.8) was designed in 1664 by Giacomo Cortesi, one of the two artists designated to paint the Joshua scene in the Gesù. Dedicated to Jan Casimir Vasa, king of Poland – it created a Polish Joshua in the Polish king much as Charles V was made into an alter Joshua in the frontispiece of the tract of the Jesuit Nicolas Smogulecki, *Sol illustratus ac propugnatus ab illustri ac generoso domino* in 1626 (Figure 3.5).[28] Joshua's battle is represented in the foreground of Borgognone's print, but the armies seen in the background are those of Vasa which defended Europe from the incursions of the Turks. Regarding the choice of Joshua, Rice has said:

> Joshua is a type for all those who wage holy war against the enemies of the faith; more specifically, it links Joshua's victory over the Amorites to Jan Casimir's campaigns against the Ottoman Turks and their Muslim allies the Tatars of Crimea.[29]

The thesis sheet lines Casimir up with Joshua on an axis and names him (on the banderole around Fame's trumpet) the Polish Joshua. What makes this further interpretation of the Joshua scene more compelling is that the Jesuits were in hot pursuit of a substantial testament the Jesuits had been left by Casimir's father at his death in 1655, earmarked for the decoration of the Gesù. In 1664, nine years after the bequest, and with a Polish nobleman coming to Rome, a thesis print couldn't hurt in the effort to grease the patronal wheels. Like many behests the Jesuits chased, this one did not bear fruit.

With several layers of meaning in play, let us address the conventionally given reason why the Joshua battle scene was not ultimately painted in the Gesù. Jesuit documents do not name the Cortese brothers around plans for the apse, but Baciccio's contract for the church left the apse fresco out. Clearly there was a separate plan for the apse fresco. Early eighteenth-century sources attest to Oliva's long-standing desire to employ Giacomo Cortese to paint the *Joshua Stopping the Sun* in the semi-dome. Giacomo, who became a Jesuit

28 See Chapter 3 in this volume by Brent Purkaple, "Appealing to Ahaz. The Jesuits and the Sundial of Ahaz in the Era of Copernican Astronomy."

29 Guglielmo Cortese and Albert Clouwet, *Thesis Print for Stanislav Andrei Lubienski*, 1664, engraving, in Rice, "Joshua and the Jesuits," fig. 15.

FIGURE 7.7 Dominique Barrière, *Joshua Stopping the Sun,* in Giovanni Domenico Roccamora, *Le cifre dell'Eucharistia*, vol. 2 (Rome: Giacomo Fei, 1670)
PHOTO: UNIVERSITY OF NOTRE DAME, HESBURGH LIBRARIES, DEPARTMENT OF SPECIAL COLLECTIONS, NOTRE DAME

FIGURE 7.8 Guglielmo Cortese and Albert Clouwet, *Thesis print for Stanislav Andrei Lubienski*, 1664, engraving, 118 x 81 cm
PHOTO: KUNSTPALAST, DUSSELDORF, REPRO LVR-ZMB

coadjutor in midlife, was a specialist in battle scenes, many explicitly Christian battle scenes. So identified was Giacomo with this subject that, as Louise Rice pointed out, a posthumous portrait shows him with a sketch easily identifiable as the Joshua scene in question.[30] His brother Guglielmo, also a painter who had painted several versions of the Joshua story, is also now known to have been involved in the commission.[31] We only know about this unexecuted plan of Joshua for the Gesù from the biographies of the Cortese published by Filippo Baldinucci, and then by Lione Pascoli, who said that the Joshua scene, for which many preparatory studies had been made, was not executed owing to the death of Giacomo around 1676:

> He had no pupils worth recalling, though he employed many young painters on his large works, and he wanted to employ many for that [project] in the tribune of the Gesù, if his death had not prevented him from starting it. Following various concepts and sketches, he had already made the design and in it he represented the miracle of Joshua stopping the sun.[32]

Hints about the brother Guglielmo's involvement by Filippo Baldinucci's collaborator Giuseppe Pinacci led Rice to conclude, justly, that the two brothers were likely to have collaborated on the work.[33] But Guglielmo died soon after, in 1679. For decades art historians have accepted the death of Giacomo in 1676 as an explanation for the cancellation of the programme – a strange conclusion given that a highly capable artist was at work on the vaults in the rest of the church. It should also be considered that the Jesuit Ludovico Bompiani, one of the champions of the project and the Joshua scene, also died in 1679.

30 Anonymous, *Triple Portrait of Jacques, Guillaume, and Jean Francois Courtois*, red and black chalk on paper, frontispiece to first of two bound volumes of drawings by Jacques and Guillaume Courtois (Istituto Centrale per la Grafica, Rome), as illustrated in Rice, "Joshua and the Jesuits," figs. 5–6.

31 Rice, "Joshua and the Jesuits," 229–233.

32 "Non lasciò scolari degni di rammentarsi, benchè di molti giovini si servisse nell'opere grandi, come di non pochi servir si voleva in quella della tribuna del Gesù, se la morte impedito non gli avesse il cominciarla. Giacchè fatto n'aveva dopo diversi pensieri, e schizzi, il disegno, e vi si vedeva delineato il miracolo di Giosuè nell'arresto del Sole." Lione Pascoli, *Vite de' pittori, scultori ed architetti moderni*, vol. 1 (Rome: Antonio de' Rossi, 1730), 121.

33 Filippo Baldinucci, *Notizie de' professori del disegno da Cimabue in qua,* vol. 6 (Florence, 1728), 425. See Rice, "Joshua at the Gesù," 229–233.

The Rejection of the Joshua Scene: Galileo and Heliocentrism

In spite of the many positive meanings that the Jesuits accrued around the figure of Joshua and the battle during which God halted the course of the Sun, there was one further meaning of the Joshua scene to both motivate and problematise its suitability for the apse. The Jesuits knew well that by 1616 this subject had become one of the inflammatory Old Testament scenes at the epicentre of the Galilean controversy waged by the Catholic Church. The literal sense of the story is that in prolonging the day by miraculously halting the usual course of the sun, Joshua and his armies were able to vanquish the enemy of Israel. But for many seventeenth-century Catholics the importance of the passage lay in its confirmation of Ptolemy's geostatic view of a cosmos ordered by a sun that moved and an earth that stood still, as real appearances suggested. With the aid of the telescope, Galileo's astronomical observations brought new levels of mechanical proof to Copernicus's heliocentric view of a cosmos with a sun that, contrary to appearances, remained still and an earth that moved around it. As Galileo's views became known, the Joshua episode which was often cited as one of the key pieces of evidence of the geostatic and geocentric view in which the Church was deeply invested, became the lynchpin of the debate. Galileo took up the Joshua passage directly in two letters of the mid-1610s. The Duchess Christina had discussed in 1613 with Galileo's disciple and friend, Benedetto Castelli, her concerns, about the challenge that the Copernican system posed for the theological investment in the geostatic view. Galileo, after Castelli had reported his own response to her, wrote to Castelli discussing the Joshua episode. That letter, subsequently expanded in the infamous Letter to Christina of 1615, marked the real beginning of Galileo's problems: in them he took on directly the theological objections to his views, directly discussing the passage in Joshua.[34] He argued, contra Catholic commentators, that it was incorrect to understand the event as the stopping of the sun alone. The entire planetary system, including the earth, would have had to have remained motionless. In Galileo's view the Joshua passage did not contradict but supported the heliocentric and dynamic model that he proposed. Galileo attempted to save the literal sense of the Old Testament text, while explaining actual appearances.[35]

---

34 He had previously been focused on the objections in physics.

35 The literature on the Joshua episode in Galileo's writings is voluminous. For orientation to the primary sources, see Stillman Drake, ed. and trans., *Discoveries and Opinions of Galileo* (New York: Doubleday, 1957); Maurice Finocchiaro, *The Galileo Affair: A Documentary History* (Berkeley: University of California Press, 1989). This author has benefited especially from Blackwell, *Galileo, Bellarmine, and the Bible*; Pietro Redondi, *Galileo Heretic*,

Much was at stake in the interpretation of this passage. For Catholics, the truth-value of Galileo's observations and mathematically derived system, were pitted against that of Scripture and the tradition of its interpretation in the writings of the Church fathers. Galileo attempted to explain the inconsistencies in the biblical passage – why the miracle may have been so described – in such a way as to *not upset* the understanding of the cosmos at the time of the writing of the Old Testament. But even in so doing, in the interpretation of this passage most notably, but also elsewhere, Galileo implicitly and explicitly challenged the Catholic view that scripture contains literal truth in scientific as well as in dogmatic matters. Galileo sought to separate these issues, and in so doing, to save himself. Ultimately, he did not, for once the epistemological framework for scripture had been so seriously questioned, his views had to be condemned. Of course, the fact that someone, not even a theologian, with no authority whatsoever would undertake a new interpretation of Scripture was also to be condemned on principle alone.[36]

The Joshua episode stood at the centre of the epistemological challenge of astronomy, mathematics, and natural history to the Catholic faith. When in 1616 the Inquisition appointed eleven theologians to examine Galileo's views, they voted unanimously that Copernicanism was philosophically and scientifically untenable and theologically heretical. And here is where the Jesuits enter. The Jesuit cardinal Robert Bellarmine was appointed to give Galileo, in a private conversation, a relatively open warning, a precept, to stop defending his heretical views. When Bellarmine died in 1621 his role in the Galileo affair was not immediately appreciated. But his warning became a critical point of contestation in Galileo's trial of 1633, following the provocative publication by Galileo of his *Dialogue Concerning the Two Chief World Systems* in 1632. One of the three charges laid at Galileo is that he had ignored Bellarmine's precept.

---

trans. Raymond Rosenthal (Princeton: Princeton University Press, 1987). For the longer durée of the position of mathematical sciences in relation to exegesis in the Society of Jesus see Volker Remmert, "'Our Mathematicians Have Learned and Verified This': Jesuits, Biblical Exegesis and the Mathematical Sciences in the Late Sixteenth and Early Seventeenth Centuries," in *Nature and Scripture in the Abrahamic Religions: Up to 1700*, ed. Jitse M. van der Meer and Scott Mandelbrote (Leiden: Brill, 2008), 665–690. I would also like to acknowledge helpful conversations over the years with Volker Remmert, Paula Findlen, Rivkah Feldhay and, most recently, Reneé Raphael, all of whom guided me to relevant literature and conveyed their own thoughts on this question.

36 Finocchiaro points out that because of the opening of interpretation of scripture by the northern Reformers, at the Council of Trent such "unbridled spirits," "relying on his own judgment" "presuming to interpret" scripture "contrary to that sense which the holy mother Church, to whom it belongs to judge of their true sense and interpretation" were condemned, and their publication forbidden. Finocchiaro, *The Galileo Affair*, 12.

The charge of heresy in the Inquisition trial boiled down to Galileo's fraudulent ignoring of Bellarmine's precept.

1670s: The State of the Heliocentrism in the Society of Jesus

Given the various investments in the Joshua scene, we must consider that the Galilean message was part of Oliva's bold vision – and that he may have been talked out of such a prominent display of a controversial topic in spite of all of the positive meanings that the Jesuits had accrued around Joshua as a figure and this event in particular. The key question to answer regards the status of the teaching of the Copernican world system around 1670. Officially, the Copernican system had been on the list of prohibited ideas included in the 1651 *Ordinatio pro Studiis Superioribus* that governed publications and teaching in the Society. The items on that list remained intact up to the Society's suppression in 1773.[37] The Jesuits continued to find new arguments, including mathematical ones, to reinforce literalist exegesis.[38] As Marcus Hellyer put it, although there was some room around the Jesuit censors and many controversial ideas were openly discussed, though called false, open adherence to Copernicanism was "impossible."[39]

And yet, the long-held view that Jesuit science was killed off by Galileo's sentence has been largely debunked. Fine grained analyses have shown that some Jesuits acknowledged Galileo's science with admiration, and hoped for the revocation of the decree against Copernicanism; in the 1660s there were Jesuits on both sides of the divide.[40] Now debate centres on whether the teaching of the new science was intended to teach the rejection of its ideas or to leave them

37 Marcus Hellyer, "'Because the Authority of My Superiors Commands:' Censorship, Physics and the German Jesuits," *Early Science and Medicine* 1 (1996): 319–354, 353; Marcus Hellyer, "The Construction of the 'Ordinatio pro Studiis Superioribus' of 1651," *Archivum Historicum Societatis Iesu* 72 (2003): 3–43.

38 Luís Miguel Carolino, "Astronomy, Cosmology, and Jesuit Discipline, 1540–1758," in *The Oxford Handbook of the Jesuits,* ed. Ines Županov (Oxford: Oxford University Press, 2019), 670–707. For a discussion of early modern biblical exegesis as it relates more specifically to Galileo's condemnation, see Jean-Robert Armogathe, "La vérité des ecritures et la nouvelle physique," in *Le Grand Siècle et la Bible,* ed. Jean-Robert Armogathe (Paris: Beauchesne, 1989), 49–60; Ernan McMullin, "Galileo on Science and Scripture," in *Galileo: Man of Science,* ed. Ernan McMullin (New York: Basic Books, Inc., 1967), 271–347.

39 Hellyer, "Because the Authority of My Superiors Commands," 354.

40 For example, Findlen, "Living in the Shadow of Galileo." I am grateful to the author for sending me this essay in manuscript form as libraries were closed during the covid-19 pandemic.

open; the extent to which controversial elements were taught around in order to preserve important scientific discoveries; and the extent to which the rejection of Copernicanism was based on scientific proofs or theology. Heilbron argued in his study of astronomy in the Church that throughout the 1660s and 1670s the Jesuits were, in myriad ways, finding ways around the Church's official position to teach Galileo's views (although not in Galileo's name).[41] Marcus Hellyer found wide discussion of Copernicus's heliocentrism by Jesuits who were required to treat his system as a "hypothesis." Teaching banned ideas as "hypotheses" meant that Copernicus's world system might explain an observed phenomenon but had no physical reality.[42] Renée J. Raphael, who studied instruction in astronomy in courses on philosophy and mathematics at the Collegio Romano from 1600–1700 noted that there was indeed an uptick in the extent to which the Copernican system was explained after 1632. Many of the texts she examined discussed Joshua 10, cited the decrees of 1616 and 1633 against Galileo's writings, and supported arguments for the earth's immobility.[43] Ultimately, though, in both disciplines in which the new science was taught, the same tripartite line of argument – observational, physical, and theological – was being invoked to reject Copernican system. There was certainly more robust argumentation, and largely on a scientific basis, instead of flat out denial. This involved more explanation and also illustration. So the ideas were not completely suppressed; but the greater circulation of them did not amount to acceptance. On the contrary, while arguments about scripture were being

---

41 John L. Heilbron, *The Sun in the Church: Cathedrals as Solar Observatories* (Cambridge, MA: Harvard University Press, 1999).

42 Since the category of hypothesis allowed for controversial ideas to be discussed, some view Jesuit science as having let Copernicanism in the back door. Marcus Hellyer ("Because the Authority of My Superiors Commands") has assessed how much 'play' there was in the system, looking in particular at the Upper Rhine Province. Renée Raphael has examined the teaching of the world systems in natural philosophy and mathematics between the 1630s and the 1660s. In addition to finding that the two world systems had migrated into natural philosophy. Renée J. Raphael, "Copernicanism in the Classroom: Jesuit Natural Philosophy and Mathematics after 1633," *Journal for the History of Astronomy* 46 (2015): 419–440.

43 For what follows see Raphael, "Copernicanism in the Classroom." I am grateful to the author for reading this essay in draft and for sending her notes on texts (published and in MS) taught at the Collegio Romano that mention Joshua 10: Gabriel Beati, *Natura in arctum coacta* (Rome, 1650), Gabriel Beati, *Sphaera Triplex* (Rome, 1662); Sylvester Mauro, *Quaestiones Philosophicae* (Rome, 1658); Domenico Marini, "Ad Primum Librum Physicorum Disputatio 11a de Principiis Rerum Naturalium Incomuni" (MS, 1660s); Octavio Cattaneo, *Cursus Philosophicus* (Rome, 1677); Archivio della Pontificia Università Gregoriana (APUG), "Physicae," Agostino de Mari, MS FC 55 (1682) (based on two student notebooks from the same course).

nuanced – when to read literally or figuratively – traditional theological views still held, including the traditional view of Joshua 10.

The work of the Jesuit Giovanni Battista Riccioli, a trained theologian and mathematician, provides one important barometer of the Jesuit appetite for Copernicanism in the middle decades of the seventeenth century and up to the decoration of the Gesù. In Riccioli's encyclopaedic publication about the new science, the *Almagestum novum* of 1651, 126 pro- and anti-Copernican arguments were evaluated.[44] But during the publication process, a shift in attitude has been detected: from a flexible and relatively open attitude to conflicting views, towards weighing the literal reading of scripture as the ultimate arbiter when evidence is ambiguous.[45] In the *Almagestum novum*, as in many other texts taught at the Collegio Romano in the seventeenth century, the Joshua episode is amply discussed. And in Riccioli's case even Bellarmine's teachings are invoked as an authority in evaluating the theological arguments around Copernicanism.[46] As Paula Findlen has pointed out, Riccioli was the first in the aftermath of Galileo's trial to conspicuously attach to an evaluation of his science the key documents connected to the Church's condemnation of him in 1633. In the *Almagestum Novum* he cast Galileo's error somewhat mildly: in Findlen's words, as "an unfortunate transgression of scriptural interpretation." Riccioli's position hardened in 1669 – just two years before the Farnese were appealed to for funds to paint Joshua in the apse – now saying that the Church's censure of Galileo was absolute because he contravened scripture.[47] In this kind of environment the Joshua scene would have affirmed the Church's position.

In contrast to Riccioli, Antonio Baldigiani acted as a mediating figure from the 1670s into the next century. As Findlen has shown in an illuminating account of his extensive behind-the-scenes efforts in the 1670s to acknowledge Galileo's science, Baldigiani corresponded privately with leading figures of the new science and those tending to Galileo's legacy, subtly paving the way for a

44 Christopher M. Graney, *Setting Aside All Authority: Giovanni Battista Riccioli and the Science Against Copernicus in the Age of Galileo* (South Bend: University of Notre Dame, 2015), 72ff. Giovanni Battista Riccioli, *Almgaestum Novum*, 2 vols. (Bologna: Ex typographia Haeredis Victorii Benatii, 1651).

45 Alfredo Dinis, "Was Riccioli a secret Copernican?," in *Giambattista Riccioli e il merito scientifico dei gesuiti nell'etá barocca*, ed. Maria Teresa Borgato (Florence: Leo Olschki, 2002), 59–62.

46 Graney, *Setting Aside All Authority*, 73.

47 Paula Findlen, "Rethinking 1633: Writing about Galileo after the Trial," in *Nature Engaged: Science in Practice from the Renaissance to the Present*, ed. Mario Biagioli and Jessica Riskin (New York: Palgrave, 2012), 213–214.

Jesuit opening to the new science in the next century.[48] Given Baldigiani's role as the Athanasius Kircher whisperer – coaxing a biography of Galileo into the *Etruscan Voyage* Kircher was preparing – he was the figure on whom Galileo's Florentine followers pinned their hopes for a Jesuit do-over in the Galileo matter. It would not have been surprising if Baldigiani himself had a voice (likely against the scheme) in the back-corridor debates about the apse fresco.

Even Jesuits who wrote and taught against Copernicanism did so increasingly using diagrams, as Renée J. Raphael has shown. One example of newly visualised discussion of the world systems at the Roman College brings us back to the Gesù. A disputation on the world system given in 1646 by professor of philosophy Ludovico Bompiani contains many astronomical diagrams, including one of the Copernican system.[49] Bompiani was an instructor at the Collegio Romano, and in 1671, as Vice-Preposito, he ran the negotiations with the Farnese, pushing the plan to move the Ignatian tomb and decorate the apse with the Joshua episode.[50] Insofar as it has been difficult to tie this plan directly to Oliva, Bompiani provides at least one direct link between the Joshua fresco project and the teaching of the geocentric World System. As we consider why Joshua Stopping the Sun could have been proposed and why it was ultimately rejected, we can say with certainty regarding the astronomical implications of the Joshua scene in 1671 that the Jesuits continued to be unwavering in arguing against Copernicanism and one of the Jesuits engaged in those disputes was directly involved in promoting the Joshua project.

Bellarmine's Tomb below the Joshua Scene

The second consideration of the context for the Joshua battle scene concerns Bellarmine's highly visible role in upholding the traditional, and literal theological reading of the Joshua passage.[51] Though his advice to treat heliocentrism

48 Findlen, "Living in the Shadow of Galileo."

49 Archivio della Pontificia Università Gregoriana (APUG), "Disputationes physicae auctore R.P. Ludovico Bompiano Anconitano Societatis Jesu excipiebat Ferdinandus Maldonatus Melitensis eiusdem Societatis," Ludovico Bompiani, MS FC 1347 (1646).

50 As Rector of the Collegio Romano for a period in the 1650s, Provincial Bompiani had been behind the designation of the Cappella della Prima Primaria in 1658 for the upper level students of philosophy at the Collegio Romano and the decorations in the 1650s by the Cortese brothers, which included scenes of battles won with the intercession of the Virgin. *Regole della congregazione Prima-Primaria nel Collegio Romano* (Rome: I. Lazzarini, 1811), 11.

51 Irving A. Kelter, "The Refusal to Accommodate. Jesuit Exegetes and the Copernican System," *Sixteenth Century Journal* 26, no. 2 (Summer 1995): 279.

as a hypothesis was considered flexible, was the representation of Joshua Stopping the Sun above his tomb significant? Would the juxtaposition have constituted praise for his role in vindicating the Catholic interpretation of that scriptural passage? The timing for a meaningful connection between Joshua and Bellarmine is uncannily right: 1671 marked fifty years since Bellarmine's death, the requisite amount of time that had to lapse before canonisation proceedings could be officially opened, though it has been said that Alexander VII had, exceptionally, lifted this restriction in 1651 to open the path to sanctity for Bellarmine.[52] But in the moment Oliva began the campaign to decorate the apse, the anniversary of Bellarmine's death must have brought the canonisation once again to his mind. The rhythm of work in gathering material for Bellarmine's canonisation in the archive of the Jesuit Postulato Generale provides some helpful context. There are some letters soliciting material for the canonisation process from right after Bellarmine's death, 1621–27, then in 1649, then starting up in 1671 and in earnest in 1675 – the same time we know Ludovico Bompiani was actively encouraging the Farnese to decorate the apse.[53] Meanwhile, in 1674, the Congregation of Rites cleared the orthodoxy of Bellarmine's writings while Daniele Bartoli, indefatigable historian of the Society, was writing a new version of Bellarmine's *vita*, which would be published in 1678 with an engraved portrait based on a newly commissioned drawing by none other than Gaulli.[54] In the *vita* Bartoli described how, in his last moments, Bellarmine said the name of Jesus thirty times, expiring with his name on his lips.[55] But Bartoli's *vita* was too little too late, as already in 1677 the letters arranging the gathering of materials ceased: in July the Sacred

52 David S. Schaff, "Cardinal Bellarmine – Now Saint and Doctor of the Church," *Church History* 2, no. 1 (1933): 46. I have been unable to confirm this date as Alexander VII was not pope in 1651.

53 Pietro Pasquale writes a few times from Capua that he is gathering material. Some letters are written directly to Oliva. Many are written to Pirro Gherardi through 1677.

54 Daniello Bartoli, *Della vita di Roberto Cardinal Bellarmino Arcivescovo di Capua della Compagnia di Giesù* (Rome: Nicolò Angelo Tinassi, 1678). Three drawings for the frontispiece portrait of Bellarmine (one with a backdrop of books and two with the dramatic drapery that was ultimately used for the engraving) have been identified. Hugh Hudson, "Giovanni Battista Gaulli: Remaking the Image of a Cardinal Saint in Seventeenth-Century Rome," *Master Drawings* 47 (2009): 70–78.

55 "Poi in voce sommessa che appena si sentiva, replicò, credo che trenta volte *Iesu, Iesu*; e placidissimamente rese quella benedetta anima al Creatore." Bartoli, *Della vita di Roberto Cardinal Bellarmino*, 255.

Congregation had voted against the canonisation based on insufficient evidence of the heroic virtue required for sanctity.[56]

Questions remain open around the connection between the Bellarmine tomb and the Joshua fresco plan. The cardinal's role in the Galileo affair is conspicuously absent from the hagiographic *vite* published in 1624 and in 1678.[57] Was the significance of Bellarmine's role in the Galileo affair clear in the 1670s? Was it considered a positive sign of Bellarmine's prudence, a virtue the Jesuits were pushing heavily for his canonisation? Or had the hagiographers discovered the Louvain lectures that have recently shown Bellarmine more sympathetic to Galileo's side?[58] Given the enduring official view of the falsity of the Copernican system, given that the geostatic view was asserted as a matter of faith after Galileo's trial, it seems unlikely that Bellarmine's involvement could have been construed as a stain on him.[59] We might also consider Stefania Tutino's interpretation of Bellarmine's position on the Galileo affair as one that was linked to his controversial argument for *potestas indirecta* (indirect papal power) – casting the papacy as reigning over an empire of souls. In her view, reading Bellarmine's theo-political thought in his position on Galileo, the emphasis shifts to Bellarmine's eye to maintaining intellectual control:

> In other words, Bellarmine was interested not so much in personally assessing Copernicanism's philosophical and scientific value, even though, we should recall, he was inclined to believe that Copernicanism had no scientific value for the reasons already discussed by Baldini. Rather, Bellarmine was interested in making sure that the Catholic Church remained the sole authority in charge of making that assessment and that such assessment was made in the most "correct" manner possible, that is to say, in a manner that would allow the Church to remain both

56 Repeated attempts to beatify and then canonise Bellarmine only succeeded in 1930. Schaff, "Cardinal Bellarmine," 46. Key to the first rejection of the cause in 1677 was the 1675 publication of Bellarmine's autobiography, penned in 1613 (before the Galileo warning). Gerard F. Giblin, "The Autobiography of St. Robert Bellarmine, S.J.," *Woodstock Letters* 89 (1960): 3. The semi-official biography of Bellarmine passes over this moment lightly. James Brodrick, *The life and work of Blessed Robert Francis cardinal Bellarmine, S. J., 1542–1621*, vol. 2 (London: Burns, Oates and Washbourne, 1928), 464.

57 Giacomo Fuligatti, *Della vita del Cardinale Roberto Bellarmino della Compagnia di Giesù* (Rome, 1624); Bartoli, *Della vita di Roberto Cardinal Bellarmino*.

58 William A. Wallace, "Galileo's Jesuit Connections and Their Influence on his Science," in *Jesuit Science and the Republic of Letters*, ed. Mordechai Feingold (Cambridge: MIT Press, 2003), 110. Robert Bellarmine, *The Louvain Lectures (Lectiones Lovanienses) of Bellarmine and the Autograph Copy of his 1616 Letter to Galileo* (Vatican City: Specola Vaticana, 1984).

59 Melchior Inchofer, *Tractatus Syllepticus* (Rome: Ludouicus Grignamus, 1633), 41.

> theologically unquestionable and philosophically up-to-date. This attitude does not make Bellarmine a defender of the principle of blind obedience to the Catholic Church at all costs. Rather, Bellarmine understood that if the Catholic Church wanted to maintain its intellectual hegemony, a degree of intellectual elasticity had to be employed, just as in the case of the pope's temporal authority. Thus, between 1616 and 1620, Bellarmine devised a strategy to make sure that while the Copernican theory was being discussed and, as he thought, proved wrong, the Catholic Church remained on top of the game, as it were.[60]

This observation helps us to pivot away from Bellarmine's particular view of the theological conflict towards his role in supporting the Church's prerogative in determining the ultimate meaning of the science to a Catholic society. Seen this way, Bellarmine's tomb acts as a visual footnote to the Joshua scene, a silent commentary that it would be the Church to decide how to interpret the scene.

With all of the accrued meanings of Joshua for the Jesuits, Bellarmine's canonisation process alone could not have motivated the fresco. But it certainly would have added another layer of meaning to the Joshua scene for the Jesuits, showing the Society's prominent place in upholding the faith.

Conclusion

Questions also remain open about Oliva's position on Copernicanism: was he a hard-liner on the challenge posed by Galileo's science to the interpretation of scripture – it seems that this is the only explanation for why he was intent on the Joshua scene. Bompiani's mediation lends support to this conclusion. On the other hand, for those like Baldigiani who acknowledged Galileo's science, the representation of the Joshua scene may have been a source of embarrassment, pointing to a Church whose traditions were out of step with its own ability to understand the problem Galileo had made such strides in resolving. Hindsight, though, is 20/20. There is ample evidence that the Jesuits were not wobbly in the teaching of Copernicanism in the context of a literalist reading of the Bible.

60 Stefania Tutino, *Empire of souls: Robert Bellarmine and the Christian commonwealth* (Oxford: Oxford University Press, 2010), 276.

Joshua 10:12–14 held multiple and highly positive meanings in the hagiography of St. Ignatius and for the Jesuit order. Burning hot amongst these in the seventeenth century was the presence of this precise episode at the centre of the tornado of the new science in the Catholic Church. Yes, this victory for the Israelites was won under Joshua's command, but the drama of the scene was the astronomical one. Arguably, this was the most important association of all, for the point of the battle was that the sun had been miraculously stopped, allowing for the victory of the Jewish people over the Amorites. The question of how to interpret the Bible, and how to interpret natural appearances was begged by the story. Given that any seventeenth-century student was most likely to encounter the Joshua passage in his astronomy classes one has to wonder whether all these Joshua images – including the Joshua Stopping the Sun that was commissioned in 1675 for the Chapel of the Presentation in St. Peters – didn't somehow tip the scales towards the question of physical reality of the geocentric world system? Could the images themselves – and here the proposed Gesù fresco – implicitly, as images, assert that one must be able to explain appearances? Compounding this layer of meaning was the presence of the tomb of the Jesuit Cardinal Bellarmine who had played a prominent role in the placement on the index of the geostatic view of Copernicus and later Galileo. By the 1670s, was this a fortuitous or a tragic coincidence? Bellarmine's hagiographers pass over it altogether. In my view, the Jesuits could not possibly have planned the scene in this very prominent site without specific and deep knowledge that this scene gave visible form to the Jesuits' controversial position on these debates, their careful parsing of the science in the course in philosophy and mathematics and their ongoing promotion of the Church's position in published texts. Given the firm teaching of the geostatic world system, it is highly unlikely that the Joshua scene could have been construed as anything other than an expression of the Jesuits' full support for the literal truth of the Bible. Whether presenting such a billboard for tradition was uncomfortable at this time, and is the reason the subject was ultimately rejected, are questions that remain to be definitively answered. It is a hypothesis, though, that cannot be ignored.

Bibliography

Pre-1800 Publications

Aguilonius, Franciscus. *Opticorum libri sex*. Antwerp: Viduam et Filios Io. Moreti, 1613.

Amman, Jost. *Bibliorum utriusque Testamenti Icones*. Frankfurt: Feyerabend, 1571.

Amman, Jost. *Eygentliche Beschreibung aller Stände*. Frankfurt, 1568.

Apian, Peter, and Gemma Frisius. *Cosmographia*. Antwerp: Gregorio Bontio, 1550.

Apian, Peter. *Horoscopion Apiani generale* [...]. Ingolstadt: Peter Apian, 1533.

Apian, Peter. *Instrument Buch*. Ingolstadt: Peter Apian, 1533.

Apian, Peter. *Introductio Geographica Petri Apiani* [...]. Ingolstadt: Peter Apian, 1533.

Apian, Peter. *Quadrans Apiani Astronomicus* [...]. Ingolstadt: Peter Apian, 1532.

Baldinucci, Filippo. *Notizie de' professori del disegno da Cimabue in qua*. Vol. 6. Florence, 1681–1728.

Bartoli, Daniello. *Della vita di Roberto Cardinal Bellarmino Arcivescovo di Capua della Compagnia di Giesù*. Rome: Nicolò Angelo Tinassi, 1678.

Beati, Gabriel. *Natura in arctum coacta*. Rome, 1650.

Beati, Gabriel. *Sphaera Triplex*. Rome, 1662.

Beham, Sebald. *Biblische Historien*. Frankfurt: Christian Egenollf, 1533.

Bellarmine, Robert. *De Ascensione mentis in Deum*. Rome: Giacomo Mascardi, 1615.

Bellarmine, Robert. *Disputationes Roberti Bellarmini Politiani, Societatis Iesu, de controversiis christianae fidei, adversus huius temporis Haereticos*. Vol. 3. Ingolstadt: David Sartorius, 1587–1593.

Bettini, Mario. *Apiaria universae philosophiae mathematicae*. Vol. 2. Bologna: Io. Baptistae Ferronij, 1642.

Biblia Iconibus artificiosissimis, quo Lectoris memoriae consulatur, tanquam emblematis quibusdam exornata. Antwerp: Johan Steels, 1541–1542.

Bobynet, Pierre. *L'Horographie ingénieuse*. Paris: François Langlois, 1647.

Bobynet, Pierre. *La cadran des cadrans*. Paris: Mathurin Henault, 1649.

Bocchi, Francesco. *Le bellezze della città di Firenze* [...]. Revised edition. Edited by M. Giovanni Cinelli. Florence: G. Gugliantini, 1677.

Borrani, Pietro. *Dissertatione sopra il retrocedimento dell'ombra solare sull'orologio di Acaz* [...]. Pistoia: Stampatore del Pubblico, 1773.

Brahe, Tycho. *Epistolarum astronomicarum*. 2nd ed. Nuremberg: Apud Levinum Hulsium, 1601.

Brentel, Georg. *Kurtzer Bericht Und Gebrauch deß beygelegten in Kupffer gezeichneten Sonnenuhr-Stöckleins*. Lauingen: Jacob Senfft, 1619.

Busch, Gabriel Christoph Benjamin. *Versuch eines Handbuchs der Erfindungen*. Vol. 6. Eisenach: Wittekinde, 1795.

Calmet, Antoine Augustin. *Commentaire littéral sur tous les livres de l'Ancien et du Nouveau Testament*. Vol. 23. Paris: Pierre Emery, 1707–1716.

Calovius, Abraham. *Biblia Testam. Veteris et Novi Illustrata*, 2nd ed. Dresden: Johann Christoph Zimmermann, 1719.

Calovius, Abraham. *Biblia Testam. Veteris Illustrata*. Vol. 4. Frankfurt am Main: Balthasar Christoph Wust, 1672–1676.

Cattaneo, Octavio. *Cursus Philosophicus*. Rome, 1677.

Celles, François Bedos de. *La gnomonique pratique ou l'art de tracer les cadrans solaires avec la plus grande precision* [...]. Paris, 1774.

Cesariano, Cesare. *Di Lucio Vitruvio Pollione de Architectura libri dece* [...]. Como: Gottardo da Ponte, 1521.

Chais, Charles. "Dissertation sur la retrogradation miraculeuse de l'ombre solaire au Cadran d'Achaz [...]." In *La Sainte Bible, avec un Commentaire Littéral, et des Notes choisies, tirée de divers Auteurs Anglois*. Vol. 6. Amsterdam: Marc Michel Rey, 1777.

Ciermans, Jan. *Disciplinae mathematicae traditae anno institutae*. Leuven: Everard de Witte, 1640.

Clavius, Christopher. *Fabrica et Usus Instrumenti ad Horologiorum Descriptionem Peropportuni* [...]. Rome: Bartolomeo Grassi, 1586.

Clavius, Christopher. *Gnominices libri octo*. Rome: Francesco Zanetti, 1581.

Clavius, Christopher. *In sphaeram Ioannis de Sacro Bosco Commentarius*. Rome: Victor Helianus, 1570.

Clavius, Christopher. *Opera Mathematica*. Vol. 5. Mainz: Johann Volmar, 1612.

Colonna, Francesco. *Hypnerotomachia. Strife of Loue in a Dreame*, translated by Robert Dallington. London: Simon Waterson, 1592.

Cornaeus, Melchior. *Curriculum philosophiae peripateticae*, Vol. 2. Würzburg: Elias Michael Zinck, 1657.

Cuilly, Jacques de, François Feuardent, and Jean de Dadré, eds. *Biblia Sacra, cum glossa ordinaria* [...] *et postilla Nicolai Lyrani*. Vol. 6. Paris: Jean-Baptiste Regnauld, 1590.

Cuilly, Jacques de, François Feuardent, and Jean de Dadré, eds. *Bibliorum Sacrorum cum Glossa ordinaria iam ante quidem a Strabo Fulgensi collecta: nunc autem novis, cum Graecorum, tum Latinorum Patrum expositionibus locupletata*. Vol. 6. Venice, 1603.

Dechales, Claude. *Cursus seu Mundus Mathematicus*. Vol. 3. Lyon: Anissoniana, 1674.

Dechales, Claude. *Cursus seu Mundus Mathematicus*. Vol. 4. Lyon: Anissonios, Joan. Posuel & Claud. Rigaud, 1690.

Den Bibel. Inhoudende het oude ende nieuwe Testament. Antwerp: Hans de Laet, 1556.

Dietrich, Veit. *Der gantz Prophet Esaias. Außgelegt durch Vitum Dietrich*. Nuremberg: vom Berg and Neuber, 1548.

Froschauer, Christoph, ed. *Die gantze Bibel der ursprünglichen Ebraischen und Griechischen waarheyt nach / auffs aller treüwlichest verteütschet*. Zurich: Christoph Froschauer, 1531.

Fuligatti, Giacomo. *Della vita del Cardinale Roberto Bellarmino della Compagnia di Giesù.* Rome, 1624.
Fuller, Samuel. *A Mathematical Miscellany in Four Parts. By a Lover of Mathematics.* Dublin: S. Fuller, 1730.
Glaser, Peter. *Sechzig kurtze Leichpredigten Von dem grossen Landtsterben zur zeit des Königs Davids* [...] *Desgleichen von der Pestilentz des Königs Hiskiae* [...]. Dresden: Bergen, 1582.
Gordon, Patrick. *Geography Anatomiz'd: or, the Geographical Grammar. Being a Short and Exact Analysis of the Whole Body of Modern Geography* [...]. London: J. and B. Sprint, 1719.
Gregory, John. *Gregorii Opuscula, or, Notes and Observations upon Passages of Scripture* [...]. 2nd ed. London: Printed for R. Royston, 1650.
Guéroult, Guillaume. *Figures de la Bible, illustrees de huictains francoys pour l'interpretation et intelligence d'icelles.* Lyon: Guillaume Rouillé, 1564.
Heissius, Joseph. *Sciatericum Achas et in eo miraculum.* Jena: Literis Krebsianis, 1689.
Henten, Johannes, ed. *Biblia, ad vetvstissima exemplaria nunc recens castigata.* Venice: Bevilacqua, 1574.
Hermann, Christian. *De Maʿalôt Aḥaz Seu de horologio Achaz* [...]. Wittenberg: Martin Schultz, 1710.
Holbein, Hans (the Younger). *Historiarum veteris instrumenti icones.* Lyon: Melchior and Caspar Trechsel, 1538.
Ibn Athari, Mashallah. *De Scientia motus orbis.* Nuremberg: Johann Weissenburger, 1504.
Inchofer, Melchior. *Tractatus Syllepticus.* Rome: Ludouicus Grignamus, 1633.
Jamnitzer, Wenzel. *Perspectiva Corporum Regularium. Das ist Ein fleyssige Fürweysung, Wie die Fünff Regulirten Cörper, darvon Plato inn Timaeo, Unnd Euclides inn sein Elementis schreibt, etc. Durch einen sonderlichen, newen, behenden und gerechten weg, der vor nie im gebrauch ist gesehen worden, gar Künstlich inn die Perspectiva gebracht* [...]. Nuremberg: Heußler, 1568.
Jode, Gerard de. *Thesaurus sacrarum historiarum Veteris Testamenti.* Antwerp: Philip Galle, 1579.
Kepler, Johannes. *Prodromus dissertationum cosmographicarum, continens mysterium cosmographicum* [...]. Tübingen: Georg Gruppenbach, 1596.
King, Josiah. *Mr. Blount's Oracles of Reason, Examined and Answered* [...]. Exeter: S. Darker, 1698.
Kircher, Athanasius. *Ars magna lucis et umbrae.* Amsterdam: Johannes Janssonius van Waesberge & Elizaeus Weyerstraet, 1671.
Kircher, Athanasius. *Ars magna lucis et umbrae.* Rome: Hermann Scheus, 1646.
Kircher, Athanasius. *Primitiae gnomonicae catoptricae.* Avignon: Jean Piot, 1635.

Koberger, Anton, ed. *Postilla super totam Bibliam (cum expositionibus Guillelmi Britonis et additionibus Pauli Burgensis et correctoriis editis a Matthias Doering)*. Nuremberg: Anton Koberger, 22 January 1481.

Lacombe, Jacques. *Dictionnaire Encyclopedique des amusemens des Sciences Mathématiques et Physiques*. Paris: Chez Panckoucke, 1792.

Lalovera, Antonio. *Veterum Geometria Promota in Septem de Cycloide Libris* [...]. Toulouse: Arnaud Colomiez, 1660.

Lapide, Cornelius a. "De horologio Achaz et egressu umbrae." In *Commentaria in Quatuor prophetas Maiores*, 339–342. Paris: Societas Minima, 1622.

Lapide, Cornelius a. *Commentaria in Quatuor prophetas Maiores*. Antwerp: Martin Nuyts, 1625.

Lempereur, Martin, ed. *La saincte Bible en Francoys, translatee selon la pure et entiere traduction de Sainct Hierome, derechief conferee et entierement revisitee selon les plus anciens et plus correctz exemplaires*. Antwerp: Martin Lempereur, 1534.

Liesveldt, Jacob van, ed. *Den Bibel met grooter neersticheyt gecorrigeert, ende op die canten gheset den ouderdom der werelt, ende hoe lange die gheschiedenissen ende historien der Bibelen elck int zijn voor Christus gheboorte gheweest zijn* [...]. Antwerp: Jacob van Liesveldt, 1535.

Line, Francis. *Explicatio horologii in horto regio Londini*. Liège: Guillaume Henri Streel, 1673.

Lobkowitz, Juan Caramuel y. *Architectura civil recta y obliqua* [...]. Vol. 3. Vigevano: C. Corrado, 1678.

Lorin, Jean. *In Acta Apostolorum commentaria*. Lyon: Horace Cardon, 1605.

Luther, Martin, and Leo Jud, trans. *Die gantz Bibel Alt vnnd Neüw Testament: Register – weiset alle Historien vnd fürnehme sprüch uber beyde Alt vnd Neüw Testament. Jtem auch mitt Zweyhundert Figuren mehr dann vor hien nie jm Truck* [...] *Verteütscht durch Doctor M. L.* [*und Leo Jud*]. Strasbourg: Wolff Köpfl, 1530.

Lyraeo, Hadriano. *De Imitatione Iesu Patientis sive de Morte et Vita*. Antwerp: Jacob van Meurs, 1655.

Mariana, Juan de. *Scholia in Vetus et Novum Testamentum*. Paris, 1620.

Mark, Johannes. *Scripturariae Exercitationes ad Quinque et Viginti Selecta Loca Veteris Testamenti*. Amsterdam: Jacobus Borstius, 1709.

Masens, Jacob. *Speculum imaginum veritatis occultae, exhibens symbola, emblemata, hieroglyphica, aenigmata ... editio nova*. Cologne: J. A. Kinckius, 1664.

Mauro, Sylvester. *Quaestiones Philosophicae*. Rome, 1658.

Mayer, Gottfried Ulrich. *Disputatione Mathematica de Sole Tempore Hiskiae Retrogrado* [...]. Wittenberg: ex officina Gerdesiana, 1719.

Mayer, Gottfried Ulrich. *Disputatione Mathematica Solem tempore Hiskiae Retrogradum Wittebergae* [...]. Wittenberg: Johann Burckhard, 1673.

Messahallah. *De Scientia motus orbis*. Nuremberg: Johann Weissenburger, 1504.

Monantheuil, Henri de. *Aristotelis Mechanica Graeca, emendate, Latina facta, & Commentaris illustrate.* Paris: Jérémie Perier, 1599.

Montucla, Jean-Étienne. *Histoire des mathématiques* […]. Vol. 1. 2nd ed. Paris: Henri Agasse, 1798.

Naxera, Emmanuel de, *In Iosue Hostilibus redimitum trophaeis Commentarii Litterales, Moralésque*. Lyon: Laurent Anisson, 1651.

Nunes, Pedro. "Tratado em defensam da carta de marear." In *Tratado da sphera com a theorica do Sol e da Lua* […]. Lisbon: Germão Galharde, 1537.

Nunes, Pedro. *Tratado da sphera com a theorica do Sol e da Lua* […]. Lisbon: Germão Galharde, 1537.

Oddi, Mutio. *De gli Horologi Solari Trattato.* Venice: Per il Ginammi, 1638.

Ozanam, Jacques. *Récréations mathématiques et physiques* […]. Vol. 4. Paris: Chez Ant. Jombert, 1778.

Ozanam, Jacques. *Recréations Mathématiques.* Vol. 2. Paris: Jean Jombert, 1694.

Pambst, Paul. *Looßbuch zu ehren der Römischen, Ungarischen und Böhmischen Künigin.* Strasbourg: Balthasar Beck, 1546.

Paradin, Claude. *Quadrins Historiques de la Bible.* Lyon: Jean de Tournes, 1560.

Parent, M. *Essais de Recherches de Mathematique et de Physique.* Vol. 3. Paris: Chez Jean de Nully, 1713.

Pascoli, Lione. *Vite de' pittori, scultori ed architetti moderni.* Vol. 1. Rome: Antonio de' Rossi, 1730.

Pèlerin, Jean, and Albrecht Glockendon. *Von der Kunnst Perspectiva.* Nuremberg, 1540.

Petri, Heinrich. *De arte atque ratione navigandi.* Coimbra, 1573.

Petri, Heinrich. *Petri Nonii Salaciensis Opera.* Basel, 1566.

Petri, Heinrich. *Petri Nonii Salaciensis Opera.* Basel: Sebastianum Henric Petri, 1592.

Pineda, Juan. *Commentariorum in Iob*. Cologne: Anton Hierat, 1600.

Pittermann, Ignatius. *Leich- und Lob-Red Der Römisch-Kayserlich, und Königlich-Katholischen Majestät, Ertz-Herzogen von Öesterreich, u.u. Karl des Sechsten Da Sr. Kayserl. Majestät Leich-Besingnuß Den 16. 17. und 18.ten Novembris 1740 In der Hof-Kirchen bey herzlichen Frauer-Gerüst gehalten worden* […]. Vienna: Heyinger, 1740.

Power, Henry. *Experimental Philosophy in Three Books.* London, 1664.

Pulci, Luigi. *La Giostra fatta in Fiorenza dal Magnifico Lorenzo de' Medici il Vecchio l'anno MCCCC–LXVIII.* Florence, 1518.

Raedt, Alardus de. *Disputatio Mathematica de Retrogressu Umbrae in Horologia Achaz* […]. Amsterdam: Jan Banningh, 1658.

Ramus, Petrus. *Defensio pro Aristotele adversus Jac. Schecium.* Lausanne: J. Probus, 1571.

Ramus, Petrus. *Prooemium mathematicum.* Paris: Wechel, 1567.

Rebenstock, Heinrich Peter. *Neuwe Biblische Figuren: Künstlich und artig gerissen durch den sinn und kunstreichen auch weitberühmte[n] Joß Amman* […]. Frankfurt: Feyerabend, 1571.

Retz, Franz von. *Defensorium inviolatae virginitatis Mariae.* Nördlingen: Friedrich Walther, 1470.

Riccioli, Giovanni Battista. *Almagestum novum.* Vol. 2. Bologna: Vittorio Benacci, 1649–1651.

Roccamora, Giovanni Domenico. *Le cifre dell'Eucharistia.* Vol. 2. Rome: Giacomo Fei, 1670.

Ryff, Walther. *Der furnembsten, notwendigsten der gantzen Architectur angehörigen Mathematischen und Mechanischen Künst eygentlicher Bericht.* Nuremberg: Johan Petreius, 1547.

Scamozzi, Vincenzo. *Dell'idea dell'architettura universale.* Vol. 2. Venice, 1615.

Scheiner, Christoph. *Rosa Ursina.* Bracciano: Andream Phaeum, 1632.

Scheiner, Christoph. *Sol elliptic.* Augsburg: Christophori Mangii, 1615.

Scheuchzer, Johann Jakob. *Kupfer-Bibel, in welcher die Physica Sacra, oder Geheiligte Natur-Wissenschafft derer in Heil. Schrifftvorkommenden Natürlichen Sachen.* Vol. 3. Augsburg: Christian Ulrich Wagner, 1731–1735.

Schmidt, Johann Andreas. *Collegii Experimentalis Physico-Mathematici Demonstrationes Singvlis Semstribus in Academia Ivlia Curiosis B.C. D. Exhibendae a Io. Andr. Schmidt D.P.P.* Helmstedt, 1700.

Schmidt, Johann Andreas. *Variorum Philosophicorum Decas* [...]. Jena: Tobias Ohrling, Bibl., 1691.

Schönberger, Georg. *Exegeses fundamentorum gnomonicorum.* Ingolstadt: Elisabeth Angermaier, 1615.

Schönberger, Georg. *Horologiorum novorum radio recto.* Freiburg im Breisgau: Johann Strasser, 1622.

Schöner, Johannes. *Opera mathematica.* Nuremberg: J. Montanus and U. Neuber, 1551.

Schöner, Johannes. *Scripta clarissimi mathematici M. Ioanis Regiomontani.* Nuremberg: J. Montanus and U. Neuber, 1544.

Serarius, Nicolaus. *In libros Regem et Paralipomenon commentaria posthuman.* Mainz: Balthasar Lipp, 1617.

Serarius, Nicolaus. *In Libros Regum et Paralipomenon Commentaria Posthuma.* Mainz: Balthasar Lipp, 1617.

Serarius, Nicolaus. *Iosue.* Vol. 2. Paris: Typographica Edmundi Martini, 1610.

Smogulecki, Nicolas. *Sol illustratus ac propugnatus ab illustri ac generoso domino.* Freiburg im Breisgau: Theodorus Meyer, 1626.

Societatis Iesu fundatoris. *Vita beati P. Ignatii Loiolae, Societatis Iesu fundatoris.* Rome, 1609.

Solis, Virgil. *Biblische Figuren des Alten und Neuen Testaments.* Frankfurt: Sigmund Feyerabend, 1560.

Spinoza, Baruch. "De prophetis." In *Tractatus Theologico Politicus*, 15–30. Hamburg: Henricus Kunraht, 1670.

Stetten, Paul von. *Kunstgewerb und Handwerks Geschichte der Reichs-Stadt Augsburg.* Augsburg, 1779.

Stöfler, Johannes. *Elucidatio fabricae ususque astrolabii* [...]. Oppenheim, 1524.

Stoy, Johann Siegmund, ed. *Bilder-Akademie für die Jugend.* Vol. 3. Nuremberg: Sixische Schriften, 1784.

Thioly, Joannes Baptista, and Petrus Taillandier. *Theses opticae, et astronomicae.* Paris, 1693.

Thümmig, Ludwig Philipp. *Phaenomenon singulare solis coelo sereno pallescentis ad rationes revocatum.* Halle: Hiiliger, 1722.

Uffenbach, Zacharias Conrad von. *Merkwürdige Reisen durch Niedersachsen, Holland und Engelland.* Vol. 3. Ulm, 1753–54.

Varenius, Bernhard. *A Compleat System of General Geography* [...] *Improved and Illustrated by Sir Isaac Newton and Dr. Jurin* [...]. London: Stephen Austen, 1734.

Varenius, Bernhard. *Geographia Generalis, in qua affectiones generales Telluris explicantur* [...]. Amsterdam: Lodewijk Elzevir, 1650.

Vaulezard, Jean-Louis. *Traicté ou usage du Quadrant analematique* [...]. Paris: au Globe céleste, 1640.

Vaulezard, Jean-Louis. *Traitté de l'origine, demonstration, construction & usage du Quadrant analematique.* Paris: J. Le Brun; au Globe céleste, 1644.

Vieri, Francesco de'. *Discorsi delle maravigliose opere di Pratolino e d'Amore.* Florence: Giovanni Marescotti, 1587.

Vitruvius, Marcus Pollio. *Di Lucio Vitruvio Pollione de Architectura libri dece* [...]. Edited by Cesare Cesariano. Translated by Cesare Cesariano, Benedetto Giovio, Bono Mauro. Como: Gottardo da Ponte, 1521.

Vitruvius, Marcus Pollio. *Vitruvius Teutsch. Nemlichen des aller namhafftigiststen und hocherfarnesten / Römischen Architecti / und Kunstreichen Werck oder Bawmeisters / Marci Vitruuij Pollionis / Zehen Bücher von der Architectur und künstlichem Bawen.* Edited and translated by Walther Ryff. Nuremberg: Johann Petreius, 1548.

Voel, Jean. *De horologiis sciothericis libri tres.* Tournon: Claude Michel & Thomas Soubron, 1608.

Vorsterman, Willem, ed. *Den Bibel. Tgeheele Oude ende Nieuw Testament met grooter naersticheyt naden Latijnschen text gecorrigeert.* Antwerp: Willem Vorsterman, 1532.

Ward, John. *The Lives of the Professors of Gresham College.* London: J. Moore, 1740.

Wicelius, Georgius. *Qvadragesimale Catholicvm. Gegrundte Auslegung aller Episteln, oder Lectionen und der Evangelien so die Kirche Gottes von alters durch die Tage der heiligen Fasten zu Lesen und Predigen Christlich und löblich verordnet hat.* Cologne: Peter Quentel, 1546.

Wilkins, John. *A discourse concerning a new planet, tending to prove, that 'tis probable our earth is one of the planets.* London: Printed by R.H for John Maynard, 1640.

Wren, Christopher, Jr. *Parentalia: or, Memoirs of the Family of the Wrens.* London: T. Osborn, 1750.

Zimmermann, Johann Jakob. *Exercitatio theoricorum Copernico-coelestium mathematico-physicotheologica.* Philadelphia, 1689.

Zimmermann, Johann Jakob. *Scriptura S. Copernizans seu potius astronomia Copernico-scripturaria bipartita.* Hamburg: Christian Reymers, 1690.

Post-1800 Publications

"Réponse." *L'Astronomie Revue d'astronomie populaire, de météorologie et de physique du globe* 4 (1885): 389–390.

Aakhus, Patricia. "Astral Magic in the Renaissance. Gems, Poetry, and Patronage of Lorenzo de' Medici." *Magic, Ritual and Witchcraft* 3, no. 2 (Winter 2008): 185–206.

d'Andrea, Francesco. "Letter to Lucantonio Porzio, 17 October 1671." Quoted in Paula Findlen, "Living in the Shadow of Galileo: Antonio Baldigiani (1647–1711), a Jesuit Scientist in Late 17th-Century Rome," in *Conflicting Duties: Science, Medicine and Religion in Rome, 1550–1750*, edited by Maria Pia Donato and Jill Kraye, 211–254. London: The Warburg Institute, 2009.

Andresen, Andreas. *Der deutsche Peintre-Graveur* […]. Vol. 1. Leipzig: Rudolph Weigel, 1864.

Andrews, Noam. "Tabula III: Kepler's Mysterious Polyhedral Model." *Journal for the History of Astronomy* 48, no. 3 (August 2017): 281–311.

Archinard, Margarida. "Les Cadrans Solaires Analemmatiques." *Annals of Science* 62, no. 3 (2005): 309–346.

Aristotle. *Meteorology* I.9. Translated by E. W. Webster. Oxford: Clarendon Press, 1908, Internet Classics Archive, n.d, http://classics.mit.edu/Aristotle/meteorology.1.i.html.

Armogathe, Jean-Robert. "La vérité des ecritures et la nouvelle physique." In *Le Grand Siècle et la Bible*, edited by Jean-Robert Armogathe, 49–60. Paris: Beauchesne, 1989.

Asselberghs, Jean-Paul. *Les tapisseries flamandes aux États-Unis d'Amérique.* Brussels: Musées Royaux d'Art et d'Histoire, 1974.

Azzi Visentini, Margherita. "Il giardino in Scandinavia tra Rinascimento e Romanticismo." In *Scandinavia. Luoghi, figure, gesti di una civiltà del paesaggio*, edited by Domenico Luciano and Luigi Latini, 133–145. Treviso: Fondazione Benetton Studi Ricerche, 1998.

Baeumker, Clemens. *Witelo. Ein Philosoph und Naturforscher des XIII. Jahrhunderts.* Münster: Aschendorff Buchhandlung, 1908.

Bailey, Gauvin A. "The Church of the Gesù in Rome and the First Phase of Decorations." In *The Holy Name: Art of the Gesù*, edited by Linda Wolk-Simon, 97–108. Philadelphia: St. Joseph's University Press, 2018.

Baldinucci, Filippo. *Notizie dei professori del disegno da Cimabue in qua.* Vol. 2. Edited by Ferdinando Ranalli and Paola Barocchi. Florence: Studio per Edizioni Scelte, 1974.

Bardtke, Hans. "Ezechias." In *Lexikon der christlichen Ikonograhie*, edited by Engelbert Kirschbaum SJ. Vol. 1. Freiburg: Herder, 1968.

Baridon, Michel. "L'islam, les jardins et la migration des savoirs en Méditerranée." *Horizons Maghrébins – Le droit à la mémoire* 45 (2001): 7–13.

Baron, Margaret E. "Hutton, Charles." In *Dictionary of Scientific Biography*, edited by Charles Gillespie, 576–577. Vol 6. New York: Charles Scribner's Sons, 1972.

Bartrum, Giulia, ed., and Gero Seelig, comp. *The New Hollstein German Engravings, Etchings and Woodcuts 1400–1700: Jost Amman.* Vol. 2. Rotterdam: Sound & Vision Publishers, 2001.

Bauer, Rotraut, and Herbert Haupt, eds. *Das Kunstkammerinventar Kaiser Rudolfs II., 1607–1611.* Vienna: Schroll, 1976.

Bellarmine, Robert. *The Louvain Lectures (Lectiones Lovanienses) of Bellarmine and the Autograph Copy of his 1616 Letter to Galileo.* Vatican City: Specola Vaticana, 1984.

Bennett, James A. "Christopher Wren: Astronomy, Architecture and the Mathematical Sciences." *Journal for the History of Astronomy* 6, no. 3 (October 1975): 149–184.

Bennett, James A. "Geometry in Context in the Sixteenth Century: the View from the Museum." *Early Science and Medicine* 7, no. 3 (2002): 214–230.

Bennett, James A. "Knowing and Doing in the Sixteenth Century: What were Instruments for?" *British Journal for the History of Science* 36, no. 2 (2003): 129–150.

Bennett, Jim. "Christopher Wren's Greshamite History of Astronomy and Geometry." In *Sir Thomas Gresham and Gresham College. Studies in the Intellectual History of London in the Sixteenth and Seventeenth Centuries*, edited by Francis Ames-Lewis, 189–197. Hampshire: Ashgate, 1999.

Bennett, Jim. "Early Modern Mathematical Instruments," *ISIS* 102 (2011): 697–705.

Bensaude-Vincent, Bernardette. "Camille Flammarion: Prestige de la science populaire," *Romantisme: Revue Du 19e Siecle* 65 (1989): 93–104.

Berninger, Ulrike. "Im Auftrag von Kaisern, Fürsten und anderen hohen Herren, der Goldschmied Wenzel Jamnitzer." In *Von nah und fern, Zuwanderer in die Reichsstadt Nürnberg*, edited by Brigitte Korn, Michael Diefenbacher, and Steven M. Zahlaus, 115–124. Petersberg: Imhof, 2014.

Bethell, Samuel L. *The Cultural Revolution of the Seventeenth Century.* London: D. Dobson, 1951.

Bieber, Margarete. "Honos and Virtus." In *American Journal of Archaeology* 49, no. 1 (1945): 25–34.

Bischoff, Michael. "Weltenträger, Kontinente, Land und Meer. Zur Ikonographie der gestochenen Titelblätter frühneuzeitlicher Atlanten." In *Kartographie der Frühen Neuzeit. Weltbilder und Wirkungen. Ergebnisse des in Kooperation mit der Kartenabteilung der Staatsbibliothek zu Berlin durchgeführten internationalen Symposiums am Weserrenaissance-Museum Schloss Brake (4.–6. April 2014)*, edited by Michael Bischoff, Vera Lüpkes, and Wolfgang Crom, 193–209. Marburg: Jonas Verlag, 2015.

Blackwell, Richard J. *Galileo, Bellarmine, and the Bible: Including a Translation of Foscarini's Letter on the Motion of the Earth.* Notre Dame, IN: University of Notre Dame Press, 1991.

Blackwell, Richard J. *Galileo, Bellarmine, and the Bible: Including a Translation of Foscarini's Letter on the Motion of the Earth.* Notre Dame, IN: University of Notre Dame Press, 1991.

Blain, A. "La retrogradation de l'ombre." *L'Astronomie: Revue d'astronomie populaire, de météorologie et de physique du globe* 4 (1885): 387–389.

Bobinger, Maximilian. *Alt-Augsburger Kompassmacher. Sonnen-, Mond- und Sternuhren, astronomische und mathematische Geräte, Räderuhren.* Augsburg: H. Rösler, 1966.

Bobinger, Maximilian. *Christoph Schissler der Ältere und Jungere.* Augsburg: Verlag Die Brigg, 1954.

Borchardt, Ludwig. *Die Geschichte der Zeitmessung und der Uhren.* Vol. 1, *Die altägyptische Zeitmessung.* Edited by Ernst von Bassermann-Jordan. Berlin: De Gruyter, 1920.

Bott, Gerhard, ed. *Wenzel Jamnitzer und die Nürnberger Goldschmiedekunst 1500–1700. Goldschmiedearbeiten, Entwürfe, Modelle, Medaillen, Ornamentstiche, Schmuck, Porträts*, exh. cat., Germanisches Nationalmuseum Nürnberg, Nuremberg. München: Klinkhardt & Biermann, 1985.

Böttiger, Johan Fredrik. *Svenska statens samling af väfda tapeter: historik och beskrifvande förteckning. Vol. 4.* Stockholm: Palmquists Aktiebolag, 1898.

Bösel, Richard. *Orazio Grassi architetto e matematico gesuita: un album conservato nell'Archivio della Pontificia Università Gregoriana a Roma.* Rome: Argos, 2004.

Boutot, Alain. "Modernité de la catoptrique de Héron d'Alexandrie." *Philosophie antique* 12 (2012): 157–196.

Brahe, Tycho. *Tychonis Brahe, Dani, Opera Omnia.* Vol. 7. Edited by John Louis Emil Dreyer. Copenhagen, 1924. Quoted in Samuel Gessner, "'Geometricus et astronomicus faber': Chr. Schissler aus Augsburg [...]," in *Weiter sehen/Seeing further*, ed. Jürgen Hamel and Michael Korey (Frankfurt am Main: Verlag Harry Deutsch, 2012), 123–154.

Brewster, David, ed. "Dialling." In *The Edinburgh Encyclopædia*, 693. Vol. 87. Edinburgh: William Blackwood, 1830.

Brodrick, James. *The life and work of Blessed Robert Francis cardinal Bellarmine, S. J., 1542–1621.* Vol. 2. London: Burns, Oates and Washbourne, 1928.

Brunon, Hervé. "'Les mouvements des eaux de l'Univers': Pratolino, jardin météorologique." In *Les Éléments et les métamorphoses de la nature: imaginaire et symbolique des arts dans la culture européenne du XVIe au XVIIIe siècle*, edited by Hervé Brunon, Monique Mosser, and Daniel Rabreau, 33–53. Bordeaux: Blake, 2004.

Brunon, Hervé. "Pratolino: art des jardins et imaginaire de la nature dans l'Italie de la seconde moitié du XVIe siècle." PhD diss., Université Paris 1 Panthéon-Sorbonne, 2001. Revised edition accessible online: TEL Multidisciplinary Theses Server. Last modified March 2, 2020. https://tel.archives-ouvertes.fr/tel-00349346.

Buchanan, Iain. "The Tapestries Purchased by King Eric XIV of Sweden in Antwerp in 1560–1561. New Documentation." In *Flemish Tapestry in European and American Collections. Studies in Honour of Guy Delmarcel*, edited by Koenraad Brosens, 93–98. Turnhout: Brepols, 2003.

Butters, Suzie. "Natural Magic, Artificial Music, and Birds at Francesco de' Medici's Pratolino." In *Sense and the Senses in Early Modern Art and Cultural Practice*, edited by Alice E. Sanger and Siv Tove Kulbrandstad Walker, 31–63. London: Routledge, 2016.

Camerota, Filippo. "Architecture and Science in Baroque Rome. the Mathematical Ornaments of Villa Pamphilj." *Nuncius* 15, no. 2 (2000): 611–638.

Camerota, Filippo. "Two new Attributions: a Refractive Dial of Guidobaldo del Monte and the "Roverino Compass" of Fabrizio Mordente." *Nuncius* 18, no. 1 (January 2003): 25 38.

Camerota, Filippo. *La linea del sole. Le grandi meridiane fiorentine.* Florence: Edizioni della Meridiana, 2007.

Carolino, Luís Miguel. "Astronomy, Cosmology, and Jesuit Discipline, 1540–1758." In *The Oxford Handbook of the Jesuits,* edited by Ines Županov, 670–707. Oxford: Oxford University Press, 2019.

Chaperon, Danielle. *Camille Flammarion: Entre astronomie et litterature.* Paris: Imago, 1998.

Charette, François. *Mathematical Instrumentation in Fourteenth-Century Egypt and Syria: the Illustrated Treatise of Najm al-Dīn al-Mißrī.* Leiden: Brill, 2003.

Chomard, Louis. "Le cadran analemmatique et la rétrogradation de l'ombre." *L'Astronomie* 20 (1906): 433–449.

Chossat, R. P. Marcel. *Les jésuites et leurs œuvres à Avignon: 1553–1768.* Avignon: F. Seguin, 1896.

Clarke, Adam. *The Holy Bible: containing the Old and New Testaments* [...]. Vol. 8. New York: Daniel Hitt and Abraham Paul, 1817.

Clément, Gilles. *Une brève histoire du jardin.* Paris: J.C. Béhar, 2011.

Clercq, Peter de. "The travel journals of Balthasar de Monconys (1608–1665) and Zacharias Conrad von Uffenbach (1683–1734)." *Bulletin of the Scientific Instrument Society* 128 (2016): 2–13.

Clifton, James. "A Lutheran Image on the Title-Page of the Last Bible without a Confessional Label." *Ephemerides Theologicae Lovanienses* 84 (2008): 69–86.

Coelen, Peter van der, and Marjolein Leesberg. "Depicting the Bible and Mapping the World: Gerard de Jode and his Legacy." In *The New Hollstein Dutch and Flemish Etchings, Engravings, and Woodcuts 1450–1700: the De Jode Dynasty, Part 1*, edited by Peter van der Coelen and Huigen Leeflang, compiled by Marjolein Leesberg, 28–93. Ouderkerk aan den Ijssel: Sound & Vision, 2018.

Coelen, Peter van der. "Thesauri en Trezoren: Boecken en bundels met oudtestamentische prenten." In *Het Oude Testament in de schilderkunst van de Gouden Eeuw*, edited by Christian Tümpel, exh. cat., Joods Historisch Museum, Amsterdam, 168–193. Zwolle: Waanders, 1991.

Cohen, H. Floris. *The Scientific Revolution: a Historiographical Inquiry.* Chicago: The University of Chicago Press, 1994.

Cohen, H. Floris. *How Modern Science Came into the World: Four Civilizations, One 17th-Century Breakthrough.* Amsterdam: Amsterdam University Press, 2010.

Collison, Robert. *Encyclopedias: their History throughout the Ages.* 2nd ed. New York: Haffner Publishing Company, 1966.

Colombo, Lino. "Refractive Sundials in Italy." Accessed February 16, 2021. https://www.academia.edu/39913916/Gli_orologi_solari_a_rifrazione_in_Italia.

Congregrazione Prima-Primaria. *Regole della congregazione Prima-Primaria nel Collegio Romano.* Rome: I. Lazzarini, 1811.

Coole, P. G. and E. Neumann. *The Orpheus Clocks.* London: Hutchinson, 1972.

Cormack, Lesley B., Steven A. Walton, and John A. Schuster, eds. *Mathematical Practitioners and the Transformation of Natural Knowledge in early Modern Europe.* Cham, Switzerland: Springer International Publishing, 2017.

Cox-Rearick, Janet. "Themes of Time and Rule at Poggio a Caiano: the Portico Frieze of Lorenzo il Magnifico." *Mitteilungen des Kunsthistorischen Institutes in Florenz* 26, no. 2 (1982): 167–210.

Cox-Rearick, Janet. *Dynasty and Destiny in Medici Art: Pontormo, Leo X and the two Cosimos.* Princeton: Princeton Unviersity Press, 1984.

Curzietti, Jacopo. *La decorazione della chiesa del SS. Nome di Gesù.* Rome: Gangemi Editore, 2011.

Dackerman, Susan, ed. *Prints and the Pursuit of Knowledge in Early Modern Europe*, exh. cat., Harvard Art Museums, Cambridge; Mary and Leigh Block Museum of Art, Northwestern University, Evanston. New Haven: Yale University Press, 2011.

Dahan, Gilbert, ed. *Nicolas de Lyre franciscain du XIVe siècle exégéte et théologien*, Collection des Études Augustiniennes – Série Moyen Âge et Temps Moderns 48. Paris: Institut des Études Augustiniennes, 2011.

Daniel, Christopher. "The Dial of Ahaz." *CLOCKS Magazine*, December 2007.

Danieluk, Robert. "Michal Boym, Andrzej Rudomina, and Jan Smogulecki: Three Seventeenth-Century Missionaries in China; a Selection of Documents from the Roman Jesuit Archives." In *Monumenta Serica*, 417–443. Vol. 59. Sankt Augustin: Monumenta Serica Insititute, 2011.

Daston, Lorraine, and Katherine Park. *Wonders and the Order of Nature, 1100–1750*. New York: Zone Books, 1998.

Davis, Charles. "Vincenzo Scamozzi architetto della luce." In *Vincenzo Scamozzi, 1548–1616*, edited by Franco Barbieri and Guido Beltramini, 33–45. Venice: Marsilio, 2003.

Davis, Herman S. "General Notes – the Dial of Ahaz." *Popular Astronomy* 7 (1899): 102–103.

Dear, Peter. "Miracles, Experiments, and the Ordinary Course of Nature." *Isis* 81, no. 4 (Dec. 1990): 663–683.

Delmotte, G. "Cadran solaire Flammarion à Rétrogradation." *L'Astronomie: Revue d'astronomie populaire, de météorologie et de physique du globe* 37 (1923): 408–409.

Dezzi Bardeschi, Marco. "Le fonti degli automi di Pratolino." In *La fonte delle fonti: iconologia degli artifizi d'acqua*, edited by Alessandro Vezzosi and Cristina Acidini Luchinat, 13–24. Florence: Aliena Editrice, 1985.

Dinis, Alfredo. "Was Riccioli a secret Copernican?" In *Giambattista Riccioli e il merito scientifico dei gesuiti nell'etá barocca*, edited by Maria Teresa Borgato, 49–77. Florence: Leo Olschki, 2002.

Dinis, Alfredo. *A Jesuit against Galileo? the Strange Case of Giovanni Battista Riccioli Cosmology*. Edited by Ricardo Barroso Batista and Alvaro Balsas. Braga: Axioma-Publicações da Faculdade de Filosofia, 2017.

Dobbel, Brian. *Augustine's Intellectual Conversion. the Journey from Platonism to Christianity*. Cambridge: Cambridge University Press, 2009.

Drake, Stillman, ed. and trans. *Discoveries and Opinions of Galileo*. New York: Doubleday, 1957.

Drinkwater, Peter I. "The Miracle of Achaz." *Bulletin of the British Sundial Society* 92.1 (1992): 24.

Dupré, Sven. "Galileo, the Telescope, and the Science of Optics in the Sixteenth Century: a Case Study of Instrumental Practice in Art and Science." PhD diss., University of Ghent, 2002.

Dupré, Sven. "The Dioptrics of Refractive Dials in the Sixteenth Century." *Nuncius* 18, no. 1 (January 2003): 39–67.

Eckhardt, Wolfgang. "Erasmus und Josua Habermel – kunstgeschichtliche Anmerkungen zu den Werken der beiden Instrumentenmacher." *Jahrbuch der Hamburger Kunstsammlungen* 22 (1977): 13–74.

Ehrenstein, Theodor. *Das Alte Testament im Bilde. Ein Illustrationswerk mit über 2000 Abbildungen von altchristlichen, mittelalterlichen und neuzeitlichen Kunstwerken*. Vienna: Kende, 1923

Ellinghaus, Julia and Volker Remmert. "Visual Worlds on Early Modern Scientific Instruments: Types and Messages." In *Reassessing Epistemic Images in the Early Modern World*, edited by Ruth Noyes, 153–174. Amsterdam: Amsterdam University Press, 2023.

Engammare, Max. "Les Figures de la Bible: le destin oublié d'un genre littéraire en image (XVIe–XVIIe siècles)." *Mélanges de l'École française de Rome* 106 (1994): 549–591.

Engelmann, Max. "Mathematische Instrumente von Wenzel Jamnitzer." *Mitteilungen aus den sächsischen Kunstsammlungen* 5 (1914): 1–11.

Engelmann, Max. "Wenzel Jamnitzers Dresdner Meßscheibe," In *Der Kunstwanderer: Zeitschrift für alte und neue Kunst, für Kunstmarkt und Sammelwesen*, edited by Adolph Donath, 311–314. Berlin, September 1919 – August 1920. https://digi.ub.uni-heidelberg.de/diglit/kunstwanderer1919_1920/0315.

Enggass, Robert. *The Painting of Baciccio: Giovanni Battista Gaulli 1639–1709*. University Park, PA: Penn State University Press, 1964.

European Cultural Heritage Online. "Historical Sources Related to Design and Construction of the Garden of Pratolino." Last modified June 2015. http://echo.mpiwg-berlin.mpg.de/content/pratolino/sources.

Exposition des Instruments et outils d'autrefois, Mars-Avril 1936, exh. cat., Musée des arts décoratifs, Paris. Paris: La Musée, 1936.

Fagiolo, Marcello. "Creare un'architettura cosmologica: l'invenzione del Pantheon e le infinite variazioni sul tema." In *Creatività, biologia, psicologia, struttura del processo creativo, Atti dei Convegni Lincei*, 245–267. Rome: Accademia Nazionale dei Lincei, 2020.

Fagiolo, Marcello. "Le due anime nelle ville della Tuscia." In *Il giardino d'Europa. Pratolino come modello nella cultura europea*, edited by Marco Dezzi Bardeschi, Alessandro Vezzosi, and Luigi Zangheri, 68–81. Milan: Mazzotta, 1986.

Fara, Amelio. *Bernardo Buontalenti.* Milan: Electa, 1995.

Farmer, Sharon. "Aristocratic Power and the 'Natural' Landscape: the Garden Park at Hesdin, ca. 1291–1302." *Speculum* 88, no. 3 (July 2013): 644–680.

Feldhay, Rivka. "Recent Narratives on Galileo and the Church: Or the Three Dogmas of the Counter-Reformation." *Science in Context* 13 (2000): 489–507.

Filson, Lily. "Magical and Mechanical Evidence: the Late-Renaissance Automata of Francesco I de' Medici." In *Evidence in the Age of the New Sciences*, edited by James A.T. Lancaster and Richard Raiswell, 177–206. Cham: Springer International Publishing, 2018.

Findlen, Paula, ed. *Athanasius Kircher: the Last Man Who Knew Everything.* New York: Routledge, 2004.

Findlen, Paula. "Living in the Shadow of Galileo: Antonio Baldigiani (1647–1711), a Jesuit Scientist in Late 17th-Century Rome." In *Conflicting Duties: Science, Medicine and Religion in Rome, 1550–1750*, edited by Maria Pia Donato and Jill Kraye, 211–254. London: The Warburg Institute, 2009.

Findlen, Paula. "Rethinking 1633: Writing about Galileo after the Trial." In *Nature Engaged: Science in Practice from the Renaissance to the Present*, edited by Mario Biagioli and Jessica Riskin, 205–226. New York: Palgrave, 2012.

Finocchiaro, Maurice, ed. and trans. *The Essential Galileo.* Indianapolis: Hackett Publishing Company, Inc., 2008.

Finocchiaro, Maurice. *The Galileo Affair: a Documentary History.* Berkeley: University of California Press, 1989.

Fischer, Bonifatius, and Robert Weber, eds. *Biblia sacra: iuxta Vulgatam versionem.* Stuttgart: Deutsche Bibelgesellschaft, 1994.

Flammarion, Camille. "Le cadran solaire a rétrogradation de l'Observatoire de Juvisy, et le miracle d'Isaie." *L'Astronomie: Revue d'astronomie populaire, de météorologie et de physique du globe* 4 (1885): 321–329.

Folkerts, Menso. "Johannes Praetorius (1537–1616) – ein bedeutender Mathematiker und Astronom des 16. Jahrhunderts." In *History of Mathematics, States of the Art*, edited by Joseph W. Dauben and Christoph Scriba, 149–169. San Diego: Academic Press, 1996.

Foucault, Michel. "The Confession of the Flesh." In *Power/Knowledge: Selected Interviews and Other Writings, 1972–1977*, edited by Colin Gordon, 194–228. New York: Pantheon Books, 1980.

Francis, Haskell, *Patrons and Painters. a Study of the Relations between Italian Art and Society in the Age of the Baroque.* London: Chatoo & Windus, 1963.

François, Wim. "Bible Production and Bible Readers in the Age of Confessionalization: the Case of the Low Countries." In *Lay Readings of the Bible in Early Modern Europe*, edited by Erminia Ardissino and Élise Boillet, 191–216. Leiden: Brill, 2020.

François, Wim. "The 'Glossed Bibles' of Jacob van Liesvelt and Willem Vorsterman (1532–1534ff.)." *Journal of Early Modern Christianity* 6, no. 2 (2019): 147–184.

Freher, Paul. *Theatrum virorum eruditione clarorum.* Vol. 2. Nuremberg: Hofmannus, 1688.

Froehlich, Karl. "An Extraordinary Achievement: the *Glossa ordinaria* in Print." In *The Bible as Book: the First Printed Editions*, edited by Kimberly van Kampen and Paul Henry Saenger, 15–22. London: The British Library, 1999.

Froehlich, Karl. "Makers and Takers: the Shaping of the Biblical *Glossa ordinaria.*" In *Biblical Interpretation from the Church Fathers to the* Reformation, 1–19. Farnham, Surrey: Ashgate, 2009.

Froehlich, Karl. "The Fate of the *Glossa ordinaria in the Sixteenth Century.*" In *Die Patristik in der Bibelexegese des 16. Jahrhunderts*, edited by David C. Steinmetz, 19–47. Wiesbaden: Harrassowitz, 1999.

Froehlich, Karl. "The Printed Gloss." In *Biblia Latina cum glossa ordinaria: Facsimile Reprint of the Editio Princeps Adolph Rusch of Strassburg 1480–81*, xii–xxvi. Vol. 1. Turnhout: Brepols, 1992.

Froehlich, Karl. "Walafrid Strabo and the *Glossa ordinaria*: the Making of a Myth." *Studia Patristica* 28 (1993): 192–196.

Frommel, Sabine. "Medici Villas as an Expression of a New Way of Life and Perception." In *The Early Modern Villa*, edited by Barbara Arciszewska, 17–34. Warsaw: Museum of King Jan III's Palace at Wilanów, 2017.

Gabriele, Mino. "Voyage de l'âme et architecture symbolique dans le *Songe de Poliphile*." *Annuaire de l'École pratique des hautes études* (EPHE). *Section des sciences religieuses* 121 (2014): 307–311.

Galilei, Galileo. *Le Opere Di Galileo Galilei, Nuova Ristampa della Edizione Nazionale*, Vol. 20. (Florence: G. Barbera, 1966).

Gatty, Mrs. Alfred, *The Book of Sun-Dials*. Edited by H. K. F. Eden and Eleanor Lloyd. 4th ed. London: George Bell & Sons, 1900.

Gaudard, Jules. *Le cadran d'Achaz et les miracles*. Lausanne: G. Bridel, 1887.

Geffroy, Mathieu Auguste. *Notices et extraits des manuscrits concernant l'histoire ou la littérature de la France qui sont conservés dans les bibliothèques ou archives de Suède, Danemark et Norvège*. Paris: Imprimerie Impériale, 1855.

Gessner, Samuel. "'Geometricus et astronomicus faber': Chr. Schissler aus Augsburg […]." In *Weiter sehen/Seeing further*, edited by Jürgen Hamel and Michael Korey, 123–154. Frankfurt am Main: Verlag Harry Deutsch, 2012.

Ghillany, Friedrich Wilhelm. *Geschichte des Seefahrers Ritter Martin Behaim, nach den ältesten vorhandenen Urkunden bearbeitet*. Nuremberg: Bauer und Raspe, 1853.

Giblin, Gerard F. "The Autobiography of St. Robert Bellarmine, S.J." *Woodstock Letters* 89 (1960): 3–30.

Gillingham, Harrold E. "The Horologium Achaz." *Connoisseur* 74 (1926): 220–222.

Gingerich, Owen, and Philip Sadler. "Christopher Schissler's Wonderful 'Bowl of Ahaz' of 1578." *Bulletin of the American Astronomical Society* 21 (1989): 1218.

Goettler, Christine. "Nomen Mirificum: Rubens' Beschneidung Jesu für den Hochaltar der Jesuitenkirche in Genua." In "Aspekte der Gegenreformation," edited by Victoria von Flemming. Special issue, *Zeitsprünge: Forschungen zur Frühen Neuzeit* 1 (1997): 796–844.

Goldstein, Bernard R. "Before the Sun in the Church." *Journal for the History of Astronomy* 32, no. 1 (2001), 73–77.

Goldstein, Bernard R. "Galileo's Account of Astronomical Miracles in the Bible: a Confusion of Sources." *Nuncius* 5, no. 1 (January 1990): 3–16.

Goldstein, Bernard R. "The Physical Astronomy of Levi ben Gerson." *Perspectives on Science* 5 (1997): 1–30.

Goldstein, Bernard R. *The Astronomy of Levi ben Gerson (1288–1344)*. New York: Springer, 1985.

Gomes Teixeira, Francisco. *História das Matemáticas em Portugal*. Lisbon: Academia das Ciências de Lisboa, 1934.

Goodwin, Gordon. "GREGORY, JOHN (1607–1646)." In *Dictionary of National Biography*, edited by Leslie Stephen and Sidney Lee, 101–102. Vol. 23. New York: Macmillan and Co., 1890.

Gosselin, Edward A. "A Listing of the Printed Editions of Nicolaus de Lyra." *Traditio* 26 (1970): 399–426.

Gouk, Penelope. *The Ivory Sundials of Nuremberg 1500–1700*. Cambridge: Whipple Museum of the History of Science, 1988.

Graney, Christopher M. *Setting Aside All Authority: Giovanni Battista Riccioli and the Science Against Copernicus in the Age of Galileo*. Notre Dame, IN: University of Notre Dame Press, 2015.

Graney, Christopher M. *Setting Aside All Authority: Giovanni Battista Riccioli and the Science Against Copernicus in the Age of Galileo*. South Bend: University of Notre Dame, 2015.

Grant, Edward. *In Defense of the Earth's Centrality and Immobility: Scholastic Reaction to Copernicanism in the Seventeenth Century*. Philadelphia, PA: The American Philosophical Society, 1984.

Grimm, Jacob and Wilhelm. "KUNSTSTÜCKLEIN." In *Deutsches Wörterbuch von Jacob und Wilhelm Grimm*. Vol. 11, column 2731. Leipzig, 1873. http://www.woerterbuchnetz.de/DWB?lemma=kunststuecklein.

Guillemin, Etienne. "De la rétrogradation de l'ombre sur le cadran solaire." *Verhandlungen der Schweizerischen Naturforschenden Gesellschaft = Actes de la Société Helvétique des Sciences Naturelles = Atti della Società Elvetica di Scienze Naturali* 60 (1878): 201–207.

Guillemin, Étienne. "Théorie, formules et épures de la rétrogradation de l'ombre." *L'Astronomie: Revue d'astronomie populaire, de météorologie et de physique du globe* 4 (1885): 329–340.

Gurlitt, Cornelius. "Aus den sächsischen Archiven. Jamnitzer und der kursächsische Hof." In *Kunstgewerbeblatt. Monatsschrift für Geschichte und Litteratur der Kleinkunst*, edited by Arthur Papst, 51–53. Leipzig, 1885. https://digi.ub.uni-heidelberg.de/diglit/kunstgewerbeblatt1885/0058.

Hartmann, Georg. *Hartmann's Practika: a Manual for Making Sundials and Astrolabes with the Compass and Rule: Written from 1518 to 1528*. Translated and edited by John Lamprey. Bellvue, CO: John Lamprey, 2002.

Hauschke, Sven. "Globen und Wissenschaftliche Instrumente. Die europäischen Höfe als Kunden Nürnberger Mathematiker." In *Quasi Centrum Europae. Europa kauft in Nürnberg 1400–1600*, edited by Hermann Maué, Thomas Eser, Sven Hauschke, and

Jana Stolzenberger, exh. cat., Germanisches Nationalmuseum Nürnberg, 385–389. Nuremberg: Germanisches Nationalmuseum, 2002.

Hauschke, Sven. "Goldschmiede als Hersteller wissenschaftlicher Instrumente und Geräte." In Goldglanz und Silberstrahl, edited by Karin Tebbe, exh. cat., Germanisches Nationalmuseum Nuremberg, 216–232. Vol. 2 of Nürnberger Goldschmiedekunst 1541–1868. Nuremberg: Germanisches Nationalmuseum, 2007.

Hauschke, Sven. "The Mathematical Instruments of Wenzel Jamnitzer (1508–1585)." In *European collections of Scientific Instruments, 1550–1750*, edited by Giorgio Strano, Stephen Johnston, Mara Miniati, and Alison Morrison-Low, 1–13. Leiden: Brill, 2009.

Hauschke, Sven. "Wenzel Jamnitzer." In *Allgemeines Künstlerlexikon*, edited by Günter Meißner and Andreas Beyer, 265f. Vol. 77. Berlin: de Gruyter, 2013.

Hauschke, Sven. "Wenzel Jamnitzers Beschreibung seines "kunstlichen und wolgetzierten Schreibtischs" und die Erfindung des Kunstschranks." *Jahrbuch der Coburger Landesstiftung* 58 (2014): 1–164.

Hayton, Darin. "Instruments and demonstrations in the astrological curriculum: evidence from the University of Vienna 1500–1530." *Studies in History and Philosophy of Biological and Biomedical Sciences* 41, no. 2 (June 2010): 125–134.

Heath, Thomas Little. *A History of Greek Mathematics.* 2nd ed. New York: Dover Publications, 1981.

Heesen, Anke te. *Der Weltkasten. Die Geschichte einer Bildenzyklopädie aus dem 18. Jahrhundert.* Göttingen: Wallstein-Verlag, 1997.

Heilbron, John L. *The Sun in the Church: Cathedrals as Solar Observatories.* Cambridge, MA: Harvard University Press, 1999.

Hellyer, Marcus. "'Because the Authority of My Superiors Commands:' Censorship, Physics and the German Jesuits." *Early Science and Medicine* 1 (1996): 319–354.

Hellyer, Marcus. "The Construction of the 'Ordinatio pro Studiis Superioribus' of 1651." *Archivum Historicum Societatis Iesu* 72 (2003): 3–43.

Hempel, Johannes. *Heilung als Symbol und Wirklichkeit im biblischen Schrifttum.* Göttingen: Vandenhoeck & Ruprecht, 1965.

Henderson, Felicity. "Taking the Moon Seriously: John Wilkins's *Discovery of a World in the Moone* (1638) and *Discourse Concerning a New World and Another Planet* (1640)." In *John Wilkins (1614–1672): New Essays*, edited by William Poole, 129–157. Leiden: Brill, 2017.

Hentschel, Klaus. "Das Brechungsgesetz in der Fassung von Snellius: Rekonstruktion seines Entdeckungspfades und eine Übersetzung seines lateinischen Manuskriptes sowie ergänzender Dokumente." *Archive for History of Exact Sciences* 55, no. 4 (2001): 297–344.

Heusinger, Christian. *Herzog Anton Ulrich-Museum Braunschweig. Die Handzeichnungssammlung.* Vol. 3, Tafelband 1. and vol. 3, Textband. Braunschweig: Herzog Anton Ulrich-Museum, 1992 and 1997.

Hibbard, Howard. *Michelangelo.* 2nd ed. Cambridge, MA: Harper & Row, 1985.

Himmelein, Volker, and Leopold, John H. *Prunkuhren des 16. Jahrhunderts. Sammlung Joseph Fremersdorf.* Stuttgart: Württembergisches Landesmuseum, 1974.

Hollstein, Friedrich. *Hollstein's German engravings, etchings and woodcuts.* Vols. 63–64. Rotterdam: Sound & Vision Publishers, 2004.

Hooykaas, Reyer, ed. and trans. *G. J. Rheticus' Treatise on Holy Scripture and the Motion of the Earth with Translation, Annotations, Commentary and Additional Chapters on Ramus-Rheticus and the Development of the Problem before 1650.* Amsterdam: North-Holland Publishing Company, 1984.

Howell, Kenneth J. *God's Two Books. Copernican Cosmology and Biblical Interpretation in Early Modern Science.* Notre Dame, IN: University of Notre Dame Press, 2002

Howson, A. G. "Charles Hutton." In *A History of Mathematics Education in England,* 59–74. Cambridge: Cambridge University Press, 1982.

Hudson, Giles M. "Torquetum." In *Instruments of Science. an Historical Encyclopaedia,* edited by Robert Bud and Deborah Jean Warner, 623–626. New York: Garland, 1989.

Hudson, Hugh. "Giovanni Battista Gaulli: Remaking the Image of a Cardinal Saint in Seventeenth-Century Rome." *Master Drawings* 47 (2009): 70–78.

Jaffé, Paul. *L'Astronomie: Revue d'astronomie populaire, de météorologie et de physique du globe* 4 (1885): 390–391.

Jensen, Kristian, ed. *Incunabula and their Readers: Printing, Selling, and Using Books in the Fifteenth Century.* London: The British Library, 2003.

Johnson-Roehr, Susan N. "The Spatialization of Knowledge and Power at the Astronomical Observatories of Sawai Jai Singh II, c. 1721–1743 CE." Ph.D. diss., University of Illinois at Urbana-Champaign, 2011.

Kaiser, Simone, and Matteo Valleriani. "The Organ of the Villa d'Este in Tivoli and the Standards of Pneumatic Engineering in the Renaissance." In *Gardens, Knowledge and the Sciences in the Early Modern Period,* edited by Hubertus Fischer, Volker Remmert, and Joachim Wolschke-Bulmahn, 77–102. Basel: Birkhäuser, 2016.

Karr Schmidt, Suzanne. "Georg Hartmann and the Development of Printed Instruments in Nuremberg." In *Prints and the Pursuit of Knowledge in Early Modern Europe,* edited by Susan Dackerman, exh. cat., Harvard Art Museums, Cambridge; Mary and Leigh Block Museum of Art, Northwestern University, Evanston, 268–279. New Haven: Yale University Press, 2011.

Karr Schmidt, Suzanne. "Georg Hartmann, *Bowl of Ahaz* Sundial." In *Prints and the Pursuit of Knowledge in Early Modern Europe,* edited by Susan Dackerman, exh. cat., Harvard Art Museums, Cambridge; Mary and Leigh Block Museum of Art, Northwestern University, Evanston, 296–297, cat. no. 69. New Haven: Yale University Press, 2011.

Karr Schmidt, Suzanne. *Altered and Adorned: Using Renaissance Prints in Daily Life,* exh. cat., Art Institute of Chicago. New Haven and London: Yale University Press, 2011.

Karr Schmidt, Suzanne. *Interactive and Sculptural Printmaking in the Renaissance.* Leiden: Brill, 2017.

Kästner, Manfred. *Die Icones Hans Holbeins des Jüngeren. Ein Beitrag zum graphischen Werk des Künstlers und zur Bibelillustration Ende des 15. und in der ersten Hälfte des 16. Jahrhunderts.* Vol. 2. Heidelberg: Esprint, 1985.

Kaye, G. R. *The Astronomical Observatories of Jai Singh.* New Imperial Series vol. 40. Calcutta, Superintendent Government Printing, 1918.

Keil, Inge and Helmut Zäh. "Tycho Brahes Aufenthalt in Augsburg (1569–1570 und 1575)." In *Beiträge zur Astronomiegeschichte*, edited by Wolfgang R. Dick and Jürgen Hamel, 45–98. Vol. 7. Frankfurt: Verlag Harri Deutsch, 2004.

Keil, Karl-August, and Inge Keil. "Astronomie am Gymnasium bei St. Anna." In *1531–1981: Festschrift 450 Jahre Gymnasium bei Sankt Anna in Augsburg,* edited by Johannes Rettenberger and Helmut Schmid, 12–23. Augsburg, 1981.

Kelin-Braslavy, Sara. "Gersonides' Use of the *Meteorology* in his Accounts of some Biblical Miracles." *Historical Studies in Science and Judaism* 10, no. 2 (July 2010): 240–313.

Kelter, Irving A. "The Refusal to Accommodate: Jesuit Exegetes and the Copernican System." *The Sixteenth Century Journal* 26, no. 2 (1995): 273–83.

Kelter, Irving A. "The Refusal to Accommodate. Jesuit Exegetes and the Copernican System." *Sixteenth Century Journal* 26, no. 2 (Summer 1995): 273–283.

Kepler, Johannes. *Gesammelte Werke.* Edited by Max Caspar and Walther von Dyck. Vol. 22. Munich: Bayerische Akademie der Wissenschaften, 1938–2017.

King, Henry Charles, and John R. Milburn. *Geared to the Stars: the Evolution of Planetariums, Orreries, and Astronomical Clocks Clocks.* Toronto: University of Toronto Press, 1978.

Klemm, Hans Gunther. *Georg Hartmann aus Eggolsheim (1489–1564): Leben und Wirken eines fränkischen Mathematikers.* Forchheim: Ehrenbürg-Gymnasium, 1990.

Knobloch, Eberhard. "Christoph Clavius. Ein Namen und Schriftenverzeichnis zu seinen *Opera Mathematica.*" *Bolletino di Storia delle Scienze Matematiche* 10 (1990): 135–189.

Knobloch, Eberhard. "Nunes and Clavius." In *The Practice of Mathematics in Portugal. Proceedings of the International Conference*, edited by Luís Saraiva and Henrique Leitão, 163–194. Coimbra: Imprensa da Universidade, 2004.

Koenigs-Erffa, Ursula. "Das Tagebuch des Sebald Welser aus dem Jahre 1577." *Mitteilungen des Vereins für Geschichte der Stadt Nürnberg* 46 (1955): 262–371.

Korey, Michael. *The Geometry of Power. Mathematical Instruments and Princely Mechanical Devices from around 1600.* Munich: Deutscher Kunstverlag, 2009.

Kosibowicz, Edward. "Un missionnaire polonais oublié: Le Père Jean Nicolas Smogulecki S.J. missionnaire en Chine au XVIIe siècle." In *Revue d'Histoire des Missions,* 335–360. Vol. 6. Paris: Les Amis des missions, 1929.

Kremer, Richard L. "How Did the Turketum (or Torquetum) Get Its Name?" In *Scientific Instruments between East and West*, edited by Neil Brown, Silke Ackermann, and Feza Günergun. Leiden: Brill, 2019. 80–107.

Krey, Philip D., and Lesley Smith, eds. *Nicholas of Lyra: the Senses of Scripture.* Leiden: Brill, 2000.

Kugler, Georg Johannes. *Die Reichskrone.* Rev. ed. Munich: Verlag Herold Wien, 1986.

Kunsthistorisches Museum Wien, *Katalog der Sammlung für Plastik und Kunstgewerbe, II. Teil: Renaissance* Vienna: Kunsthistorisches Museum Wien, 1966.

Kuntz, Paul G. "The Hierarchical Vision of St. Robert Bellarmino," In *Acta Conventus Neo-Latini Turonensis*, edited by Jean-Claude Margolin, 959–977. Paris: J. Vrin, 1980.

La Cotardiere, Philippe de, and Patrick Fuentes. *Camille Flammarion.* Paris: Flammarion, 1994.

Landi, Fabrizia. *Le temps revient: il fregio di Poggio a Caiano.* Florence: Landi, 1986.

Lapide, Cornelius a. *The Great Commentary.* Translated by Thomas W. Mossman. Edinburgh: John Grant, 1908.

Lattis, James M. *Between Copernicus and Galileo: Christoph Clavius and the Collapse of Ptolemaic Cosmology.* Chicago: University of Chicago Press, 1994.

Leitão, Henrique. "A Brief Note on the Power of Mathematics. Pedro Nunes and the Retrogradation of Shadows." In *Proceedings of Recreational Mathematics Colloquium V G4G (Europe)*, edited by Jorge Nuno Silva, 45–52. Lisbon: Ludus, 2017.

Lempertz. "Tischuhr von Hans Gruber aus Nürnberg. Auction 1096, Lot 1607." Accessed February 6, 2021. https://www.lempertz.com/de/kataloge/lot/1096-3/1607-tischuhr-von-hans-gruber-aus-nuernberg.html.

Levy, Evonne. "Giovanni Battista Gaulli, *Adoration of the Name of Jesus* [Rome, Galleria Spada]." In *Saint, Site and Sacred Strategy: Ignatius, Rome and Jesuit Urbanism*, edited by Thomas Lucas, 195 (cat. no. 122). Rome: Biblioteca Vaticana, 1991.

Levy, Evonne. "The Honor of the altar, the fate of cults." In *The Holy Name: Art of the Gesù*, edited by Linda Wolk-Simon, 145–165. Philadelphia: St. Joseph's University Press, 2018.

Levy, Evonne. "The Institutional Memory of the Roman Gesù: Plans for Renovation in the 1670s by Carlo Fontana, Pietro da Cortona and Luca Berrettini." *Römisches Jahrbuch der Bibliotheca Hertziana* 33 (1999/2000): 373–426.

Levy, Evonne. *Propaganda and the Jesuit Baroque.* Berkeley: University of California Press, 2004.

Liisberg, Bering. *Urmagere og ure i Danmark.* Copenhagen: Martius Truelsen, 1908.

Littré, É. "De quelques phénoménes naturels donnés ou pris dans la Bible comme miraculeux." *La Philosophie positive* 22 (1879): 147–149.

Lohne, Johannes. "Thomas Harriot (1560–1631), the Tycho Brahe of Optics." *Centaurus* 6 (1959): 113–121.

Lohne, Johannes. "Zur Geschichte des Brechungsgesetzes." *Sudhoffs Archiv für Geschichte der Medizin und der Naturwissenschaften* 47, no. 2 (June 1963): 152–172.

Loyola, Ignatius of. "Letter to the Brothers in Gandia, 29 July 1647." In *Personal Writings*, edited and translated by Joseph A. Munitiz and Philip Endean, 184. London: Penguin, 1996.

Luther, Martin. *The Babylonian Captivity of the Church, 1520*. Edited by Erik H. Herrmann and Paul W. Robinson. Minneapolis, MN: Fortress Press, 2016.

Lynch, Peter. "Pedro Nunes and the Retrogression of the Sun." *Bulletin of the Irish Mathematical Society* 81 (Summer 2018): 23–30.

MacDougall, Bonnie G. "Jantar Mantar: Architecture, Astronomy, and Solar Kingship in Princely India." *The Cornell Journal of Architecture* 5 (1996): 16–33.

Mader, Günter, and Laila Neubert-Mader. *Jardins italiens*. Translated by Aude Virey-Wallon. Fribourg: Office du Livre, 1987.

Mancha, José Luis. "The Latin Translation of Levi Ben Gerson's Astronomy." In *Studies in Medieval Astronomy and Optics*, 1–25. Burlington: Ashgate, 2006.

Marconi, Paolo. "Una chiave per l'interpretazione dell'urbanistica rinascimentale: la cittadella come microcosmo." In *Quaderni dell'Istituto di Storia dell'Architettura. Serie XV. Fascicoli 85–90*, 53–94. Rome: Facoltà Di Architettura Università Di Roma, 1968.

Marianeschi, Edmondo, and Nicola Severino. "La Retrogradazione dell'Ombra." In *Atti del VI° Seminario di Gnomonica, 30 Sett. – 2 Ott.* San Benedetto del Tronto, 1994, 7–36.

Martin, Philippe, and Fabienne Henryot. *Dom Augustin Calmet; un itinéraire intellectuel.* Paris: Riveneuve éditions, 2008.

Maué, Hermann, Thomas Eser, Sven Hauschke, and Jana Stolzenberger, eds. *Quasi Centrum Europae. Europa kauft in Nürnberg 1400–1600*, exh. cat., Germanisches Nationalmuseum Nürnberg. Nuremberg: Germanisches Nationalmuseum, 2002.

Maunder, Edward Walter. *The Astronomy of the Bible. An Elementary Commentary on the Astronomical References of Holy Scripture.* London: T. S. Clark & Co., 1908

Mayaud, Pierre-Noël. *Le conflit entre l'Astronomie Nouvelle et l'Écriture Sainte aux XVIe et XVIIe siècles*, 6 vols. Paris: Champion, 2005

Maylender, M. *Storia delle Accademie d'Italia.* Bologna, 1926–1930.

McMullin, Ernan. "Galileo on Science and Scripture." In *Galileo: Man of Science*, edited by Ernan McMullin, 271–347. New York: Basic Books, Inc., 1967.

Meißner, Günter, and Andreas Beyer, eds. *Allgemeines Künstlerlexikon.* Vol. 76. Berlin: de Gruyter, 2013.

Melanchthon, Philip. *Orations on Philosophy and Education*, edited by Sachiko Kusukawa. Cambridge: Cambridge University Press, 1999.

Melion, Walter S. "Scripture for the Eyes: Bible Illustration in the Sixteenth-Century Low Countries." In *Scripture for the Eyes: Bible Illustration in Netherlandish Prints of the Sixteenth Century*, edited by James Clifton and Walter S. Melion, exh. cat.,

Museum of Biblical Art New York; Michael C. Carlos Museum, Emory University, 1–100. London and New York: D. Giles and Museum of Biblical Art, 2009.

Mende, Matthias, Rainer Schoch, and Anna Scherbaum. *Albrecht Dürer. Das druckgraphische Werk.* Vol. 3. München: Prestel, 2004.

Mersmann, Jasmin. "Moving Shadows, Moving Sun. Early Modern Sundials Restaging Miracles." *Nuncius* 30, no. 1 (January 2015): 96–123.

Mielke, Hans. "Antwerpener Graphik in der 2 Hälfte des 16. Jahrhunderts: Der *Thesaurus veteris et novi Testamenti* des Gerard de Jode (1585) und seine Künstler." *Zeitschrift für Kunstgeschichte*, 38 (1975): 29–83.

Milanesi, Gaetano. Editor's note to *Le vite de' più eccellenti pittori, scultori ed architettori*, by Giorgio Vasari, edited by Gaetano Milanesi, 593. Vol. 2. Florence: Sansoni, 1878.

Mills, Allan. "The "dial of Ahaz," and Refractive Sundials in General. Part I: Scaphe Dials." *Bulletin of the Scientific Instrument Society* 44 (1995): 21–24.

Mills, Allan. "The "dial of Ahaz," and Refractive Sundials in General. Part II: Horizontal Planar Dials." *Bulletin of the Scientific Instrument Society* 45 (1995): 25–29.

Morrall, Andrew. "Craftsmen and the Courts. Some Strategies of Production and Sale of Works of Art and Science in late 16th Century Augsburg." In *The Artist between Court and City (1300–1600). L'artiste entre la cour et la ville. Der Künstler zwischen Hof und Stadt,* edited by Dagmar Eichberger and Phillipe Lorentz, 220–245. Petersberg: Michael Imhoff Verlag, 2017.

Morrall, Andrew. "Virgil's Flute: the Art and Science of 'Antique Letters' and the Origins of Knowledge." In *The Primacy of the Image in Northern European Art, 1400–1700. Essays in Honor of Larry Silver,* edited by Debra Taylor Cashin, Henry Luttikhuizen and Ashley D. West, 193–206. Leiden: Brill, 2017.

Morrison, Joseph. "The Sun Dial of Ahaz." *Popular Astronomy* 6 (1898): 537–549.

Mosley, Adam. "Objects of Knowledge: Mathematics and Models in Sixteenth-Century Cosmology and Astronomy." In *Transmitting Knowledge. Words, Images and Instruments in Early Modern Europe*, edited by Sachiko Kusukawa and Ian Maclean, 193–216. Oxford: Oxford University Press, 2006.

Mosley, Adam. *Bearing the Heavens. Tycho Brahe and the Astronomical Community of the Late Sixteenth Century.* New York: Cambridge University Press, 2011.

Müller, Kathrin. *Visuelle Weltaneignung: Astronomische und kosmologische Diagramme in Handschriften des Mittelalters.* Göttingen: Vandenhoeck & Ruprecht, 2008.

Multhauf, Robert P., and David Davies. *A Catalogue of Instruments and Models in the Possession of the American Philosophical Society.* Philadelphia: American Philosophical Society, 1961.

Murray, Luke. *Jesuit Biblical Studies after Trent: Franciscus Toletus and Cornelius a Lapide.* Göttingen: Vandenhoeck & Ruprecht, 2019.

Nardi, Antonio. "Un galileiano eccentrico: il gesuita François Milliet Dechales tra Galileo e Newton." *Archives Internationales d'Histoire des Sciences* 49, no. 142 (1999): 32–74.

Natali, Antonio, Alessandro Nova, and Massimiliano Rossi, eds. *La Tribuna del principe.* Florence: Giunti, 2014.

Nauta, Doede. *Biografisch lexicon voor de geschiedenis van het Nederlands protestantisme.* Vol. 6. Kampen: Kok, 2006.

Navarro Morales, Maria Elisa. "The Astronomer's Palace: an Instrument for the Observation of the Sky." *Rupkatha Journal on Interdisciplinary Studies in Humanities* 3, no. 1 (2011). Available online at https://rupkatha.com/V3/n1/02Astronomer%27s-Palace.pdf, accessed July 29, 2020.

Nayyar, Reshma. "*Joshua Stopping the Sun* and Ignatius of Loyola at Il Gesù in Rome." *Journal of Jesuit Studies* 3 (2016): 211–237.

Nazé, Yaël. *L'astronomie des anciens.* Paris: Belin, 2009.

Neudörffer, Johann. *Des Johann Neudörfer Schreib- und Rechenmeisters zu Nürnberg Nachrichten von Künstlern und Werkleuten daselbst aus dem Jahre 1547 nebst der Fortsetzung des Andreas Gulden nach den Handschriften und mit Anmerkungen.* Edited by G. W. K. Lochner. Vienna: Baumüller, 1875.

Nunes, Pedro. *Obras.* Edited by Henrique Leitão. Vol. 6. Lisbon: Academia das Ciências de Lisboa and Fundação Calouste Gulbenkian, 2002.

Ozanam, Jacques. *Recreations in Mathematics and Natural Philosophy* [...]. Translated by Charles Hutton. Vol. 4. London: G. Kearsley, 1803.

Pancani, Ugo. "L'orologio di Villa Mediciea di Poggio a Caiano." Accessed July 25, 2020. http://www.avrvm.it/lorologio-di-villa-medicea-di-poggio-a-caiano/.

Pantin, Isabelle. "New Philosophy and Old Prejudices: Aspects of the Reception of Copernicanism in a Divided Europe." *Studies in the History and Philosophy of Science* 30, no. 2 (1999): 237–262.

Parisot, Jean-Paul. "La rétrogradation de l'ombre dans les cadrans solaires analemmatiques." *Journal for the History of Astronomy* 16, no. 1 (February 1985): 43–48.

Pecchiai, Pio. *Il Gesù di Roma.* Rome: Società Grafica Romana, 1952.

Pedretti, Carlo. "Il Codice di Benvenuto di Lorenzo Golpaja." In *Studi Vinciani. Documenti, Analisi, e Inediti leonardeschi*, 23–41. Geneva: Librairie E. Droz, 1957.

Pellecchia, Linda. "Stepping up: Observations on the Renaissance Staircase in Florence." *Opus Incertum* 2, no. 4 (2008): 42–49.

Perlus, Barry. *Celestial Mirror. The Astronomical Observatories of Jai Singh II.* New Haven: Yale University Press, 2020.

Peterson, Ivars. "Turning Back Time: an Antique Sundial Simulates a Biblical Miracle." *Ivars Science News* 137, no. 6 (February 1990): 91.

Pilz, Kurt. "Jost Amman 1539–1591." *Mitteilungen des Vereins für Geschichte der Stadt Nürnberg* 37 (1940): 201–252.

Pilz, Kurt. *600 Jahre Astronomie in Nürnberg.* Nuremberg: Verlag Hans Carl, 1977.

Plaßmeyer, Peter. “Christoph Schissler: the Elector's Dealer.” In *European Collections of Scientific Instruments, 1550–1750*, edited by Giorgio Strano, Stephen Johnston, Mara Miniati, and Alison D. Morrison-Low, 15–25. Leiden: Brill, 2009.

Plaßmeyer, Peter, and Christoph Emmendörffer, eds. *Weltenglanz: Der Mathematisch-Physikalische Salon Dresden zu Gast im Maximilianmuseum Augsburg.* Berlin and Munich: Deutscher Kunstverlag, 2009.

Pomel, Fabienne, ed. *Cloches et horloges dans les textes médiévaux. Mesurer et maîtriser le temps.* Rennes: Presses Universitaires de Rennes, 2012.

Raphael, Renée J. “Copernicanism in the Classroom: Jesuit Natural Philosophy and Mathematics after 1633.” *Journal for the History of Astronomy* 46 (2015): 419–440.

Redondi, Pietro. *Galileo Heretic.* Translated by Raymond Rosenthal. Princeton: Princeton University Press, 1987.

Reeves, Marjorie. *Joachim de Fiore and the Prophetic Future.* New York: Harper and Row, 1977.

Remmert, Volker. “‘Our Mathematicians Have Learned and Verified This’: Jesuits, Biblical Exegesis, and the Mathematical Sciences in the Late Sixteenth and Early Seventeenth Centuries.” In *Nature and Scripture in the Abrahamic Religions: Up to 1700*, edited by Jitse M. van der Meer and Scott H. Mandelbrote, 665–691. Leiden: Brill, 2008.

Remmert, Volker. “‘Whether the Stars Are Innumerable for Us?’: Astronomy and Biblical Exegesis in the Society of Jesus around 1600.” In *The Word and the World: Biblical Exegesis and Early Modern Science*, edited by Kevin Killeen and Peter J. Forshaw, 157–173. New York: Palgrave Macmillan, 2014.

Remmert, Volker. “Analyse des Imageries des Instruments Scientifiques de L'Époque modern (XVIe–XVIIe siècle),” Images des Mathématiques, CNRS, last modified May 29, 2019, https://images.math.cnrs.fr/Analyse-des-imageries-des-instruments-scientifiques-de-l-epoque-moderne-XVIe.html.

Remmert, Volker. “On Picturing the Past: Arithmetic and Geometry as Wings of the Mind.” *The Mathematical Intelligencer* 31, no. 3 (June 2009): 42–47.

Remmert, Volker. “Picturing Jesuit Anti-Copernican Consensus: Astronomy and Biblical Exegesis in the Frontispiece of Clavius' *Opera mathematica* (1612).” In *The Jesuits II: Cultures, Sciences, and the Arts, 1540–1773*, edited by John W. O'Malley, Gauvin Alexander Bailey, Steven J. Harris, and T. Frank Kennedy, 291–313. Toronto: University of Toronto Press, 2006.

Remmert, Volker. “Tycho Brahes Nase, oder: Atlas und Herkules und die visuelle Legitimierung der neuen Astronomie im 17. Jahrhundert.” *Zeitschrift für Kunstgeschichte* 66 (2003): 177–206.

Remmert, Volker. “Visuelle Strategien zur Konturierung eines jesuitischen Wissensreiches.” In *Le monde est une peinture. Jesuitische Identität und die Rolle der Bilder*, edited by Elisabeth Oy-Marra, Volker Remmert. Berlin: Akademie Verlag, 2011.

Remmert, Volker. *Picturing the Scientific Revolution: Title Engravings in Early Modern Scientific Publications.* Philadelphia: St. Joseph's University Press, 2011.

Remmert, Volker. *Widmung, Welterklärung und Wissenschaftslegitimierung: Titelbilder und ihre Funktionen in der wissenschaftlichen Revolution.* Wiesbaden: Harrassowitz, 2005.

Rey Pailhade, J. de, A. Rome, and Aug. Favot, "Le Cadran solaire du Lycée de Jeunes Filles de Grenoble." *Bulletin de la Société de Statistique, des sciences naturelles et des arts industriels du département de l'Isère*, 4th series, no. 15 (1920): 213–310.

Ribouillault, Denis. "Le Salone de la Villa d'Este à Tivoli: un théâtre des jardins et du territoire." *Studiolo* 3 (2005): 65–94.

Ribouillault, Denis. "Sundials on the Quirinal: Astronomy and the Early Modern Garden." In *Gardens, Knowledge and the Sciences in the Early Modern Period*, edited by Hubertus Fischer, Volker Remmert, and Joachim Wolschke-Bulmahn, 103–134. Cham: Spring International Publishing, 2018.

Ribouillault, Denis. Forthcoming. *Gardens of the Heavens*.

Rice, Louise. "Joshua and the Jesuits: the Battle for the Apse of the Gesù." In *The Holy Name: Art of the Gesù*, edited by Linda Wolk-Simon, 225–268. Philadelphia: St. Joseph's University Press, 2018.

Richard, Jules-Marie. *Inventaire-sommaire des archives départementales antérieures à 1790. Pas-de-Calais.* Vol. 1. Arras: Impr. de la Société du Pas-de-Calais, 1878.

Richter, Fleur. *Die Ästhetik Geometrischer Körper in der Renaissance.* Stuttgart: Verlag Gerd Hatje, 1995.

Ritvo, Lucille B. "Hartmann, Georg," Complete Dictionary of Scientific Biography. Accessed March 17, 2021. https://www.encyclopedia.com/science/dictionaries-thesauruses-pictures-and-press-releases/hartmann-georg.

Roguet, D. "Le cadran solaire de l'Observatoire Flammarion de Juvisy et les Cadrans Solaires en general." *L'Astronomie: Revue d'astronomie populaire, de météorologie et de physique du globe* 26 (1912): 441–464.

Rohr, René R. J. "Les cadrans solaires universels de Jacques Ozanam." *Centaurus. International Magazine of the History of Mathematics, Science, and Technology* 29, no. 3 (September 1986): 165–177.

Rohr, René R. J. *Sundials. History, Theory and Practice*. Revised edition. New York: Dover Publication, 1996.

Rohr, René R.J. "The universal sundials of Jacques Ozanam." *Antiquarian Horology* 18 (1989): 171–177.

Rosier, Bart. *The Bible in Print: Netherlandish Bible Illustration in the Sixteenth Century.* Vol. 2. Leiden: Foleor, 1997.

Rousseau, Claudia. "Cosimo I de' Medici and Astrology: the Symbolism of Prophecy." PhD diss., Columbia University, 1983.

Sachse, Julius F. "Horologium Achaz." *Proceedings of the American Philosophical Society* 34, no. 147 (January 1895): 21–30.

Sadler, Philip M. "An Ancient Time Machine: the Dial of Ahaz." *American Journal of Physics* 63, no. 3 (1995): 211–216.

Salmas, Anne-Claire. "La mesure du temps de la journée (II). Modules et fonctionnement des horloges à ombre tardives et des cadrans solaires." *Bulletin de l'Institut français d'archéologie orientale* 114 (2014): 419–446.

Sambin de Norcen, Maria Teresa. "*Ut apud Plinium*: Giardino e paesaggio a Belriguardo nel Quattrocento." In *Delizie in villa. Il giardino rinscimentale e i suoi committenti*, edited by Francesco Ceccarelli and Gianni Venturi, 65–90. Florence: Olschki, 2008.

Sauer, Christine. "Die Memorabilien in der Stadtbibliothek." In *Wunderkammer im Wissensraum. Die Memorabilien der Stadtbibliothek Nürnberg im Kontext städtischer Sammlungskulturen*, edited by Christine Sauer, 17–95. Wiesbaden: Harrassowitz, 2021.

Savoie, Denis. "Cadran de Georg Hartmann." *Cadran Info* 43 (May 2021): 11–14.

Savoie, Denis. "Le cadran solaire de hauteur de Wenzel Jamnitzer de l'Observatoire de Paris." *Cadran Info* 36 (October 2017): 136–151.

Savoie, Denis. "Le cadran solaire du barrage de Castillon." *L'astronomie* (February 2010): 12–19.

Savoie, Denis. "Les cadrans solaires à corniche." *Cadran Info* 19 (May 2009): 71–80.

Savoie, Denis. "Les cadrans solaires du Château de Denainvilliers." *Bulletin* ANCAHA 97 (Summer 2003): 79–82.

Savoie, Denis. *La Gnomonique.* Paris: Les Belles Lettres, 2001.

Savoie, Denis. *Recherches sur les cadrans solaires.* Turnhout: Brepols, 2014.

Schaaf, William L. "Jacques Ozanam." In *Dictionary of Scientific Biography*, edited by Charles Coulston Gillispie, 263–265. Vol. 10. New York. Charles Scribner's Sons, 1974.

Schaff, David S. "Cardinal Bellarmine – Now Saint and Doctor of the Church." *Church History* 2, no. 1 (1933): 41–55.

Schaff, Phillip, and Jakob Herzog. "Clarke, Adam." In *The New Schaff-Herzog Encyclopedia of Religious Knowledge*, 126. Vol. 3. Grand Rapids: Baker, 1953.

Schaller, Andrea. *Prunkuhren der Renaissance. Württembergisches Landesmuseum Stuttgart.* Stuttgart: Württembergisches Landesmuseum, 2001.

Scholz-Hänsel, Michael. *Eine spanische Wissenschaftsutopie am Ende des 16. Jahrhunderts. Die Bibliotheksfresken von Pellegrino Pellegrini im Escorial.* Münster: Lit, 1987

Schroeder, Horst. *Der Topos der Nine Worthies in Literatur und bildender Kunst.* Göttingen: Vandenhoeck & Ruprecht, 1971.

Schuchard, Margaret, ed. *Bernhard Varenius (1622–1650).* Leiden: Brill, 2007.

Sellers, Ian. "Clarke, Adam." In *Oxford Dictionary of National Biography*, edited by H. C. G. Matthew and Brian Harrison. Oxford: Oxford University Press, 2004.

Sellers, Ian. *Adam Clarke, controversialist: Wesleyanism and the historic faith in the age of Bunting.* St Columb Major: Wesley Historical Society, 1976.

Severino, Nicola. "Georg Hartmann, Il primo grande gnomonista della Rinascenza." Last modified December 2008. https://www.academia.edu/39912162/George_Hartmann_Il_primo_grande_gnomonista_della_Rinascenza.

Sharma, Virendra Nath. "Míshra Yantra of the Delhi Observatory." *Indian Journal of History of Science* 29, no. 3 (1994): 477–487.

Sloley, R. W. "Primitive Methods of Measuring Time: With Special Reference to Egypt." *The Journal of Egyptian Archaeology*, 17, no. 3–4 (Nov. 1931): 166–178.

Smith, Jeffrey Chipps. *Nuremberg. a Renaissance City, 1500–1618*. Austin: University of Texas Press, 1983.

Smith, Lesley. *The 'Glossa Ordinaria': the Making of a Modern Bible Commentary*. Leiden: Brill, 2009.

Smith, Webster. "Pratolino." *Journal of the Society of Architectural Historians* 20, no. 4 (December 1961): 155–168.

Smolarski, Dennis C. "The Jesuit Ratio Studiorum, Christopher Clavius, and the Study of Mathematical Sciences in Universities." *Science in context* 15, no. 03 (2002): 447–457.

Solomon, Xavier F. "The Lost Monument to Cardinal Robert Bellarmine." In *The Holy Name: Art of the Gesù*, edited by Linda Wolk-Simon, 125–144. Philadelphia: St. Joseph's University Press, 2018.

Sprat, Thomas. *The History of the Royal Society*. Edited by J. I. Cope and H. W. Jones. London: Routledge & Keagan Paul, 1959.

Spurr, John. ""Rational Religion" in Restoration England." *Journal of the History of Ideas*, 49, no. 4 (1988): 563–585.

Steche, Richard. "Astronomische Instrumente von W. Jamnitzer." *Anzeiger für Kunde der deutschen Vorzeit* 24, no. 2 (February 1877): 55.

Steche, Richard. "Zwei Radierungen von Wenzel Jamitzer." *Mitteilungen aus dem Germanischen Nationalmuseum* (1891): 36–39.

Stockler, Garção, and Francisco de Borja. *Ensaio Historico sobre a Origem e Progressos das Mathematicas em Portugal*. Paris: Officina de P. N. Rougeron, 1819.

Stolberg, Michael. *Gelehrte Medizin und ärtzlicher Alltag in der Renaissance*. Berlin: De Gruyter, 2020.

Syndram, Dirk, Martina Minning, and Jochen Vötsch, eds. *Die kurfürstliche sächsische Kunstkammer in Dresden. Das Inventar der Kunstkammer 1587*. Dresden: Sandstein, 2010.

Taub, Liba. "Introduction: Reengaging with Instruments," *ISIS* 102 (2011): 689–696.

Taub, Liba. "On Scientific Instruments," *Studies in History and Philosophy of Science* 40 (2009): 337–343.

Taylor, René. "Architecture and Magic. Considerations on the 'Idea' of the Escorial." In *Essays in the History of Architecture Presented to Rudolf Wittkower*, edited by Douglas Fraser, Howard Hibbard and Milton J. Lewine, 81–109. London: Phaidon, 1967.

Thieme, Ulrich, and Felix Becker, eds., *Allgemeines Lexikon der bildenden Künstler von der Antike bis zur Gegenwart*, vol. 18 (Leipzig: E.A. Seemann, 1925).

Thunø, Erik. *The Apse Mosaic in Early Medieval Rome: Time, Network, and Repetition.* Cambridge: Cambridge University Press, 2015.

Thürigen, Susanne. *Turm, Spiegel, Buch. Astronomische Tischuhren in Süddeutschland zwischen 1450 und 1650*. Berlin/Boston: De Gruyter, 2022.

Truitt, Elly Rachel. *Medieval Robots: Mechanism, Magic, Nature, and Art.* Philadelphia: University of Pennsylvania Press, 2015.

Turner, Anthony J. "A Biblical Miracle in a Renaissance Sundial." *Bulletin of the Scientific Instrument Society* 61 (1999): 11–14. .

Tutino, Stefania. *Empire of souls: Robert Bellarmine and the Christian commonwealth.* Oxford: Oxford University Press, 2010.

Valleriani, Matteo. "Ancient Pneumatics Transformed during the Early Modern Period." *Nuncius* 29, no. 1 (January 2014): 127–173.

Valleriani, Matteo. "Pratolino: the History of Science in a Garden." Accessed July 24, 2020. http://pratolino.mpiwg-berlin.mpg.de/.

Van Dorp, Jaap. "The Prayer of Isaiah and the Sundial of Ahaz (2Kgs 20:11)." In *Psalms and Prayers*, edited by Bob Becking and Eric Peels, 253–265. Leiden: Brill, 2007.

Van Kasteren, John P. "Cornelius Cornelii a Lapide." In *The Catholic Encyclopedia*, edited by Charles G. Herbermann, Edward A. Pace, Condé B. Pallen, Thomas J. Shahan, and John J. Wynne, 377. Vol. 4. New York: Robert Appleton Company, 1908.

Van Miert, Dirk. *Humanism in an Age of Science. the Amsterdam Athenaeum in the Golden Age, 1632–1704.* Translated by Michiel Wielema and Anthony Ossa-Richardson. Leiden and Boston: Brill, 2009.

Ventrone, Paola, ed. *Le Temps Revient, 'L Tempo si Rinuova; Feste e Spettacoli nella Firenze di Lorenzo il Magnifico.* Milan: Silvana, 1992. An exhibition catalogue.

Vermij, Rienk. *The Calvinist Copernicans: the Reception of the New Astronomy in the Dutch Republic, 1575–1750.* Amsterdam: Koninklijke Nederlandse Akademie van Wetenschappen, 2002.

Vieri, Francesco de'. *Trattato nel quale si contengono i tre primi libri delle Metheore, nuovamente ristampati, et da lui coretti con l'aggiunta del quarto Libro.* Florence: Giorgio Marescotti, 1582.

Vigouroux, Fulcran. *Les Livres Saints et la Critique Rationaliste.* Vol. 5. Paris: Roger & Chernoviz, 1891.

Vincent, Gilbert. "Construction géométrique simplifiée d'un cadran solaire analemmatique et rappel de quelques propriétés." *Observations & Travaux* 63 (2006): 1–5.

Volwahsen, Andreas. *Cosmic Architecture in India. the Astronomical Monuments of Maharaja Jai Singh II.* Munich: Prestel, 2001.

Waard, Cornelius de. "Le manuscrit perdu de Snellius sur la réfraction." *Janus* 39 (1935): 51–73.

Wallace, William A. "Galileo's Jesuit Connections and Their Influence on his Science." In *Jesuit Science and the Republic of Letters*, edited by Mordechai Feingold, 99–126. Cambridge: MIT Press, 2003.

Wardhaugh, Benjamin, ed. Introduction to *The Correspondence of Charles Hutton (1737–1823)* […], xiv–xxxv. Oxford: Oxford University Press, 2017.

Watanabe-O'Kelly, Helen. *Court Culture in Dresden. From Renaissance to Baroque.* Basingstoke: Palgrave, 2002.

Weil, Mark. "The Devotion of the Forty Hours and Roman Baroque Illusions." *Journal of the Warburg and Courtauld Institutes* 37 (1974): 218–248.

Westman, Robert S. "The Melanchthon Circle, Rheticus, and the Wittenberg Interpretation of the Copernican Theory." *Isis* 66 (1975): 165–193.

Westman, Robert S. "Three Responses to the Copernican Theory: Johannes Praetorius, Tycho Brahe and Michael Maestlin." In *The Copernican Achievement*, edited by Robert S. Westman, 285–345. Berkeley: University of California Press, 1975.

Wilkes, John, ed. "How the Shadow of a Style, on a Sun-Dial, might go backwards without a miracle is treated." In *Encyclopedia* Londinensis, 300. Vol 10. London: J. Adlard, 1811.

Wilkins, John. *The Mathematical and Philosophical Works of the Right Reverend John Wilkins* […]. Vol. 1. London: Printed by C. Whittingham, for Vernor & Hood, Cuthell & Martin and J. Walker, 1802.

Willers, Johannes, Gerhard Bott, and Wolfgang Pülhorn. *Focus Behaim-Globus*, exh. cat., Germanisches Nationalmuseum Nürnberg. Vol. 2. Nuremberg: Verlag des Germanischen Nationalmuseums, 1992.

Winkler, Louis. "Pennsylvania German Astronomy and Astrology X: Christopher Witt's Device." *Pennsylvania Folklife* 24, no. 2 (Winter 1974–1975): 36–39.

Wolfangel, Doris. *Dr. Melchior Ayrer (1520–1579)*. Würzburg: Baehr, 1957.

Wreede, Liesbeth Cornelia de. "Willebrord Snellius (1580–1626), a Humanist Reshaping the Mathematical Sciences." PhD diss., University of Utrecht, 2007.

Wright, Michael T. "Archimedes, Astronomy, and the Planetarium." In *Archimedes in the 21st Century: Trends in the History of Science*, 125–141. Cham: Springer International Publishing, 2017.

Wunderlich, Herbert. *Das Dresdener 'quadratum geometricum' aus dem Jahre 1569 von Christoph Schissler der Ä., Augsburg.* Berlin: VEB Deutscher Verlag der Wissenschaften, 1960.

Wunderlich, Herbert. *Kursächsische Feldmeßkunst, artilleristische Richtverfahren und Ballistik im 16. Und 17. Jahrhundert, Beitrag zur Geschichte der praktischen Mathematik, der Physik und des Artilleriewesens in der Renaissance unter Zugrundelegung von Instrumenten, Karten, Hand- und Druckschriften des Staatlichen Mathematisch-Physikalischen Salons Dresden.* Berlin: Deutscher Verlag der Wissenschaften, 1977.

Wyss, Robert Ludwig. "Die neuen Helden: eine ikonographische Studie." *Zeitschrift für schweizerische Archäologie und Kunstgeschichte* 17 (1957): 73–106.

Zangheri, Luigi. "Pratolino. La grande macchina del cosmo." In *Il giardino e la memoria del mondo*, edited by Giuliana Baldan Zenoni-Politeo and Antonella Pietrogrande, 43–52. Florence: Leo S. Olschki, 2002.

Zangheri, Luigi. *Pratolino. Il giardino delle meraviglie.* Vol. 1. Florence: Gonelli, 1979.

Zentralinstitut für Kunstgeschichte und Städtische Kunstsammlungen Augsburg. *Welt im Umbruch. Augsburg zwischen Renaissance und Barock.* Vol. 2. Augsburg: Augsburger Druck- und Verlagshaus, 1980.

Zimmermann, Heinrich, ed. "Das Inventar der Prager Schatz- und Kunstkammer vom 6. Dezember 1621. Nach Akten des k. und k. Reichsfinanzarchivs in Wien." *Jahrbuch der Kunsthistorischen Sammlungen des Allerhöchsten Kaiserhauses* 25 (1905): XIII–LXXV.

Zinner, Ernst. *Deutsche und niederländische astronomische Instrumente des 11.–18. Jahrhunderts.* Munich: C. H. Beck, 1979.

Zinner, Ernst. *Geschichte und Bibliographie der astronomischen Literatur in Deutschland zur Zeit der Renaissance.* 2nd ed. Stuttgart: Hiersemann, 1964.

Zuber, Mike A. "Copernican Cosmotheism: Johann Jacob Zimmermann and the Mystical Light." *Aries* 14, no. 2 (2014): 215–245.

Zuber, Mike A. "Johann Jacob Zimmermann and God's Two Books: Copernican Cosmology in Lutheran Germany around 1700." In *Knowing Nature in Early Modern Europe,* edited by David Beck, 83–99. London: Routledge, 2015.

Manuscripts

Archivio della Pontificia Università Gregoriana (APUG), "Ad Librum Physicorum Disputatione de Principiis Rerum Naturalium in Comuni," Giovanni Filippo de Marini, MS FC 2161 (1660s).

Archivio della Pontificia Università Gregoriana (APUG), "Disputationes physicae auctore R.P. Ludovico Bompiano Anconitano Societatis Jesu excipiebat Ferdinandus Maldonatus Melitensis eiusdem Societatis," Ludovico Bompiani, MS FC 1347 (1646).

Archivio della Pontificia Università Gregoriana (APUG), "Physicae," Agostino de Mari, MS FC 55 (1682).

Archivum Romanum Societatis Iesu (ARSI), Rom. 143/II, fols. 359^{r}–^{v} (correspondence, 1675).

Bayerische Staatsbibliothek München, "Collectio figurarum," Georg Hartmann, MS Rar. 434 (1600s). Accessed March 17, 2021. https://daten.digitale-sammlungen.de/~db/0002/bsb00028171/images/.

Biblioteca degli Intronati, "Libro degli artifizii spiritali over di fiato d'Herone Alessandrino," Oreste Vannoci Biringucci, MS L. VI. 44 (1582).

Biblioteca Marciana, "Libro di macchine diverse," Benvenuto della Volpaia, MS. 5363, formerly cod. It. IV.41 (ca. 1520).

Biblioteca Nazionale Centrale, "Breve discorso intorno all'arte dell'alchimia," Francesco de' Vieri, MS. Magl. XVI 78 (sixteenth century).

Biblioteca Nazionale Centrale, "Della natura del voto di Erone Alessandrino," Bernardo Davanzati, MS Rari 223, formerly MS Palat. 1166 (1582). Reedited as Bernardo Davanzati, Della natura del voto di Erone Alessandrino. Volgarizzamento inedito di Bernardo Davanzati. Florence: Stamperia del Monitore, 1862.

Biblioteca Apostolica Vaticana, Barb. 5341, fol. 207v.

Biblioteca Riccardiana, MS Riccardiana 2312, fol. 108v. Accessed online May 10, 2021. http://echo.mpiwg-berlin.mpg.de/MPIWG:9MHXWTWK.

Bibliothèque Nationale de France, "Psautier de Paris," MS. gr. 139, fol. 446v. Available online since March 9, 2015. https://gallica.bnf.fr/ark:/12148/btv1b10515446x/f916.item/.

Hauptstaatsarchiv Dresden, "Ein klein kunststücklein zu dem heiligen Christ und einem Newen Jahresgeschenk," Loc. 8679, fols. 44–45 (1572).

Hauptstaatsarchiv Dresden, "Inventar über des Kurfürsten von Sachsen und Burggrafen zu Magdeburg Kunstkammer," 10009 Kunstkammer, Sammlungen und Galerien, Nr. 001 (1587).

National Art Library of the Victoria & Albert Museum, "Ein gar Kunstlicher und wolgetzierter Schreibtisch sampt allerhant kunstlichen silbern und vergulten newerfunden Instrumenten so darin zufinden. Zum gebrauch der Geometrischen und Astronomischen, auch andern schönen und nützlichen Kunsten [...]," Wenzel Jamnitzer, 2 vols., MSL/1893/1600 and MSL/1893/1601 (1585).

Staats- und Stadtbibliothek Augsburg, "Thomas Rucker an den Kaiser," Cod. Halder 2, Cod. MS 231 (1574).

Württembergische Landesbibliothek Stuttgart, [Description of Pratolino, 1600], Heinrich Schickhardt, MS Cod. hist. qt. 148b, fol. 49r–53v. Quoted in Luigi Zangheri, Pratolino. Il giardino delle meraviglie, vol. 1 (Florence: Gonelli, 1979), 221–225.

Index

www.ingramcontent.com/pod-product-compliance
Lightning Source LLC
Chambersburg PA
CBHW071249180926
51613CB00002B/176
* 9 7 8 9 0 0 4 4 7 1 8 8 7 *